THE AGE OF THE SOYBEAN:
AN ENVIRONMENTAL HISTORY OF SOY
DURING THE GREAT ACCELERATION

THE AGE OF THE SOYBEAN:
AN ENVIRONMENTAL HISTORY OF SOY
DURING THE GREAT ACCELERATION

Edited by
Claiton Marcio da Silva and Claudio de Majo

© 2022
The White Horse Press, The Old Vicarage, Main Street, Winwick,
Cambridgeshire, UK

Set in 11 point Adobe Caslon Pro and Lucida Sans
Printed by Lightning Source

British Library Cataloguing in Publication Data
A catalogue record for this book is available from the British Library

ISBN   978-1-912186-64-8 (PB)
          978-1-912186-65-5 (ebook)

doi: 10.3197/63800040695086.book

Cover and interior layout by Stefania Bonura Graphics Web & Books
Image re-use may be subject to restrictions.

CONTENTS

LIST OF FIGURES

The Age of the Soybean

List of figures

CONTRIBUTOR BIOGRAPHIES

Editors

Claiton Marcio da Silva is an Associate Professor at the Federal University of the Southern Frontier (UFFS), Santa Catarina, Brazil and a CNPq Productivity Researcher. He began studying the modernisation of Brazilian agriculture in 2002, completing his MA in history at the Federal University of Santa Catarina (UFSC) with a thesis on rural youth in Southern Brazil. At Casa de Oswaldo Cruz (COC/Fiocruz), his doctoral dissertation explored the role of Nelson Rockefeller's American International Association for Economic and Social Development (AIA) in introducing the US extension service model in Latin America. Recently, he has been exploring the global environmental history of soybeans and their impacts on traditional populations in the Global South.

Claudio de Majo is a doctoral candidate at the Rachel Carson Center for Environment and Society (Ludwig-Maximilians University, Munich). His research interests include global environmental history, the environmental history of Europe and Latin America, and the relation between the commons and ecology. He is also interested in cutting-edge methodologies such as evolutionary history and neo-materialism. Claudio's research has received funding from the Andrea von Braun Stiftung for Interdisciplinary Studies, the Leibniz Institute of European History (IEG) and the Amerika-Institut (LMU Munich), where he also works as a lecturer. He is deputy editor of *Global Environment: A Journal of Transdisciplinary History*.

Chapter 1

Samira Peruchi Moretto is Professor at the Federal University of Fronteira SUL (UFFS) Santa Catarina, Brazil and a CNPq Productivity Researcher 2. She completed an MA (2010) in History and her Doctorate (2014) both at the Universidade Federal de Santa Catarina (UFSC), Santa Catarina, Brazil. She advises master and doctorate students at the Graduate Program of History at Federal University of Fronteira Sul and at the Universidade Federal de Santa Catarina. She researches and advises on Environmental History, domestication and introduction of plant species, deforestation, reforestation, landscape transformations, biodiversity conservation and history of the Brazilian Republic, with emphasis on the period from 1964 to recent times.

Eunice Sueli Nodari is a Full Professor at the Universidade Federal de Santa Catarina (UFSC), Santa Catarina, Brazil and a CNPq Productivity Researcher 1D. She completed her MA (1992) in History at the University of California at Davis (CA), USA and her Doctorate at the Pontifícia Universidade Católica do Rio Grande do Sul (PUC/RS). She advises master and doctorate students at the Graduate Program of History and at the Interdisciplinary Doctoral Program in Human Sciences. For more than twenty years, Professor Nodari has been doing research on environmental history, with focus on transformation of the landscape, the devastation of Araucaria Forests, agriculture and migration and environmental disasters. She currently coordinates an international project on Environmental history of vitiviniculture in the Americas.

Rubens Onofre Nodari is Full Professor at the Universidade Federal de Santa Catarina (UFSC), Santa Catarina, Brazil and a CNPq Productivity Researcher 1A. He completed his M.Sc. (1980) in Plant Breeding at the Universidade Federal do Rio Grande do Sul and his Ph.D. in Genetics (1992) at University of California, Davis (CA), USA. In the Graduate Program on Plant Genetic Resources he advises M.Sc. and Ph.D. students and carries out research on plant genetics and breeding, conservation and use of plant genetic resources, organisation of the genetic variability of natural and experimental populations, biohazards and biosafety of genetically modified organisms and plant genomics/transcryptomics/proteomics.

Chapter 2

Brian Lander is assistant professor of history at Brown University in the USA, where he is also a fellow of the Institute at Brown for Environment and Society. He has published on various topics in the environmental history of premodern China, including agriculture, water control, politics, and animals. He is the author of *The King's Harvest: A Political Ecology of China from the First Farmers to the First Empire* (Yale University Press, 2021).

Contributor biographies

Thomas David DuBois is a historian of modern China, and Distinguished Professor in Beijing Normal and Hebei Universities. He is the author of three books on religion and law, including *Empire and the Meaning of Religion in Northeast Asia* (Cambridge 2017), which was centred on China's Northeast. His current research on China's food industries is based on years of fieldwork in farms, dairies, and restaurants across China, and has thus far produced about a dozen journal articles. He is currently finishing a book that recreates China's long food history as a series of seven banquets.

Chapter 3

Rhuan Targino Zaleski Trindade holds a bachelor's and a master's degree in History from the Federal University of Rio Grande do Sul (UFRGS) and a Ph.D. in History from the Federal University of Paraná (UFPR). He currently works as an adjunct at the Centro-Western State University (Unicentro), Irati Campus, lecturing in the bachelor's program. The author has developed studies on immigration, colonisation and Polish ethnicity in Brazil, especially during the interwar period. He has authored several articles, and book chapters on several research themes, such as the introduction of soybeans in the state of Rio Grande do Sul, rural and peasant history, as well as on Polish-Brazilian relationships during the Vargas Era, with a focus on national and ethnic identities.

Chapter 4

Ernst Langthaler is Professor of Social and Economic History at the Johannes Kepler University Linz, Austria, and Head of the Institute of Rural History in St. Pölten, Austria. He was Visiting Professor at the Universities of Innsbruck and Vienna and Visiting Fellow at the Rachel Carson Center for Environment and Society at the University of Munich. He is editorial board member of several journals and book series (e.g., Rural History Yearbook) as well as Secretary of the European Rural History Organisation (EURHO). His research focuses on agricultural and food history since 1850 from regional to global scales. His current book-length project deals with agro-food globalisation since the 1870s through the lens of soy.

Chapter 5

Matilda Baraibar Norberg (Ph.D.) is a researcher and lecturer at the Department of Economic History and International Relations (Stockholm University, Sweden). She is also a member of the advisory board of the South American Institute for Resilience and Sustainability Studies (SARAS). Her research area is within a historically informed political economy, with particular focus on agrarian change, food systems, development and sustainability. Her latest book is *The Political Economy of Agrarian Change in Latin America: Argentina, Paraguay and Uruguay* (Springer International Publishing, 2020).

Chapter 6

Cassiano de Brito Rocha holds a Master's degree in Social Sciences and Humanities from the Universidade Estadual de Goiás. He currently works as Monitoring Manager of Infrastructure and Economic Development Projects in the Government of the State of Goiás. His research focuses on the history of agriculture with an emphasis on the expansion of soy agricultural frontiers in the Cerrado.

Kárita de Jesus Boaventura has a Ph.D. in Environmental Sciences from the Postgraduate Program in Natural Resources of the Cerrado of the State University of Goiás. She works as a History Teacher in the public primary education system of the State of Goiás. She is a member of the Research Group 'Environmental History of the Cerrados' of the CNPq Research Directory Group.

Giovanni de Araujo Boggione has a Ph.D .in Remote Sensing from the National Institute for Space Research (INPE). He is currently a professor at the Federal Institute of Education, Science and Technology of Goiás (IFG) where he works in Geoprocessing, Digital Processing of Satellite Images and Geographic Database. He is also pursuing post-doctorate training in the Graduate Program in Society, Technology and Environment at the Evangelical University of Goiás, under the supervision of Sandro Dutra e Silva. He is a member of the CNPq Research Directory Group 'Environmental History of the Cerrados'.

Sandro Dutra e Silva holds a Ph.D. in History from the University of Brasília (UnB). He is a professor of Brazilian environmental history at the State University of Goiás and the Evangelical University of Goiás, Brazil.

He currently serves as editor-in-chief of *Historia Ambiental Latinoamericana y Caribeña* (HALAC), the journal of the Latin American and Caribbean Society for Environmental History (SOLCHA). His main research interests include the environmental history of the Cerrado and the agricultural frontier in Central Brazil. He is a Research Productivity Fellow of the National Council for Scientific and Technological Development (CNPq).

Chapter 7

Eduardo Relly is currently Post-Doc Researcher at Friedrich-Schiller-University in Jena, Germany, and works in the Collaborative Research Centre TRR 'Structural Change of Property'. He completed his Ph.D. Studies in History at Free University Berlin (2019) with a dissertation on the transfer of environmental knowledge between Germany and Brazil under the background of commons privatisation and settler colonialism policies. He has been active mostly in Brazilian and German academia, with research positions or academic stays in Bielefeld (BGHS, University of Bielefeld), Munich (Rachel Carson Centre, LMU München) and São Leopoldo-Porto Alegre (Universidade Vale do Rio dos Sinos, UNISINOS). He has lately been dealing with the topic of propertisation of genetic resources of the biodiversity under the guidelines of the Convention on biological diversity and the Nagoya Protocol.

Claudio de Majo is a doctoral candidate at the Rachel Carson Center for Environment and Society (Ludwig-Maximilians University, Munich). His research interests include global environmental history, the environmental history of Europe and Latin America, and the relation between the commons and ecology. He is also interested in cutting-edge methodologies such as evolutionary history and neo-materialism. Claudio's research has received funding from the Andrea von Braun Stiftung for Interdisciplinary Studies, the Leibniz Institute of European History (IEG) and the Amerika-Institut (LMU Munich), where he also works as a lecturer. He is deputy editor of *Global Environment: A Journal of Transdisciplinary History*.

Chapter 8

Larissa de Lima Trindade has a degree in Accounting from the Federal University of Santa Maria, Brazil (2006), a master's degree in Administration from the Federal University of Santa Maria-BR (2009) and a joint Doctorate (2016) in Human Sciences from the Interdisciplinary Postgraduate Program in Human Sciences at the Federal University of Santa Catarina and Kent State University, Ohio, USA (2015/2016). She was a Capes/Fulbright Scholarship. She is currently an adjunct professor at the Federal University of Fronteira Sul, in Chapecó, Santa Catarina State - Brazil. She has experience in Administration, Accounting and Public Policy, working mainly on the themes: sustainability, environmental responsibility in public and private organisations, environmental public policies, environmental governance and finance.

Rodrigo Fortunato de Oliveira is an Environmental and Sanitary Engineering undergraduate student at Federal University Fronteira Sul. He was a scientific initiation scholarship holder in projects focused on the environmental area. His work focuses on forest hydrology and integrated water resources management, but he has also worked with statistics teaching and management of solid waste cooperatives.

Joshua Filla is a research staff member with Kent State University's Center for Public Policy and Health. He and his colleagues have produced literature and professional publications related to water pollution policy, including policy issues related to the nutrient enrichment of water bodies in the United States. His work also focuses on other aspects of public health policy and management, including care coordination and risk reduction, local health department management, and intergovernmental collaboration. He received his Master's in Public Administration and Public Policy from Kent State University in 2011.

John Hoornbeek serves as Director of the Center for Public Policy and Health and Professor of Health Policy and Management at Kent State University (KSU) in the United States. His areas of research include environmental and water policy, alternative approaches to health services administration, intergovernmental relations, and public management. Dr. Hoornbeek has published approximately twenty articles in peer reviewed scholarly journals, multiple chapters in edited books, a single-authored book on American water pollution policy published by the State University of New York Press, and an array of professional reports sponsored by government and non-profit organisations. He has also held professional positions at the federal, state, and local levels of government in the US. Dr Hoornbeek earned his Bachelor's degree at Beloit College, his Master's degree at the University of Wisconsin – Madison, and his Doctorate at the University of Pittsburgh.

Mutlaq Albugmi is a Ph.D. candidate at the College of Public Health at Kent State University (KSU), where he has also worked as a Graduate Research Assistant for the Center for Public Policy and Health. He earned his Master's degree from KSU, after working in the healthcare industry in Saudi Arabia prior to moving to the United States. He has worked in multifaceted environments such as hospitals and universities, where he has been responsible for conducting activities such as research, policy analysis, and program evaluation. His areas of research and teaching interest are in health policy, health system reform, and health economics.

Chapter 9

Enrique Antonio Mejia is a doctoral candidate at the Department of Economic History and International Relations at Stockholm University and is a member of the Resilience Research School at Stockholm Resilience Centre. He works within the project 'Unequal Exchange and Agrofood Globalization: Nitrogen, Soybeans and Latin America', funded by the Jan Wallanders and Tom Hedelius Foundation as well as the Tore Browaldhs Foundation. His research interests lie in the nexus of economic history, ecological economics and political ecology.

Chapter 10

Anna Zeide is an Associate Professor of History and Director of the Food Studies Program at Virginia Tech. She is the author of *Canned: The Rise and Fall of Consumer Confidence in the American Food Industry* (University of California Press, 2018); a co-editor of *Acquired Tastes: Stories About the Origins of Modern Foods* (MIT Press, 2021); and the author of the forthcoming *US History in 15 Foods* (Bloomsbury Press, forthcoming 2023).

Chapter 11

Janina Priebe holds a Ph.D. in History of Science and Ideas. She is a researcher in environmental history at Umeå University, Sweden. Her work centres on visions of natural resource development in the twentieth century in the northern hemisphere, and cultural and historical perspectives on natural resources. She works within interdisciplinary research projects on forestry and climate action, and energy transformation in northern rural areas. She is currently the Arctic Five Chair in Environmental History at Umeå University (2022–2024), and the assistant programme leader for Future Forests, an interdisciplinary forest research and communication platform between the Swedish University of Agricultural Sciences, Umeå University, and the Forestry Research Institute of Sweden.

Chapter 12

Jo Klanovicz is professor of history at the Universidade Estadual do Centro-Oeste do Parana (Unicentro), Brazil, and CNPq Productivity Researcher . In 2017, he was visiting researcher at Freie University, Berlin. Since 2013 he has overseen the Center for Environmental History, Culture & Technology (CHAT-Unicentro). He advises masters and doctoral students in the Graduate Program in Community Development (Unicentro), and in the Graduate Program in History (Universidade do Estado de Santa Catarina, UDESC). His research interests include agro-environmental history, Latin American environmental history, history of commodities, and environmental humanities. Currently, he coordinates a project on Agro-Environmental History of Apple Orchards and Climate Change, and a project on a multispecies history of bees and beekeeping. He is chief editor of *Esboços: Histórias em Contextos Globais*.

Chapter 13

Evelien de Hoop is based at the transdisciplinary Athena Institute (Vrije Universiteit Amsterdam, NL). With a background in Science and Technology Studies, she works with scholars from diverse disciplines and societal stakeholders to address contemporary health and sustainability challenges. Empirical foci include sustainable Dutch countrysides, Indian farmers' engagements with biofuels, and inclusive food systems in deprived urban areas across Europe. She mobilises research on the historical and transnational entanglements of contemporary challenges, with a particular focus on knowledge politics, to render her contemporary work historically-sensitive and pluralise ways of thinking about the future. She recently edited *Historicising Entanglements: Science, Technology and Socio-ecological Change in the*

Contributor biographies

Postcolonial Anthropocene (special issue of *Global Environment* 15 (2) (2022), with Aarthi Shridhar, Claiton M. da Silva and Erik van der Vleuten).

Erik van der Vleuten teaches at Eindhoven University of Technology and researches the connected histories of technology and socioecological change in an infrastructured world. Earlier work includes *Europe's Infrastructure Transition: Economy, War, Nature* (2015, with Per Högselius and Arne Kaijser); *Engineering the Future, Understanding the Past: A Social History of Technology* (2017, with Ruth Oldenziel and Mila Davids); and *Historicising Entanglements: Science, Technology and Socio-ecological Change in the Postcolonial Anthropocene* (special issue of *Global Environment* 15 (2) (2022), with Evelien de Hoop, Aarthi Shridhar and Claiton M. da Silva). His current research and education focus on diverse and contested sustainability histories and futures of distant-yet-connected regions across the globe. See https://www.eindhovenhistorylab.nl/erik-van-der-vleuten/

Chapter 14

Richa Kumar is Associate Professor of Sociology and Policy Studies at the Indian Institute of Technology, Delhi in the Department of Humanities and Social Sciences. Her research and teaching interests are in the Sociology of Agriculture, Food and Nutrition, Science and Technology Studies, and Rural and Agrarian Policy. She is a co-author of the *State of Rural and Agrarian India Report 2020* (http://www.ruralagrarianstudies.org/state-of-rural-and-agrarian-india-report-2020/) and her current research is on the impact of monoculture farming on farm systems, the environment and human health (see https://www.ted.com/talks/richa_kumar_our_food_and_our_future). Her book, *Rethinking Revolutions: Soyabean, Choupals and the Changing Countryside in Central India* was published by Oxford University Press in 2016. She is a recipient of the inaugural Elizabeth Adiseshiah Memorial Award (2019) and the New India Fellowship (2010). She completed her Ph.D. in the Science, Technology and Society Program at the Massachusetts Institute of Technology, Cambridge, USA.

Chapter 15

Vimbai Kwashirai is a Zimbabwean, Oxonian and Marie Curie scholar. He researches on the interface of eco-economic themes in Zimbabwe and Africa. Author of *Green Colonialism in Zimbabwe* and other works, he has published widely on ecological, socio-economic and political developments in Zimbabwe. His forthcoming book with Cambridge University Press is on election violence, human rights, politics and power in Zimbabwe. He has taught at several universities in Zimbabwe, the United Kingdom and Germany, winning several academic awards and fellowships in Europe. Based in Switzerland, his Marie Curie Horizon 2020 research analyses drought mitigation strategies in modern southern Africa, 1965–2022.

Chapter 16

Jose Muzlera is director of the Bachelor's Degree in Social Sciences at the National University of Quilmes (UNQ), Researcher of the Center for the Study of Rural Argentina at the National University of Quilmes (CEAR-UNQ) and Researcher of the National Council for Scientific and Technical Research (CONICET). Together with Alejandra Salomón, he is the editor of the *Diccionario del Agro Iberoamericano*. His research, always taking the Argentinean Pampean region as a territorial frame, has been linked to the dynamics of inheritance and social reproduction of farmers and contractors, to the effects of late modernity in the socio-productive framework and, in recent years, to the effects of the agribusiness model on the quality of life of those who inhabit the agro-territories.

PREFACE

Antônio Inácio Andrioli

WINNER OF THE 2020 BAVARIAN PRIZE FOR NATURE CONSERVATION

The construction of the global economic system has been going on for centuries. The emergence and development of world trade; the prioritisation of industrialisation in Global North countries; colonialism and semi-colonialism; the export of capital, its collapse and its complete reintegration into the world market are the most significant features. In colonial times, there were already attempts to integrate intercontinental regions into Empires, for which the definition of 'world domination' is more appropriate. This form of globalisation – conducted under the catchwords of 'freedom' and 'market economy' – subjugates nature to the form of merchandise and in the interest of profit and economic exploitation.

Soy production for export, and the effects thereof, illustrates well the context of the globalisation of capital in agriculture. Particularly since the end of the 1990s, Europe has increased the value of soy as a cheap source of protein for intensive animal husbandry, leading to a rapid increase in its importation. Furthermore, in order to cultivate large areas at low cost, the adoption of transgenics is being advocated, with consequences continuing to be evaluated only from a technical point of view, ignoring environmental, social and biosecurity problems.

The increasing dependence of the southern hemisphere on northern hemisphere countries has decisive political importance for the global debate on soy cultivation. The international division of labour is maintained and deepened. Developing countries concentrate on exporting raw materials, while industrialised countries deal in manufactured goods. Large multinational corporations seek to integrate 'as yet unaccounted for' natural resources and their economic potential into the capitalist market economy, exploiting these to accumulate capital. The mainstream agrarian structure continues to exist, with the Global South providing the raw materials while profits and economic power remain in the North.

THE AGE OF THE SOYBEAN: 1–3 **doi:** 10.3197/63800040695086.preface

Antônio Inácio Andrioli

This is about the exploitation of developing countries by new means. The monetary crises and indebtedness of the poorest countries lead them to a dead-end, increasing the dependence of the periphery on the economic centres. In the 'globalised' world economy, existing asymmetries between rich and developing countries, based on industrialised countries' high standard of living, are aggravated, externalising social and ecological costs. These dynamics generate new dependencies and the deepening of disparate power relations, both between countries and between actors within the developing countries themselves, concerning unequal access to the means of production and vital resources. Both the 'debt trap' and the resulting financial dependency of countries play a crucial role in the subjugation and growing fragility of developing countries' national economies.

Governments and companies in industrialised countries, as well as international organisations such as the World Bank and the International Monetary Fund, point out that, precisely because of their wealth of natural resources, developing countries have the chance to attract investments and increase their exports, thus improving their trade balance and honouring their foreign debts. In this scenario, soy cultivation is increasingly present in monoculture, using large areas and marginalising peasants and traditional populations, substantially impacting people's food sovereignty. Soy has been transformed into global merchandise, a commodity, expanding worldwide with support from governments and multinational corporations and leaving behind a trail of social and environmental destruction.

The monoculture of soy allows us to analyse the productive dynamics of broad economic sectors that benefit from it. It is a matter of producing protein at a low cost to supply an increasingly dependent world market. This aim requires subsidies, notable financing credits, the construction of an extensive storage system, marketing and transport structures in producing countries, and a growing industrialisation network in importing countries. The soybean's most famous derivatives, such as lecithin, bran and biodiesel, are responsible for the global supply of food markets. But its primary use has remained the same for decades: animal feed. How to measure its effects? How to analyse its history? How to assess its future?

The present book, divided into five parts, seeks to answer some of these questions, deepening the academic debate about a real problem with increasing global significance. The volume's intended challenge is to contribute to constructing a comprehensive global environmental history of soybean cultivation, with particular emphasis on the Great Acceleration period. This approach allows us to understand the main economic incentives that led to the worldwide expansion and commodification of soy. Particular emphasis is given to the resulting global food chains, the consequences of the increasingly destructive use of soil, the biotechnological manipulation of seeds, deforestation, the destruction of native forests, soils and biomes and the multiple hazards posed by pesticides (e.g. water contamination and health effects).

By dealing with different periods of soy production and its environmental impacts, the authors, writing in several corners of the planet, manage to convey crucial empirical and methodological contents to build a comprehensive analytical framework for assessing the main dilemmas of soybean environmental impacts in a globalised environment. The book's main reference is sustainability and its narratives, the idea of natural resources as a vehicle for a possible alternative future for humankind.

The book's conclusions inspire new questions that could spur further research on the environments and populations affected by international soybean trade. A crop with immense potential for food that could be a substantial substitute for growing worldwide meat consumption has become one of the greatest environmental and social villains. Its form of cultivation and growing incorporation into the logic of capital show how productive forces can be brought into play towards the destruction of nature.

INTRODUCTION: WRITING A GLOBAL ENVIRONMENTAL HISTORY OF SOYBEAN FARMING DURING THE GREAT ACCELERATION

Claudio de Majo and Claiton Marcio da Silva

The soybean is far more than just a versatile crop whose derivatives serve the protein needs of a meatless diet. One of the world's most important commodities, soy represents the embodiment of mechanised industrial agriculture and is one of the main actors behind the socioeconomic, political, and ecological transformations of industrial farming in several world regions. Snowballing soybean expansion has mobilised different social actors, with scientific research and the free market playing a decisive role in the grain's nutritional and industrial ubiquity. Extending this argument, we could say that we live in a world dependent not only on fossil fuels but also on soy. At the core of this dependency lies the combination of the market and scientific research that has allowed soy to increase productivity, adapt to less fertile soils and be used in a wide range of products: from oils and animal feed to mattresses and cars. Soybeans have been adapted to areas previously unthinkable, such as the cold regions of Russia and Great Britain and the North American state of North Dakota. The advances obtained by this combination of scientific research and market might in future mitigate the effects of harmful activities linked to soy. For example, it could potentially increase its contribution to global protein demand, balancing the impacts of the global livestock industry.

Following this line of thought, the problem is not soybeans themselves, but the high social and environmental price paid for the rapid expansion of the crop in the last sixty years – a 1,200 per cent total production increase since 1961.[1] This expansion has been at the expense of territories previously

---

1   See H. Ritchie and M. Roser, 'Soy', *Our World in Data*. Available at https://ourworldindata.org/soy (Accessed 29 June 2022).

THE AGE OF THE SOYBEAN: 5–15 **doi:** 10.3197/63800040695086.intro

dedicated to polyculture or cereals, and it continues apace, advancing over savannah and forest ecosystems. As recently as 2019, Eliseu Guarani-Kaiowá, a Brazilian indigenous leader, declared that 'our blood irrigates the soya consumed in Europe',[2] directly linking the expansion of soybean monocultures to the eviction and murder of indigenous peoples and other traditional populations. Currently, soy is the world's third most important commodity in production and export, totalling 350 million tonnes harvested annually (250 million of which are almost equally shared by Brazil and the United States, the world's leading producers). Although soy is an essential asset in the diets of vegans and vegetarians, less than twenty per cent of global production is destined to direct human consumption (mainly cooking oil, milk, and tofu). The remaining amount primarily serves as animal feed, fuelling the global livestock industry (about 77 per cent) and, to a lesser extent, for other industrial purposes (e.g., biofuels, lubricants, etc.).[3]

Considering the dramatic changes brought by soy expansion, this book discusses the global history of this expansion from an environmental perspective, through various theoretical-methodological lenses. The main chronological focus is a judgement call: the post-World War II epoch – what John McNeill and Peter Engelke have defined as the Great Acceleration.[4] In several world regions, this process has proved indissolubly linked to the rise of mechanised industrial agriculture with socio-economic, political and ecological consequences. Due to their versatility and resilience, soybeans have played an essential role in regional dietary regimes and rotational agricultural techniques. Since the 1950s, soybean farming has become one of the most valuable agricultural commodities on the global market.[5] As a result, several countries have heavily invested in biotechnologies to convert soybeans into cash crops. In particular, with the crop's adaptation to tropical climates since

---

2   R. Belincanta, '"Nosso sangue irriga a soja consumida na Europa", denuncia comitiva indígena em Roma', *RFI* 21 Oct. 2019: https://www.rfi.fr/br/brasil/20191021-nosso-sangue-irriga-soja--consumida-na-europa-denuncia-comitiva-indigena-em-roma (Accessed 29 June 2022).

3   Ritchie and Roser, 'Soy'.

4   On the Great Acceleration, see J. R. McNeill and P. Engelke, *The Great Acceleration. An Environmental History of the Anthropocene since 1945* (Cambridge, MA: Belknap Press of Harvard University Press, 2014). On the role of soybean during the Great Acceleration, see C.M. da Silva and C. de Majo, 'Genealogy of the Soyacene: The tropical bonanza of soyabean farming during the Great Acceleration', *International Review of Environmental History* **7** (2) (2021): 65–96.

5   On the historical appreciation of soybean's qualities see I. Prodöhl, 'Versatile and cheap: A global history of soy in the first half of the twentieth century', *Journal of Global History* **8** (3) (2013): 461–82. On the role of soy as an international commodity, see e E. Langthaler, 'The soy paradox: The Western nutrition transition revisited 1950–2010', *Global Environment* **11** (1) (2018): 79–104.

the 1970s, several countries of the Global South (especially Latin American ones) have specialised in expanding monocultural production. As a result, soy has rapidly replaced cereal fields, orchards and other spaces typically used for family agriculture and small-scale production in only a few decades. Moreover, soybean cultivation occurs at various scales; it is easier to find large farms in the top ten producing countries – Brazil, the United States, Argentina, China, India, Paraguay, Canada, Russia, Ukraine and Bolivia.[6] However, the advance of large soybean farms is also discernible in smaller producers such as Mexico, Mozambique, and Colombia. In these countries, the conjunction between favourable market prices and soybean's versatility and resilience has led local farmers to replace crops such as lentils, corn and vines, as well as animal breeding. The conjunction of these factors has turned Global South countries with extensive agricultural frontiers and the United States into the planet's ideal territories for soy production. Together, Argentina, Bolivia, Brazil, Paraguay, Uruguay, India, and Mexico produce more than half of global soya yields. Brazil and Argentina yield almost 160 million metric tons (MMT) combined. The other half is primarily occupied by the United States and China, among the world's top five producers.[7]

Controversially, soybean monocultures producing cash crop commodities have brought both positive and negative economic and socio-environmental consequences. On the one hand, soy commodities have brought economic prosperity to several peripheral territories, generating crucial sources of wealth for countries that have invested in social programmes, such as Brazil and Bolivia. The high dollar price against local currencies has encouraged the export of primary products such as soy and beef since the early 2000s. For economists and politicians, this investment choice towards agriculture and cattle ranching has relieved the effects of the 2008/2009 world economic crisis, generating revenues that are then invested in heating the domestic market. Revenues from the soybean sector, and from agribusiness in general, have also been used to sustain income transfer programmes and social policies in leading producing countries such as Brazil and Argentina.

On the other hand, the proliferation of soybean monocultures has ge-nerated controversial effects for society and ecology. As soy has swelled the

---

6   The Science Agriculture, *Top 10 Biggest Soybean Producers in the World*: https://scienceagri.com/top-10-biggest-soybean-producers-in-the-world/ (Accessed 29 June 2022).

7   For further information, see J. Karunga, '10 countries with largest soybean production', *World Atlas* (30 August 2018): www.worldatlas.com/articles/world-leaders-in-soya-soybean-production-by-country.html (Accessed 29 June 2022).

Claudio de Majo and Claiton Marcio da Silva

pockets of big landowners, the rise of agricultural monopolies has determined the forced eviction of landless populations and the dismantlement of their traditional lifestyle. Since its approval in Latin America during the 1990s, GM soy has overwhelmingly dominated the agricultural sector, increasing farmers' dependence on seed-producing multinational agribusinesses and affecting the biodiversity of entire bioregions. Although countries like India have maintained their cultivation from non-transgenic seeds, other activities leading to socio-environmental damage are generated by large-scale cultivation, such as use of pesticides or high consumption of water resources. While the dramatic consequences illustrated above are relatively well-known among environmental organisations and policymakers, they are only part of a complex global network that links western markets with powerful geoeconomic entities such as China and the European Union, while at the same time side-lining traditional and alternative forms of production.

Over recent years, the rapidly changing global and environmental dynamics characterising soybean farming have prompted scholarly attention. From East to West, soy as a research topic has mobilised researchers of various trends and approaches. In recent years, essential monographs on the subject have been published, such as Christine DuBois' *The Story of Soy* (Reaktion Books, 2018), which addresses the transformation of the soy crops from their first domestication efforts in the Far East to their ubiquitous presence on the global food-chain in the present. Equally important, in *Rethinking Revolutions: Soyabean, Choupals and the Changing Countryside in Central India* (Oxford University Press, 2016), Richa Kumar addresses, from extensive ethnographic and documentary research, the changes brought about by the advance of soybean in Central India. On Latin America, Matilda Baraibar Norberg's *The Political Economy of Agrarian Change in Latin America* (Palgrave McMillan, 2020) addressed the cases of Argentina, Uruguay and Paraguay, demonstrating the different actors that came together in the composition of the food chain connecting Latin America's Southern Cone to the global market, mainly for beef and soybeans. More specifically referring to Argentina, Pablo Lapegna's *Soybeans and Power: Genetically Modified Crops, Environmental Politics, and Social Movements in Argentina* (Oxford University Press, 2016) and Amalia Leguizamón's *Seeds of Power: Environmental Injustice and Genetically Modified Soybeans in Argentina* (Duke University Press, 2020) address the contestable rise of GM soy crops, a springboard for the introduction of GMOs all over Latin America.

While several recent publications have successfully unveiled the key historical drivers of the rise of soybean in the global food chain, the environmental impacts of this transition are an ongoing process that will continue permeating our society for years to come. The growing dependency of global markets' on the numerous soybean varieties yielded at different latitudes will undoubtedly determine the continuation of such international exploits. This volume is thus timely in tackling the challenge of exploring the biological and environmental transformations determined by the rise of soy in global food chains. Recent research efforts have begun to outline the first sets of thematic guidelines and potential historical narratives linked to the rise of this leguminous crop on the world stage. In attempting to move beyond the one-sided term *Soylandia*, these studies have proposed concepts such as *Sojización* and *Soyacene* to convey the complex political and techno-scientific entanglements and power relations brought by soybean's planetary expansion.[8] Yet, this comprehensive analytical framework still lacks the empirical corpus to assess the multiple economic, social, technological and environmental facets. The present book addresses this significant gap, proposing a wide-ranging global environmental history of soybean farming, mainly centred upon the Great Acceleration.

Building on the growing corpus mentioned above, this book represents the first attempt to address these challenges, proposing one of the most significant efforts to date at an environmental history of soybeans. Its pages gather a wide range of researchers of various nationalities, addressing broad regions and the dispersion, increased demand for, production and consumption of soy, with attendant socio-environmental effects. In unpacking the multi-scalar histories of soy farming, emphasising impacts on ecology and society, the research team that collaborated on this project includes environmental, social and economic historians, STS scholars, anthropologists, public health and policy managers and geneticists. Such a wide range of expertise has significantly enriched the traditional historical approaches with ethnographic and scientific contents.

As a result, the chapters in this book cover a wide geographical range, examining soybean histories in countries including Japan, China, India, Zimbabwe, Brazil, Argentina, Uruguay, Paraguay, Germany, the Nether-

---

8   On *Sojización*, see Matilda Baraibar Norberg, *The Political Economy of Agrarian Change in Latin America: Argentina, Paraguay and Uruguay* (Cham: Palgrave Macmillan, 2020). On the *soyacene*, see C.M. da Silva and C. de Majo, 'Towards the Soyacene: Narratives for an environmental history of soy in Latin America's Southern Cone', *Historia Ambiental Latinoamericana y Caribeña (HALAC)* **11** (1) (2021): 329–56.

lands and the United States; and simultaneously addressing both national and transnational or world/global histories. The volume is divided into five parts, addressing different moments of soy production and its environmental impacts. Through this vast selection of empirical and methodological content, the contributors strive to add historiographical nuance to previous scholarly efforts while at the same time hoping to inspire related research agendas.

Part I provides three comprehensive historical overviews of soybean cultivation before the Great Acceleration, mainly focused on Asia and the Americas. While chronologically situated outside our intended framework, these are necessary to help readers situate early soybean farming experiences in time and space and understand the multiple environmental challenges. Despite it being virtually impossible to retrace the complex genealogy of soy in the limited space of a few book chapters, the nuanced analyses featured in this volume allow us to partly solve what is perhaps the most complex task for any historian – namely what Marc Bloch would define as the 'myth of origins'. In the first chapter, Samira Peruchi Moretto, Eunice Sueli Nodari and Rubens Nodari trace a general overview of soybean's biogenetic characteristics, contributing to its worldwide success. Drawing from complex bio-genetic notions, they speculate on the origins of soy and the possible factors that allowed this resilient legume to adapt to different geographical and climatic contexts. Such an ambitious reconstruction is continued in the second chapter, where Brian Lander and Thomas DuBois propose a *longue durée* case study on soybean farming and its multiple nutritional uses in China, allegedly the cradle of soy domestication and the first region of the world where it became a staple crop. Intriguingly, most of the agricultural and environmental challenges brought by the expansion of soybean farming in the Chinese context will return in other geographical and historical contexts. The third chapter, authored by Rhuan Targino Zaleski Trindade, closes the circle, interlinking early soybean farming experiences in southern Brazil with Polish farmers who migrated overseas between the late nineteenth and early twentieth centuries. The adoption of this Asian crop among Polish communities, led by intellectual Ceslau Biezanko, constituted one of the first steps towards the rise of the crop in South America, today the world's primary production hub.

Part II discusses the economic dynamics that led to the global expansion of soybean during the Great Acceleration, emphasising global food chains. In the fourth chapter, Ernst Langthaler sets the tone of this discussion, looking at the process of agro-food globalisation witnessed by an essential

Introduction

crop commodity such as soy since the mid-twentieth century. Drawing from a vast set of empirical data, the author outlines the different phases that characterised soybean commodity networks from the 1950s to the present and their relations with dramatic global socioeconomic and environmental transformations. Similarly, in the fifth chapter, Matilda Baraibar Norberg looks at the main economic drivers that led to the progressive marketisation of soybean in Latin America – what several scholars from Spanish-speaking countries have defined as *sojización*. Using Karl Polanyi's seminal book, *The Great Transformation,* as a litmus test, the author explores the multiple historical market shifts and continuity that propelled *sojización*. While Baraibar Norberg primarily focuses on national market dynamics in Hispanic countries, the sixth chapter, authored by Cassiano de Brito Rocha, Kárita de Jesus Boaventura, Giovanni de Araujo Boggione and Sandro Dutra e Silva, looks at the dynamics of soy commodification in Brazil. In particular, it focuses on the economic region of Matopiba, the country's last soybean frontier belonging to the Cerrado savanna bioregion. Drawing from empirical sources and remote sensing detections, the chapter analyses land utilisation patterns linked to the soybean frontier expansion over the last fifty years and its projected future growth.

The following three sections complement the global reach of the previous one, delving into localised histories and experiences linked to the acceleration of soybean farming since the mid-twentieth century. The third part provides a fine selection of the different ecological milieus affected by the rising age of soybeans. In Chapter Seven, Eduardo Relly and Claudio de Majo explore the unknown igneous geographies of soybean in several bio-regions of Latin America. Although fire played a crucial role in developing millennial coevolutionary human-nature interactions in many of these territories, it also became one of the main tools propelling the tropicalisation of soybean farming. The result was an indissoluble relationship between the rise of soybean monocultures and fire-based forest clearing and destruction. While fire contributed to the unfolding of the age of soybeans, water supplies were progressively affected by the same phenomenon. As shown by the eighth chapter, authored by Larissa de Lima Trindade, John Hoornbeek, Mutlaq Albugmi, Joshua Filla and Rodrigo Fortunato de Oliveira, the massive growth of soybean has provoked dramatic water contamination issues both in Brazil and the US. Drawing from a large corpus of scientific literature, the authors also explicitly link the risks of water contamination with adverse health effects on the rise. The section closes with chapter nine, where Enrique

Claudio de Majo and Claiton Marcio da Silva

Antonio Meija discusses the impact of intensive soybean farming on the soil nitrogen cycle of the Argentinean Pampas. Adopting an original theoretical framework, the chapter looks at the often-unaccounted dynamics of ecological exchanges permeating the global nitrogen trade and their hidden socioeconomic costs.

Just as natural milieus have historically borne the mark of intensive soybean farming, this leguminous crop has dramatically reshaped nutritional habits and technologies. Part IV addresses these transformations in several world regions throughout the last decades. In Chapter Ten, Anna Zeide illustrates the multiple uses of soybeans in Northern America from the 1950s, often unrecognised in contemporary scholarship. As the author demonstrates, the incredible versatility of soy allowed it to become a mainstay of American diets, utilised as a core of different food production processes and animal nutrition. Meanwhile, countercultural movements also increasingly adopted soy as a meat and dairy substitute. While soy entered American diets through the back door, the eleventh chapter demonstrates how scientific soybean imaginaries influenced public opinion and production choices in the European continent. By focusing on the example of Germany, Janina Priebe brings to light scientific visions and socio-technical imaginaries linked to soy throughout the twentieth century. As the author demonstrates, these imaginaries provided the breeding ground to transform soy into an age-defining commodity, inaugurating a trend whose social, economic and environmental developments would accelerate from the 1950s onwards. Moving to the southern hemisphere, Jo Klanovicz's chapter provides a fascinating example of the biotechnological innovations that allowed the subtropical regions of southern Brazil and Paraguay to become world-producing leaders. However, as Klanovicz demonstrates, such leadership comes with a price, as local soybean agro-landscapes entail radical shifts in labour, land, capital and technology relations. The immediate consequence of this revolutionary set of transformations is perhaps best reflected in present debates opposing advocates of genetically modified soybean breeds and defenders of organic farming. However, such discussions do not belong just to the southern hemisphere. As the chapter authored by Erik van der Vleuten and Evelien de Hoop demonstrates, sustainability debates have permeated public discourse over the last five decades in the Netherlands. Drawing from a vast array of studies, the authors argue for the historiographic need to incorporate discussions addressing the relationship between soybean consumption at a

Introduction

regional level (both for human diet and animal feeding) and the different sustainability narratives that arose simultaneously.

The fifth and final part of the volume delves into the regional environmental histories of soybean farming over three continents. The fourteenth chapter, authored by Richa Kumar, looks at the ambivalent effects of soybean farming in the central Indian region of Malwa in the state of Madhya Pradesh, the nation's production core where soybeans are already a consolidated cash crop. Adopting a mixed historiographic and ethnographic approach, Kumar demonstrates the increasingly pervasive role of soybean monocultures in the Indian domestic commodity market and their localised socio-environmental downsides, shedding light on a context that conventional narratives have often omitted. Similar issues are also indirectly raised by Vimbai Kwashirai in his chapter on Zimbabwe. As the author notes, so far, soybean farming experiences in Zimbabwe uniquely depart from other experiences. The adoption of soybean as a subsistence rather than a cash crop has broadly benefited local farmers, improved their daily diets and significantly ameliorated crop rotation regimes, especially for nutrient-demanding staple crops such as maize. However, as Kwashirai notes, such an eminently positive experience is also justified by the relatively slow uptake of soybean crops over the last century. This scenario leaves the potential consequences of soybean marketisation still unclear. Much more tangible are the socio-environmental implications of soybean farming in the Buenos Aires Pampean region, as described by José Muzlera. Through a mixed methodology adopting historical records and structured interviews, the chapter describes the socio-environmental impacts of the dominant agribusiness model. Looking at its specific effects on local economic activities and land tenure patterns, Muzlera draws a nuanced picture of this productive paradigm's dynamics and how it affects individual and collective wellbeing.

Although each one deals with the history of soybean farming in its own unique way, the chapters together convey a comprehensive historical picture of the social, environmental and economic implications of the rise of soy on the world stage. Nonetheless, further scholarly endeavours could, of course, potentially enrich the literature on the topic, adding depth and nuance to the themes tackled here. Possible future research lines would tackle the relationship between the growth of soybean farming and another global behemoth: the livestock industry. While the multiple interrelations between these two industrial sectors are relatively well-known, systematic historical

14

Claudio de Majo and Claiton Marcio da Silva

reconstructions, both locally and globally, have only been partly undertaken.[9] Another potentially engrossing point would concern the changing energy inputs fuelling the soybean farming industry (e.g., oil, gas, coal) and the proportional outputs generated by the production of biofuels. These two research trajectories could potentially converge in studies linking soybean industrial products and patterns of land grabbing and environmental contamination at a local level.

Overall, this volume lays vital foundations for understanding and tackling global soy cultivation issues. Its contributions offer a historically-grounded analysis of some of the most distinctive historical transitions that have made soybean one of the world's major commodities and of the socio-environmental consequences of such expansion. Besides stimulating further scholarship and new research lines, as indicated above, this volume aspires to provide intellectual grounding for more nuanced discussions about the challenging future of commodity markets and modern agriculture, two sectors where soy has played – and will undoubtedly continue to play – a significant role.

ACKNOWLEDGEMENTS

We would like to acknowledge the institutions that supported this project, which originated from a research project linked to the Environmental History Laboratory of the Federal University of the Southern Frontier (UFFS) and funded by the Brazilian National Council for Scientific Research (CNPq). In 2019, we had the opportunity to gather part of the group during the 3rd World Congress of Environmental History in Florianopolis (Brazil). Gradually, new collaborators joined the project, contributing to our volume with new chapters and resources. Therefore, besides the UFFS, this book includes researchers from the Federal University of Santa Catarina (UFSC), the State University of the Midwest of Paraná (Unicentro), the State University of Goiás (UEG), and the Evangelical University of Goias (Unievangélica), all in Brazil. Other institutions include the Rachel Carson Center for Environment and Society of the Ludwig-Maximilians University of Munich and

9   For a historical perspective on Uruguay, Argentina, and Paraguay, see Baraibar Norberg, *The Political Economy of Agrarian Change in Latin America*. For a discussion of this relation today in Brazil, see Rebecca Lima Albuquerque Maranhão et al., 'The spatiotemporal dynamics of soybean and cattle production in Brazil', *Sustainability* **11** (7) (2019): 2150; and Raoni Rajão et al., 'The rotten apples of Brazil's agribusiness', *Science* **369** (6501) (2020): 246–48.

Introduction

the Friedrich Schiller University of Jena in Germany, the Johannes Kepler University of Linz in Austria, the University of Basel in Switzerland, the Eindhoven University of Technology and Vrije Universiteit of Amsterdam in the Netherlands, Stockholm University and Umeå University in Sweden, Kent State University, Virginia Tech and Brown University in the USA, Beijing Normal University in China, the Indian Institute of Technology in Delhi, India, and the National University of Quilmes in Argentina. We also thank the Federal University of the Southern Frontier (UFFS), CNPq and the Research and Innovation Support Foundation of Santa Catarina State (FAPESC), the Johannes Kepler University of Linz, the Eindhoven University of Technology, Stockholm University, Umeå University, Virginia Tech and Brown University in the USA for their financial support to the open-access publication of this volume. All the contributors to the volume worked with commitment, participating in a dynamic international team of scholars from all over the world. Finally, we thank Sarah Johnson and James Rice at The White Horse Press for believing in our project and making it possible, lending their dedication and professionalism to the cause.

Part I

Soybeans Before the Great Acceleration

CHAPTER 1.

SOYBEAN AS A CRITICAL GENETIC RESOURCE: DOMESTICATION, DISSEMINATION AND INTRODUCTION OF *GLYCINE MAX*

Samira Peruchi Moretto, Eunice Nodari and Rubens Nodari

The Soybean (Glycine max)

Cultivated soybean (*Glycine max* (L.) Merr.) is an important crop as a main source of dietary protein and oil for animals and human beings. In 2018, the global production of soybeans reached 347 million tons, 87 per cent of this in the Americas.[1] There is a consensus about the wild type from which the cultivated soybean was domesticated (*Glycine soja* Sieb. & Zucc.). The species *G. soja* is naturally distributed throughout East Asia, which includes China, Korea, Japan and part of Russia.[2] The most robust evidence suggests that domestication occurred in the Northern part of East China.

Vavilov was a Russian botanist and one of the first geneticists to describe diversity of plant type and forms of cultivated populations of domesticated species. In addition, he used the high diversity of type and forms of the plants as a criterion to define the species' centre of origin, because the domestication process in this location has numerous morphological types intermediate between derivatives and wild types, and because wild populations continue to hybridise with cultivated populations.[3] According to Wang et al., scientists have been developing many studies in the last few decades using different data resources, including the distribution of wild soybean, and historical records and archaeological findings.[4] Thus, different hypotheses as to the centre of

---

1   FAO, *World Food and Agriculture – Statistical Yearbook 2020* (Rome: FAO, 2020).

2   T. Hymowitz, 'Soybeans', in N.W. Simmonds (ed.), *Evolution of Crop Plants* (Oxford: Willey, 1976), pp. 159–82.

3   N.I. Vavilov, *Five Continents* (Rome: International Plant Genetic Resources Institute, 1997).

4   L. Wang, F. Lin, L. Li, W. Li, Z. Yan, W. Luan, R. Piao, Y. Guan, X. Ning, L. Zhu, Y. Ma, Z. Dong, H. Zhang, Y. Zhang, R. Guan, Y. Li, Z. Liu, R. Chang and L. Qiu, 'Genetic diversity center

THE AGE OF THE SOYBEAN: 19–28 **doi:** 10.3197/63800040695086.ch01

Samira Peruchi Moretto, Eunice Nodari and Rubens Nodari

origin of soybean have been proposed. These include the Yellow River valley, the lower reaches of the Yellow River, North China, South China, and even multiple centres of origin simultaneously.

Soybean is a plant belonging to the *Fabaceae* family, which also comprises plants such as beans, lentils and peas. The word soy originates from an oriental name for soy sauce.[5] There are many collections of cultivated, semi-domesticated and wild types of soybean accessions conserved in germplasm banks across the world. The latest survey done by FAO indicated that there were 227,944 accessions, 14 per cent stored in the Chinese Germplasm Bank.[6] The US National Plant Germplasm System (NPGS) maintains a collection with more than 9,000 species.[7] Among them are more than 35,000 accessions of both *G. max* and *G. soja*. In Brazil, the National Centre of Genetic Resources and Biotechnology (CENARGEN/EMBRAPA) conserves *ex situ* 370,066 accessions from more than 2,330 species.[8] The genus with the highest number of preserved accessions at CENARGEN is *Glycine* with approximately 55,000 accessions.[9]

The diversity found in domesticated soybean is the result of over 3,000 years of cultivation in which Chinese farmers selected more than 20,000 landraces (defined as cultivars that predate scientific breeding). The extensive range in phenotype embodied in landraces today is the result of the slow spread of soybean throughout geographically diverse Asia (China first, then Korea and Japan), the continual occurrence of natural mutations in the crop and both conscious and unconscious selection for local adaptation.[10]

of cultivated soybean (Glycine max) in China – New insight and evidence for the diversity center of Chinese cultivated soybean', *Journal of Integrative Agriculture* **15** (11) (2016): 2481–87.

5   T. Hymowitz and C.A. Newell. 'Taxonomy of the genus Glycine, domestication and uses of soybeans', *Economic Botany* **35** (3) (1981): 272–88.

6   FAO, *The State of the World's Plant Genetic Resources for Food and Agriculture*, Second Report (Rome: FAO, 2010).

7   ARS, *Center for Agricultural Resources Research* (US: USDA, 2021). https://www.ars.usda.gov/ plains-area/fort-collins-co/center-for-agricultural-resources-research (accessed 12 Sep. 2021).

8   EMBRAPA, *Recursos Genéticos e Biotecnologia* (Brazil: EMBRAPA, 2021) https://www.embrapa. br/recursos-geneticos-e-biotecnologia (accessed 12 Sep. 2021).

9   *Recursos Genéticos Vegetais para a Alimentação e a Agricultura no Brasil. Relatório Nacional - 2012 a 2019*, (Brasília, DF: Embrapa, 2021).

10   T.E. Carter, T. Hymowitz and R.L. Nelson, 'Biogeography, local adaptation, Vavilov, and genetic diversity in soybean', in D. Werner (ed.), *Biological Resources and Migration* (Berlin, Heidelberg: Springer, 2004), pp. 47–59.

1. Soybean as a Critical Genetic Resource

About ninety per cent of soybean yield is used for oil production.[11] Besides its importance as a source of protein and oil, soybean also contributes to nitrogen fixation in the soil.[12] The characteristics of the grain depend on the variety, and the size may be from 1 to 3.5 centimetres.[13]

Soybean can also be processed into various forms of food, from soy sauce to tofu. Immature green beans and sprouts also eaten. Seeds contain 18–23 per cent oil and 39–45 per cent protein,[14] which are used in various forms and products. Most of the meal is used as a high protein animal feed, mainly to produce eggs, poultry and pork.[15] Currently the main producers of soybean are the United States, Brazil, China, Argentina and India.

Domestication and Dissemination

Studying the domestication of a plant species is a dynamic process that goes beyond borders and points to the importance of humans in enacting landscape changes. Assuming that the environment is constantly changing and that humans are part of it, continuous recombination is possible. Genetic resource management practices were developed in this context. Plant domestication is a very vivid example of this association. The mapping of plant species is very complex, because they happened in dominant societies, which left us records. According to Hymowitz and Shurtleff, historical and popular literature concerning soybean is replete with factual inaccuracies that keep recycling from one publication or website to another without documentation.[16]

Considering that the investigation of the records from past societies is part of the process of identifying the domestication of plants, Hymowitz and Shurtleff[17] bring evidence against the statement that the book *Pen Ts'ao Kong Mu* presents the first written record on soybean. It is common to find references to Emperor Shennong, considered the Father of Agriculture, as

11   J.G. Vaughan and P.A. Judu, *The Oxford Book of Health Foods*, Current Online Version 2009 (Oxford University Press, 2003).

12   G. Chung and R.J. Singh. 'Broadening the genetic base of soybean: A multidisciplinary approach', Critical Reviews in Plant Sciences **27** (5) (2008): 295–341.

13   T. Sorosiak, 'Soybean', in K.F. Kiple and K.C. Ornelas (eds), *The Cambridge World History of Food* (Cambridge University Press, 2000), pp. 422–27.

14   Hymowitz, *Soybeans*, p. 60.

15   Ibid.

16   T. Hymowitz and W.R. Shurtleff. 'Debunking soybean myths and legends in the historical and popular literature', *Crop Science* **45** (2) (2005): 473–76.

17   Hymowitz and Shurtleff, *Debunking Soybean Myths*.

the author of the book, in which soybean appears among the five main grains of Chinese civilisation. However, according to Hymowitz and Shurtleff,[18] the myths about Emperor Shennong are attributable to Han historians, since other relevant researchers do not mention their stories. Furthermore, six different publication dates are found, all between 2838 and 2383 BCE, which differs from the dating system that existed in China until 841 BCE.

The study developed by Ping-Ti Ho,[19] based on archaeological, botanical, historical and philological evidence, pointed out that the origins of Chinese agriculture are different from those of other agricultural systems of the Old World. One of the differences highlighted by Ho is the absence of leguminous plants rich in protein, even though the existence of wild species of soybean, especially in the north of the Yangtze, indicates that the plant is indigenous to China. The lack of protein is also implied as one of the reasons for the malnutrition presented in skeletons. In addition, the importance of hunting for protein in a society in which the agriculture of other plants had already dominated for millennia leads to the fact that domestication of soybean was belated, or domesticated types were not widespread or the issue was not enough scrutinised. Yet, carbon dating research by Hymowitz and Shurtleff indicates that, contrary to the idea that soybean was one of the oldest crops grown by humans, at least thirty other crops were domesticated before soybean.[20] The domestication of soybean was thought to have taken place around 3,100 years ago. However, recent molecular evidence suggests that in fact it was probably between 9,000–5,000 years ago.[21]

As for its geographical context, pollen profile studies and geographic evidence suggest that soybean was first domesticated in the low plains of North China. Both Vavilov and Harlan indicate the eastern north region of China as the main centre of domestication of soybean.[22] The plant 'usually requires a long growing season with a plentiful water supply'.[23] The time that elapsed before the successful domestication of soybeans indicates that the

18   Ibid.
19   P.T. Ho. 'The loess and the origin of Chinese agriculture', *The American Historical Review* **75** (1) (1969): 1–36.
20   Hymowitz and Shurtleff, *Debunking Soybean Myths*.
21   E.J. Sedivy, F. Wu and Y. Hanzawa. 'Soybean domestication: the origin, genetic architecture and molecular bases', *New Phytologist* **214** (2) (2017): 539–53.
22   N. Vavilov, *Studies on the Origin of Cultivated Plants* (Leningrad, 1926) and J.R. Harlan, *Crops and Man* (Madison: American Society of Agronomy, 1975), p. 214.
23   Ho, *The Loess*, p. 28.

ancient Chinese agricultural system was the result of a long process of trial and error.[24] The migration of soybean from the northern to southern China and to Korea, Japan and southeast Asia countries probably took place between 3100 and 2300 BP.[25] According to Thomas Sorosiak,[26] the dissemination of soybean to other Asian regions was influenced by the missions of Buddhist priests in 1440 BP. Soldiers, merchants and travellers also contributed to its introduction.

Soybean's potential aroused European attention by 1712, through the publications of botanist Engelbert Kaempfer, who lived in Japan two decades before.[27] The author also reported that, in Austria, Prof. Frederick Haberlandt had attempted to expand soybean in the region, with no success. Despite that, soybeans remained relatively unknown in Europe until the end of the seventeenth century, when travellers started to describe the bean in their diaries and letters. In the eighteenth century, scientific interest motivated the introduction of the soybean in Europe, in botanical gardens and experimental stations[28]. Like most of foreign plants, soy was grown for botanical and taxonomy purposes, in botanical gardens in the Netherlands, Paris and United Kingdom. The first records of agricultural production in Europe were in Croatia, Romania, Czechoslovakia and Austria.[29] However, it was after its introduction in the Americas that soy became a major actor on the agricultural world stage.

Soy in the Americas

Soy first reached North America in 1765, with seeds shipped from England into Savannah, Georgia, by Samuel Bowen.[30] Cultivated grains were processed into soy sauce, vermicelli (soybean noodles) and a sago powder (substitute for that made from sweet potatoes), which were then exported to England. Since then, the United States of America has imported soybean germplasm directly from China. In 1908, soy was imported from Northeast China, and Manchuria as well, where it rapidly became the region's cash

24   Ibid.

25   Hymowitz, *Soybeans*, p. 160.

26   Sorosiak, *Soybean*.

27   Hymowitz, *Soybeans*, p. 160.

28   Ibid., p. 161.

29   V. Đorđević, 'Carte Blanche: Soybean, the Legume Queen', *Legume Perspectives* 1 (2013): 4

30   T. Hymovitz and J.R. Harlan. 'Introduction of soybean to North America by Samuel Bowen in 1765', *Economic Botany* **37** (4) (1983): 371–79.

crop.[31] As a result, soybean became a cash crop first in the US, and later in other American countries. The fundamental change from staple food status in China was provided by a set of national and transnational actors with distinct interests in the global and national spread of soybeans.[32] Among the botanical expeditions to Asian countries organised by the Bureau of Plant Industry of United States Department of Agriculture (USDA) agency, the one carried out by P. Howard Dorse and William J. Morse, from 1929 to 1931, collected 4,500 different types of soybean out of 9,000 distinct specimens for agricultural experiments.[33]

In 1922, German botanist Hermann Bollmann separated lecithin from soy oil, propelling the large-scale production of soy-derived oils.[34] This aroused the interest of countries such as the United States. Overall, the combination of agricultural technologies on the rise, the search for species with high commercial value and plenty of physical space turned the US into one of the largest world producers of the twentieth century. As global soy demand continued to increase over the next few decades, its uses multiplied. In the following years, experiments evaluating the nutritional value of soybeans multiplied, driven by the influential experiments of George Washington Carver at the Tuskegee Institute (Alabama), together with Lafayette Mendel and Thomas Osborne. In the early 1930s, the western world showed a growing interest in researching and breeding varieties towards soybean improvement, mainly in the United States, whose production rates were still lower than those of China, Japan and Korea. The mass spread of soybean cultivation in the United States was characterised by a wide variety of crops from different regions of Asia. In 1939, Morse and Cartter described 108 soybean cultivars in the United States.[35]

The first soybean cultivation reference in Brazil dates back to 1882, when some genotypes were experimentally introduced in the State of Bahia by

31   I. Prodöhl, 'Versatile and cheap: a global history of soy in the first half of the twentieth century', *Journal of Global History* **8** (3) (2013): 461–82, at 462

32   Ibid.

33   Ibid., 476.

34   G.R. List, *The History of Lipid Science & Technology* (AOCS Lipid Library, 2021). https://lipidlibrary.aocs.org/resource-material/the-history-of-lipid-science-and-technology/hermann-bollmann-(1880-1934)-bruno-rewald-(1882-1947)-heinrich-buer-(1875-1962)-stroud-jordan-(1885-1947)-percy-julian-(1899-1975)-joseph-eichberg-(1906-1997) (accessed 12 Sep. 2021).

35   Ibid.

agronomist Gustavo Dutra.[36] Dutra referred to soy as 'Chinese Beans'.[37] Nevertheless, there are also reports that Japanese people who migrated to São Paulo reintroduced accessions of soybean and tried to cultivate the grain. The Japanese brought some varieties of soybeans to Brazil from 1908. However, it was only in 1914 that acclimatisation was successful in the municipality of Santa Rosa, state of Rio Grande do Sul.[38] At the time, the main use of soybeans in Brazil was as forage. In the 1930s, studies on soybean started being developed in the former Colony Phytotechnical Experimental Station.[39]

Until the 1960s, the Brazilian southern states were the main soybean producers of the country, with Santa Catarina and Paraná also beginning to grow the crop from the 1950s. At first, the introduction of this crop was due to migration of farmers from Rio Grande do Sul to the west and Vale do Rio do Peixe regions, and later it was motivated by its use as green manure in coffee plantations.[40] In the same decade, Rio Grande do Sul had three different regions leading in soybean production: the region of Missões, first for self-consumption and later for export too; the Alto Uruguai region, as pig manure; and the Planalto Médio, linking soybean farming to the process of agricultural mechanisation already ongoing in wheat crops.[41]

In other tropical regions of Brazil, soybean cultivation was allowed by technological innovations, which included experiments with different strains, adapting the cultivation to lower latitudes. Many research centres were involved in this process, with governmental and private efforts, such as Instituto de Pesquisa Agropecuária do Sul (IPEAS), Centro Nacional de Pesquisa da Soja (CNPSo), also called Embrapa/Soja, and many others. With the development of new cultivars, the 1970s marked the expansion of soy to the Cerrado,[42] while the already-mentioned regions of Rio Grande

36   M.A.D. Queiróz, C.O. Goedert and S.R.R. Ramos (eds), *Recursos genéticos e melhoramento de plantas para o nordeste brasileiro* (Brasília: Embrapa, 1999), p. 129.

37   L.P. Bonetti, 'Distribuição da soja no mundo', in S. Miasaka and J.C. Medina (eds), *A soja no Brasil* (Campinas: Seção de divulgação do Instituto de Tecnologia de Alimentos, 1981), pp.1–6.

38   Ibid.; F.J.B. Reifschneider, G.P. Henz, C.F. Ragassi, U.G. Anjos and R.M. Ferraz, *New Perspectives on the History of Brazilian Agriculture* (Brasília, DF: Embrapa, 2012), p. 75.

39   F. Teresawa, J.M. Teresawa and M.M. Teresawa. 'FT Sementes and the expansion of soybeans in Brazil', in F.L. Silva et al. (eds), *Soybean Breeding* (New York: Springer Publisher, 2017).

40   Ibid.

41   O.A. Conceição, A expansão da soja no Rio Grande do Sul – 1950–75. (Ph.D. Thesis, Fundação de Economia e Estatística (RS), 1984).

42   Teresawa et al., *FT Sementes.*

do Sul were already consolidated soybean producers.[43]

Why has soy become the most economically important crop in Brazil? The success of genetic improvement carried out in Brazil can be indicated by the value of the average genetic gain, which was close to 0.9 per cent per year. In addition to improving the productive potential per se, two other major contributions of soybean genetic improvement in Brazil can be highlighted.[44] The first was the adaptation of soybeans to low latitudes (for crops in the Amazon region) through the introduction of genes for a 'long juvenile period' in Brazilian germplasm.[45] This was the starting point for the spread of soy culture to the Brazilian Cerrados. The second contribution, supporting the first, was the various works in improvement of genetic resistance to the most expressive diseases of the crop.[46] However, soy cultivation in Brazil is still responsible for more than a third of all pesticides consumed in the country.[47]

Soybean farming also gained importance in Argentina from the 1970s. At first, soybean crops in the country were located in the humid Pampas region, as with wheat. However, as in Brazil, it later expanded to other agricultural areas, reaching much of northern Argentina.[48] The Chaco region of Argentina experienced a very intense expansion of soybean production in the 1990s, with the development of transgenic seeds. In addition to the environmental and social impact, the expansion of soy has also led to a decrease in more traditional crops.[49] Figure 1 shows the main soybean production areas of Brazil and Argentina.

Interestingly, phenotypic analysis of modern Chinese and North American cultivars follows the same diversity patterns.[50] Pedigree analyses of Latin American breeding programmes, although incomplete, show that

---

43   M. Gerhardt. 'Uma história ambiental da modernização da agricultura: o norte do Rio Grande do Sul', *Revista História: Debates E Tendências* **16** (1) (2016): 166–80.

44   F.R. Ferreira and S.P. Silva Neto, 'Exemplos Bem-Sucedidos de Introdução e Aclimatação de Plantas no Brasil', in R.F.A. Veiga and M.A. Queiróz (eds), *Recursos Fitogenéticos: a base da agricultura sustentável no Brasil* (UFV, 2015), pp. 148–59.

45   E.E. Hartwig and R.A.S. Kiihl. 'Identification and utilization of a delayed flowering character in soybean for short-day conditions', *Field Crops Research* **2** (1979): 145–51. L.A. Almeida and R.A.S Kiihl, 'Melhoramento da soja no Brasil – desafios e perspectivas', in G.M.S. Câmara (ed.), *Soja: Tecnologia da Produção* (Piracicaba: USP-ESALQ, 1998), pp. 40–54.

46   Ferreira and Silva Neto, *Exemplos Bem-Sucedidos.*

47   Ibid.

48   S. Gómez Lende. 'El modelo sojero em Argentina (1996–2014), um caso de acumulación por desposesión', *Mercator* **14** (3) (2015): 7–25.

49   A. Zarrilli. '¿Una agriculturización insostenible? La provincia del Chaco, Argentina (1980–2008)', *Historia Agraria* **51** (2010): 143–76.

50   Carter, Hymowitz and Nelson, *Biogeography, Local Adaptation.*

1. Soybean as a Critical Genetic Resource

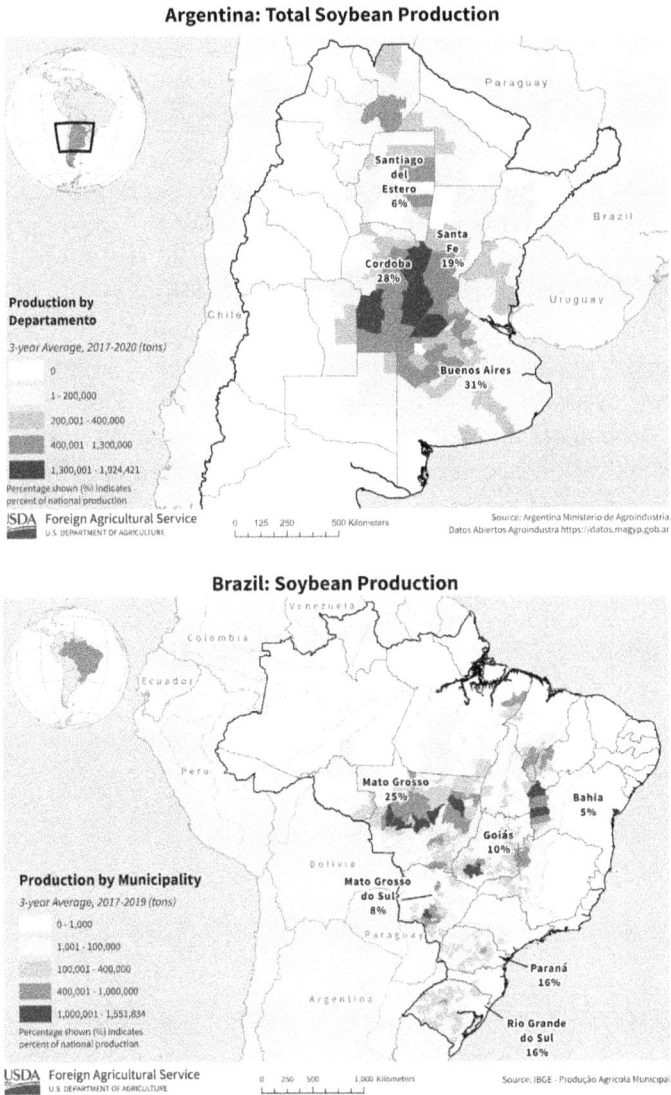

Figure 1.

Zones of soybean production in Brazil and Argentina.

Sources: USDA, *Brazil – Crop Production Maps* (USA, Foreign Agricultural Service, 2021) https://ipad. fas.usda.gov/rssiws/al/crop_production_maps/Brazil/Municipality/Brazil_Soybean.png (accessed 11 Sept. 2021); USDA, *Southern South America Crop Production Maps* (USA, Foreign Agricultural Service, 2021) https://ipad.fas.usda.gov/rssiws/al/crop_production_maps/ssa/AR_Delegation/Argentina_ Total_2017_20_Soybean.png (accessed 11 Sept. 2021).

28

Samira Peruchi Moretto, Eunice Nodari and Rubens Nodari

these programmes are derived primarily from a subset of North American breeding stock and are thus likely to be less diverse than the North American breeding programme. They argue that, though conscious breeding choices, the high economic costs of breeding and historical factors can be used to explain the reduced diversity in breeding programmes outside of China compared to within. It is important to note that these results, obtained from modern breeding programmes, are consistent with (1) Vavilov's principle of crop domestication, which states that genetic diversity will be greatest at the centre of domestication (China in the case of soybean); and (2) the concept of Darwinian genetic drift which can be used to infer that genetic relatedness or uniformity will increase within breeding populations that are derived from relatively few founding members.[51] A precaution gleaned from the observed trend in diversity is that all soybean breeding programmes outside China, regardless of the phenotypic superiority of their genetic breeding materials, should be examined to determine the adequacy of genetic diversity.[52]

51   Ibid., p. 47.
52   Ibid.

CHAPTER 2.

A HISTORY OF SOY IN CHINA: FROM WEEDY BEAN TO GLOBAL COMMODITY

Brian Lander and Thomas David DuBois

Introduction

In under a century, soybeans have become one of the world's predominant crops, transforming economies and ecologies around the globe. Soy has spread so rapidly in the Americas that it feels to many like a new crop, but today's global soybean was created through millennia of cultivation and processing in East Asia.[1] Generations of farmers domesticated wild beans and developed varieties suited to specific conditions and uses. They also created techniques to transform soybeans into food products: cooking oil, a variety of condiments and the endlessly versatile tofu. For most of their history, soybeans were grown on small farms and often on marginal land. Only in the twentieth century did farmers around the world begin to grow soy as a large-scale monoculture and only in the past few decades has it become one of the world's most traded commodities.

Today much of world's soybean crop is fed to livestock, but the main importance of soy in China was historically as food for humans. Instead of eating unprocessed soybeans, people turned them into more digestible foods and learned to press them to extract the oil, which is now a key part of industrial soy processing. The story of how people learned to convert soybeans into tasty high-protein foods is relevant for thinking about the ecological impact of modern agriculture. Soybeans are sometimes vilified because they occupy so much of the world's land, but most of those beans are fed to animals that are later fed to humans, an extremely inefficient way to produce food. It is to

---

1   Global soybean production ballooned from 27 million tonnes in 1961 to 353 million tonnes in 2017, and most of that expansion has been in the Americas. For useful visualisations, see H. Ritchie and M. Roser, 'Soy', at the *Our World in Data* https://ourworldindata.org/soy (Accessed 28 July 2021).

THE AGE OF THE SOYBEAN: 29–47 **doi:** 10.3197/63800040695086.ch02

Brian Lander and Thomas David DuBois

support its growing meat industry that China has recently emerged as the world's largest consumer of soybeans. Yet, while the practice of producing meat by feeding soy to farmed livestock is extremely inefficient, Asian soy foods are very efficient sources of food and have the potential to help humans live more sustainably as our populations continue to grow.

Domestication

The soybean (*Glycine max*) was domesticated from a wild species (*G. soja*) whose range extends from Afghanistan in the west to Japan in the east, and from southern Siberia in the north to the subtropics of the Yangzi River valley. Like many members of the bean family, soy plants have a symbiotic relationship with bacteria that allows their roots to ingest atmospheric nitrogen. Because of this, the plants add nitrogen to soil, giving farmers a strong incentive to rotate them with other crops. Soybeans are annuals, dropping their seeds and dying at the end of each growing season, and they grow vigorously in disturbed habitats, a trait commonly described as 'weedy'. These types of plants established themselves in areas on the fringes of ancient human settlements and as weeds with other cultivated crops, and were relatively easy to cultivate for their seeds, thus encouraging people to find a use for them. While wild soybeans tend to grow as vines that flourish in sunny open areas (they are still common weeds in China's farms), the domestication process transformed the plants into separate erect stems that stop growing once they produce bean pods. Since people tended to collect the beans in intact pods, the process of domestication selected the varieties whose pods stayed closed.[2]

Soybeans were domesticated in what is now North China by farmers whose main crops were varieties of millet.[3] Soybeans probably began their career in agriculture not as valued crops, but as weeds that could be eaten if necessary. The earliest excavated soybeans come from sites over 8,000 years old, a time when stable agricultural communities were just beginning to form. As far as archaeologists can tell, these soybeans are no different from

2   G.A. Lee et al., 'Archaeological soybean (*Glycine max*) in East Asia: Does size matter?', *PloS One* **6** (11) (2011): 1–12; For the bigger picture, see G.W. Crawford, 'Domestication and the origins of agriculture in China', in D. Bekken, L. Graumlich, and G. Feinman (eds), *China: Visions through the Ages* (Chicago: The University of Chicago Press, 2018), pp. 45–63.

3   Lee et al., 'Archaeological soybean in East Asia'; G.A. Lee et al., 'Plants and people from the early Neolithic to Shang periods in North China', *Proceedings of the National Academy of Sciences* **104** (3) (2007): 1087–92.

2. *A History of Soy in China*

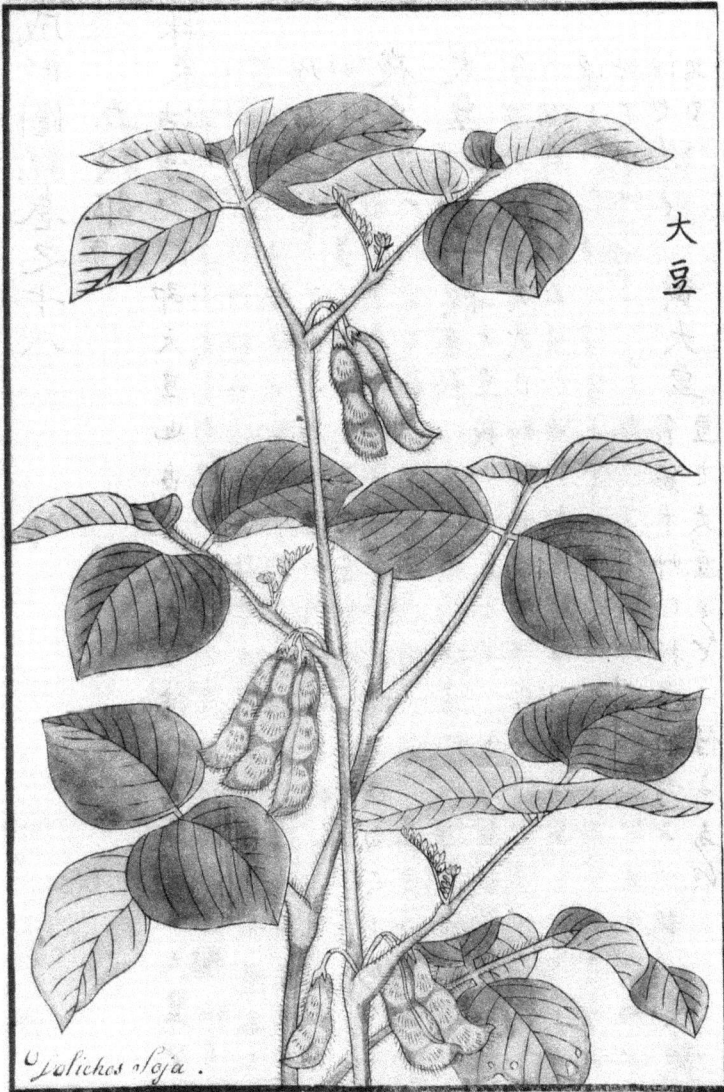

Figure 1.

Image from Sōhan and S. Shigehi (eds), *Seikei Zusetsu* (Illustrated Explanations of the Forms of Things) courtesy of Leiden University Libraries (Ser.1042 vol. 18, page 002) http://hdl.handle.net/1887.1/ item:938295 From the collection of Philipp Franz von Siebold (1796–1866), who presumably wrote 'Dolichos Soja' at the bottom.[4]

4   On this book, see S.A. Chatterjee and T. van Andel, 'Lost grains and forgotten vegetables from Japan: The Seikei Zusetsu agricultural catalog (1793–1804)', *Economic Botany* **73**(3)(2019): 375–89.

the wild form. Only a few millennia later is there evidence that the beans had begun to change in form in response to cultivation pressure and perhaps intentional selection, something that may have occurred independently in several areas of East Asia. At some point farmers began to select beans that were larger and oilier, and by 4,000 years ago they were planting cultivars with higher oil content.[5] Domestication also involved selection for plants that produced more and longer-lasting leaves, perhaps an unintended consequence of selecting varieties that produced more beans, though people could also eat the leaves.[6]

The development of improved varieties was one reason farmers gradually began to grow more soybeans. Another was their beneficial effect on soil fertility. As human populations increased in the first millennium BCE, people could no longer leave their fields fallow for years before planting them again, and planting beans helped them keep fields productive even when they were being used more intensively.[7] Similarly, as populations grew in core farming regions there was less and less land left for domestic or wild animals and meat became less common in people's diets. The reduced availability of animal protein was surely a main reason people grew more soybeans. Modern soybean cultivars are often composed of forty per cent protein and twenty per cent oil, and they produce between four and eighteen times as much usable protein per unit of land as milk, eggs or meat.[8] People did not know about protein, but they surely could tell that eating beans regularly made them feel healthier than just eating millet and vegetables. Soybeans were not highly regarded because they are hard to digest, but they increased in popularity as people learned to turn them into sprouts, tofu and various sauces. The later introduction of Buddhism may have encouraged the spread of vegetarian cuisine, though many people rarely ate meat simply because of the scarcity of animals to eat.[9]

5   Y. Zong et al., 'Selection for oil content during soybean domestication revealed by x-ray tomography of ancient beans', *Scientific Reports* **7** (1) (2017): 43595.

6   A. Togashi and S. Oikawa, 'Leaf productivity and persistence have been improved during soybean (*Glycine max*) domestication and evolution', *Journal of Plant Research* **134** (2) (2021): 223–33.

7   This is argued, plausibly but with little evidence, in B. Zhang and Z.M. Fan (eds), *Zhongguo nongye tongshi: Zhanguo Qin Han juan* (Beijing: Zhongguo nongye, 2007).

8   F. Simoons, *Food in China: A Cultural and Historical Inquiry* (Boca Raton: CRC Press, 1991), p. 71; C.W. Wrigley, H. Corke and C. Walker, *Encyclopedia of Grain Science* (Elsevier Academic Press, 2004), vol. 3, p. 142.

9   J. Kieschnick, 'Buddhist vegetarianism in China', in Roel Sterckx (ed.), *Of Tripod and Palate: Food, Politics and Religion in Traditional China* (New York: Palgrave Macmillan, 2005), pp. 186–212.

2. A History of Soy in China

Agricultural books from ancient China contain the oldest writings on soybeans. The earliest of these is a partially extant farming manual written by Fan Shengzhi about 2,100 years ago. Its passage on soybeans begins by arguing that the main attraction of soybeans was their ability to flourish in adversity: 'From soybeans a good crop can be easily secured even in adverse years, so it is natural that the ancient people grew soya as a provision against famine.'[10] Given the irregular nature of East Asia's monsoon, a crop that would reliably produce food provided an important safety net, even if it was not especially palatable. Fan recommends that a farmer should plant about half an acre of soybeans per member of the family, and says that they could be sown anytime between spring and early July, which made them a particularly versatile crop. He notes that the soil does not need to be well ploughed and the seeds should not be planted deep. Fan also provides instructions for planting in shallow pits with manure, a labour-intensive method that surely produced good yields.[11]

The sixth-century *Essential Techniques for the Common People* [*Qimin yaoshu*], a manual from north China on how to raise and process various domestic plants and animals, often mentions soybeans. It suggests that, since soybeans grow well in poor soils, farmers should plant them there and reserve their better fields for more desirable crops like millet, wheat and rice. It also states that soy only needs to be hoed twice in a season. Regarding varieties, the *Essential Techniques* says there are two main types of soybeans, black and white, but it quotes an earlier text that mentions three varieties, one of which had edible leaves. A thousand years later, the canonical Chinese medical encyclopaedia *Herbal Compendium* [*Bencao gangmu*] states that soybeans come in several colours, including black, white, yellow, brown, blue-green and spotted. It records that the black ones could be eaten, used as medicine or fermented into a seasoning. The yellow ones could be made into bean curd, pressed to get oil or processed into sauce. It then states that the other species can only be used for making bean curd or cooked and eaten.[12] Being a medical manual, the *Compendium* frequently mentions the use of

The word 'tofu' is the Japanese pronunciation of the term, which has become standard in English. In Mandarin it is pronounced 'doufu'.

10   S.H. Shih, *On 'Fan Sheng-Chih Shu' an Agriculturistic Book of China Written by Fan Sheng-Chi in the First Century B.C.* (Beijing: Science Press, 1959), pp. 18–19.

11   Ibid., pp. 19, 35.

12   S. Li, *Compendium of Materia Medica (Bencao Gangmu)*, 6 vols (Beijing: Foreign Language Press, 2006), p. 2370; S. Li, *Bencao gangmu*, ed. Y. Wang (Beijing: Renmin weisheng, 2005), ch. 24, p. 1499.

soybeans as medicine for both people and livestock. Virtually every part of the plant was used as medicine, including the leaves, sprouts, flowers and beans, sometimes fermented. It also mentions that soy should be avoided by people with some illnesses.[13] It mentions alcohol with beans soaked in it.[14] Apparently, soybeans were even used to dye hair.

Processing

Soybeans are now valued as a 'flex' crop, meaning that they can be used for a wide variety of things. Many of the basic processing techniques were pioneered in China, innovations probably spurred by the necessity of subsisting on nutritious but somewhat indigestible soybeans in years when other crops failed. Dried soybeans are unusually wholesome, containing various essential nutrients in addition to their high protein and fat content. However, raw soybeans contain proteins that are toxic to humans as well as carbohydrates that cause flatulence, and have a 'beany' flavour that many find unappealing. Beginning in ancient times, people in China learned that sprouting and boiling the beans solves many of these problems and began to develop a wide variety of soy foods. The classic history of soy processing is Hsing-Tsung Huang's volume of *Science and Civilisation in China*.[15]

Raw soy milk is the basis of many soy foods. It is made by boiling beans, grinding them into a paste and then filtering through a rough cloth. The resulting juice is commonly consumed hot for breakfast in China and has a more beany flavour than the more processed soy milk available in grocery stores. To make tofu, a coagulating agent such as calcium sulphate is added to the raw soy milk. This process creates fresh unpressed tofu, a soft pudding that is also a common breakfast food. Wrapped in cloth and pressed to remove water, this soft product becomes more solid forms of tofu. The more water is removed, the harder the tofu becomes. Some tofu products can be cured or dried and stored for long periods, and many are additionally fermented, brined or smoked, both for preservation and for flavour. Tofu has been made for at least a thousand years in China, and perhaps twice that long.[16]

---

13   Li, *Compendium of Materia Medica*, p. 1907.

14   Ibid., p. 2367; Li, *Bencao gangmu*, ch. 48, p. 2605.

15   H.T. Huang, *Science and Civilisation in China 6.5: Fermentations and Food Science* (Cambridge: Cambridge University Press, 2000).

16   Ibid., pp. 292–378; R.P. Hommel, *China at Work; an Illustrated Record of the Primitive Industries of China's Masses, Whose Life Is Toil, and Thus an Account of Chinese Civilization* (Cambridge, Mass.: MIT Press, 1969), pp. 105–09.

2. A History of Soy in China

A Spanish missionary who lived in the southeastern province of Fujian in the mid-seventeenth century considered tofu one of the marvels of China:

> The delicacy that is widely used, common, inexpensive and abundant in all China, and which is eaten by everyone in that empire, from the emperor to the most ordinary Chinese person – the emperor and lords as a treat and the ordinary people for sustenance and necessity – is called tofu, which is bean paté. I did not see how they make it; they make milk from the beans, curdle it like cheese as large as a millstone, five or six fingers thick. The whole mass is as white as fresh snow. It has everything you can wish for. It can be eaten raw, but is normally cooked, and prepared with vegetables, fish, and other things. On its own it is bland, but cooked in the aforementioned way it is good, and is excellent fried in cow butter. They also dry and smoke it, mixing in caraway seeds, which is even better. It is incredible what vast quantities of it are purchased and eaten in China, and hard to conceive that there is such a large quantity of beans. A Chinese person who has tofu, vegetables, and rice does not need anything else to work. Nor do I think that there is anyone who cannot get it because for a quarter one can buy twenty ounces or even more.[17]

This passage makes clear just how common tofu had become. Its description of tofu being consumed by the elite was no exaggeration, since there was by this time a well-developed culture of tofu connoisseurship that is recorded in culinary texts.[18]

Today's practice of separating soybeans into oil and protein-rich solids also goes back centuries in China. Like the seeds of brassicas and other plants, soybeans were crushed, heated and pressed to make oil for cooking and a variety of other uses.[19] Once the oil had been squeezed out, what remained was a tightly pressed 'bean cake' that was valued as fertiliser because it was high in protein and other nutrients. Bean cakes also made excellent feed for pigs, as did byproducts from the production of tofu and other soy-based foods. Although often considered inferior for tofu or oil, black soybeans were a hardy crop that served as backup in case other crops failed and could in any case be fed to livestock.[20]

Soybeans were also fermented to make a variety of condiment pastes that

17   D. Fernández Navarrete, *Tratados historicos, politicos, ethicos y religiosos de la monarchia de China* (Madrid: en la Imprenta Real por Iuan Garcia Infançon, 1676), pp. 347–48. This is our translation (with help from Augusto Garcia-Agundez), drawing on the English translation An Account of the Empire of China (London: H. Lintot, J. Osborn, 1732), pp. 278–79. The seeds identified as caraway were more likely fennel.

18   M. Brown, 'On bird's nests and bean curds: Reflections on the rise of tofu connoisseurship', presented at Harvard University in 2018. This was the source of the Fernández Navarrete quote above.

19   F. Bray, *Science and Civilisation in China 6.2: Agriculture* (Cambridge: Cambridge University Press, 1984), pp. 293–97, 430–33, 518–26; Hommel, *China at Work*, pp. 85–93.

20   Bray, *Science and Civilisation in China 6.2*, p. 514.

Brian Lander and Thomas David DuBois

天工開物 中卷

六八

南方榨

Figure 2.

This woodblock print depicts an oil press made from a tree trunk with a slot hollowed in the middle. Boiled and crushed beans were wrapped in cloth or placed inside rings of metal or woven bamboo and then stacked in the slot (these are the rectangles in the cylinder). Above these, workers inserted wedges and used the suspended log to hammer them in between blocks of wood, creating enough pressure to crush the beans, whose oil drained into the bowl below.[21]

are still widely consumed in East Asia, as well as the liquid soy sauce, varieties of which are now used globally.[22] During the sixteenth and seventeenth centuries, European ships arrived along the coasts of Asia and, although they did not bring back soybeans for widespread cultivation in Europe, they did bring back the word 'soy'. Processed soy products had spread across Southe-

21 We are grateful to the National Library of France for putting this 1637 edition of Song Yingxing's *Tiangong kaiwu* online at https://gallica.bnf.fr/ark:/12148/btv1b52505781g/f274.item. This is p. 68 of the middle of three books. Courtesy of gallica.bnf.fr and the Bibliothèque National de France. Image edited for clarity. For an English translation see Sung Y.H., E.T.Z. Sun, and C.C. Sun, *Tien-Kung Kai-Wu: Chinese Technology in the Seventeenth Century* (University Park: Pennsylvania State University Press, 1966), pp 214–21. For photographs and explanations of this equipment, see Hommel, *China at Work*, pp. 89–93.

22 Huang, *Science and Civilisation in China 6.5*, pp. 346–78.

ast Asia along with Chinese migrants by the time Europeans arrived in the region, and the name 'soya' comes not from the beans themselves but from the closely related Chinese and Japanese words for soy sauce.[23]

While the large-scale trade in soybeans began in the twentieth century, it should be noted that soybeans have been traded for over two millennia in China. At least as early as the third century BCE, the Yellow River valley was quite commercialised and merchants traded in bulk commodities including soybeans and soybean condiments.[24] At the same time, increasingly powerful states collected soybeans as tax, and fed them to their horses.[25] By the 1700s, merchant houses conducted a brisk overland and coastal trade in soybeans, bean cake, tofu products and a variety of soy-based condiments.[26]

Manchuria and the Growth of a Global Market

By the dawn of the twentieth century, China's population was well over four hundred million, and farmers throughout China were growing soybeans. Soy was widely cultivated as a subsidiary crop, planted in fallow land, or around the edges of productive fields. In his famous surveys of over 16,000 farms, agronomist John Lossing Buck discovered that farmers in most of China's agricultural regions devoted less than five per cent of their land to soybeans. Moreover, the farmers only ate a small portion of their soybeans themselves. They sold most of their yellow soybeans for processing into food products like bean paste or oil, while protein-rich black soybeans were often used for animal feed. For agrarian households, soybeans generally represented less than five per cent of total calories consumed, though they were an important source of protein. The nitrogen-rich green waste, including stalks and husks leftover from processing, was frequently combined with planted

---

23   Oxford English Dictionary Online, accessed March 2021. After receiving soybeans from East Asia, Southeast Asians also developed their own soy-based foods such as tempeh.

24   Huang, *Science and Civilisation in China 6.5*, pp. 333–78; W.H. Nienhauser (ed.), *The Grand Scribe's Records XI: Memoirs of Han China, Part IV* (Bloomington: Indiana University Press, 2019), pp. 292–93 (*Shiji* ch. 129).

25   A.F.P. Hulsewé, *Remnants of Ch'in Law* (Leiden: Brill, 1985), p. 42; J.W. Dauben, 'Suan Shu Shu: A book on numbers and computations', *Archive for the History of Exact Sciences* 62 (2008): 91–178 at 137, 142; A.J. Barbieri-Low and R.D.S. Yates, *Law, State, and Society in Early Imperial China* (Leiden: Brill, 2015), pp. 923, 934; M. Korolkov, 'Empire-building and Market-making at the Qin Frontier' (Ph.D. Dissertation, Columbia University, 2020), pp. 258, 577–78.

26   R. von Glahn, *The Economic History of China: From Antiquity to the Nineteenth Century* (Cambridge: Cambridge University Press, 2016).

Brian Lander and Thomas David DuBois

seeds as fertiliser.[27]

The shift to large-scale soybean monocultures, an epochal transition in the history of soy, first occurred in Manchuria. Bounded by Russia, Korea and Mongolia, this vast region has a temperate climate with colder winters than the rest of China. The homeland of the country's ruling Manchu dynasty (1644–1911), Manchuria had been largely closed to Chinese migration since the seventeenth century. It was only during the nineteenth century that waves of new migrants from crowded north China began gradually expanding the agrarian frontier northward into the provinces of Jilin and Heilongjiang.[28]

Manchuria would play a pivotal role in the soybean story for two reasons. The first is that Manchuria had the open spaces, famous black soils, and 'something approaching the optimum climate' for the extensive cultivation of soybeans.[29] Unlike the marginal cultivation of soybeans on millions of family farms, this crop was for export. As early as the eighteenth century, Manchuria's seaports had already begun exporting soybeans to coastal cities like Shanghai and Amoy, where the beans were processed into condiments or pressed for oil. From there, the bean cake was sold locally as fertiliser, or reshipped to markets in southern China and Southeast Asia.[30] In this way, the extensive cultivation of northern soybeans became a counterpart to intensive southern crops such as rice and sugar cane. Although the rest of China produced more soybeans overall, ideal growing conditions and low population density made Manchuria a far better export producer.

The second reason Manchuria was so important in the history of soybeans is that explosion of global interest in soybeans coincided with a race to develop Manchuria itself. In 1908, the Harbin-based businessman R.M. Kabalkin sent the pioneer shipment of 'the miracle bean' from Vladivostok to the British port of Hull, sparking a rush of European and American interest

---

27   J.L. Buck, *Land Utilization in China: A Study of 16,786 Farms in 168 Localities, and 38,256 Farm Families in Twenty-two Provinces in China, 1929–1933*, (Nanjing: University of Nanking, 1937), pp. 236–37; 404–10, 417.

28   A. McKeown, 'Global migration, 1846–1940', *Journal of World History* **15** (2) (2004): 155–89. For the political and ecological chronology of this shift, see J. Reardon-Anderson, 'Land use and society in Manchuria and Inner Mongolia during the Qing Dynasty', *Environmental History* **5** (4) (2000): 503–30. For a similar story along the border with Inner Mongolia, see Y. Wang, 'Irrigation, commercialization, and social change in nineteenth-century Inner Mongolia', *International Review of Social History* **59** (2) (2014): 215–46.

29   G.F. Deasy, 'The soya bean in Manchuria', *Economic Geography* **15** (3) (1939): 303–10.

30   C.V. Piper and W.J. Morse, *The Soybean* (New York: McGraw-Hill, 1923), p. 5; J.R. Stewart, 'The soya bean and Manchuria', *Far Eastern Survey* **5** (21) (1936): 221–26.

in soybeans.[31] This new global attention not only created new markets, it also sparked investment in scientific research to develop soybeans into plastics, soap, fuel and, of course, food products. It was during this time that British and German researchers began working to improve traditional processes like pressing to capture more oil, with higher nutrition and better taste.[32] As soybeans moved from Manchuria to the world they were increasingly seen and classified as an industrial crop.[33]

At this same time, Manchuria became the centre of intense geopolitical struggle. China, Russia and Japan jockeyed for power by developing infrastructure such as railroads to establish spheres of influence. After Japan emerged victorious from war with Russia in 1905, its influence over the region grew, as did the region's economy. Soybeans were a primary driver of this growth: between 1906 and 1921, soybean production grew from 600,000 to 4.5 million tons.[34] The Japanese-owned railway which ran from the northern city of Harbin to the deep-water port of Dalian (Dairen in Japanese) became a vast highway of soybean exports, facilitated by a whole infrastructure of banking, commodity exchanges and buyers and more or less state-linked companies like Mitsui Bussan.[35] Little of Manchuria's soybean crop was destined for direct human consumption. Some was processed into bean paste or pressed into oil. Chinese buyers used the raw oil for cooking, European ones processed it into products like soap, some of which was then sold back to China. When possible, European buyers preferred to buy whole beans, which shipped without spoiling and could be profitably processed in mills on the continent. Most of Manchuria's soybean exports consisted of

---

31 Kabalkin registered his Sino-Anglo Orient Trading Company in hopes of creating a stable trade with European markets but was subsequently squeezed out by the Japanese trade. Our thanks to Dr. Dan Ben-Canaan for this important insight on the Harbin industry.

32 I. Prodöhl, 'Versatile and cheap: A global history of soy in the first half of the twentieth century', *Journal of Global History* **8** (3) (2013): 461–82.

33 S. Wen. 'From Manchuria to Egypt: Soybean's global migration and transformation in the 20th century', *Asian Journal of Middle Eastern and Islamic Studies* **13** (2) (2019): 176–94.

34 Piper and Morse, *The Soybean*, p. 7.

35 H. Mizuno and I. Prodöhl, 'Mitsui Bussan and the Manchurian soybean trade: Geopolitics and economic strategies in China's Northeast, ca. 1870s–1920s', *Business History* (2019); Y. Enatsu, 'The role of private companies in the expansion of Japan's interests in Manchuria in the 1920s: The case of the Toa Kangyo Company (Tōa kangyō kabushiki kaisha)', *Chinese Business History* **15** (2) (2005). M. Hiraga and S. Hisano, 'The first food regime in Asian Context? Japan's capitalist development and the making of soybean as a global commodity in the 1890s–1930s' (Kyoto: Asian Platform for Global Sustainability & Transcultural Studies; AGST Working Paper Series No.2017-03); R.E. Wells. 'The Manchurian Bean: How the Soybean Shaped the Modern History of China's Northeast, 1862–1945' (Ph.D. Thesis, University of Wisconsin-Madison, 2018).

Piles of Beans awaiting shipment.

Figure 3.

This image from the port of Dalian shows bags of beans stacked into pyramidal piles. The round discs in the centre are bean cakes. These soy products had been carried by the South Manchurian Railway (note the rail cars marked SMR) to the coast whence they would presumably be shipped out. Source: SMR postcard in the collection of Brian Lander.

bean cake to fertilise Japanese rice fields, a substantial transfer of nutrients from a frontier to the imperial metropole.[36]

Elsewhere in China, the reputation of soybeans enjoyed a renaissance as the key to improving the national diet. Following the latest trends in dietary science, political reformers encouraged consumption of protein to strengthen the character and physical strength of the people. Unable to provide meat or dairy like the Western powers, China would instead switch to bean products, such as tofu and bean milk.[37] Soy held an almost magical appeal for

36   D. Wolff. 'Bean there: toward a soy-based history of northeast China', *South Atlantic Quarterly* **99** (1) (2000): 242–52.

37   J.C. Fu, *The Other Milk: Reinventing Soy in Republican China* (Seattle: University of Washington, 2018). F. Sabban. 'The taste for milk in modern China (1865–1937)', in J.A. Klein and A. Murcott (eds), *Food Consumption in Global Perspective*. Consumption and Public Life (London: Palgrave Macmillan, 2014). T. D. DuBois, 'China's dairy century – making, drinking and dreaming of milk', in R. Kowner, G. Bar-Oz, M. Biran, M Shahar and G. Shelach (eds), *Animals and Human Society in Asia: Historical and Ethical Perspectives* (London: Palgrave Macmillan, 2019), pp. 179–212.

political reformers like anarchist Li Shizeng, who in 1908 combined social idealism with scientific training in biology and chemistry to build the first soy processing plant in France. More than just providing cheap food to the poor nations of the world, Li's Caséo-Sojaïne was also a social experiment in improving the moral character of the young Chinese men sent there to work and study.[38] The call to build the strength of the Chinese nation on a foundation of cheap vegetable protein increased in intensity after Japan carved off Manchuria in 1931 and invaded China proper in 1937. With China's government forced by the war to seek refuge in the mountainous Southwest, a new generation of dietary reformers saw in the humble soybean a solution to feeding both troops and refugees.

Soybeans and Socialism

In 1945, as Japan's empire crumbled, Chinese communist forces seized Manchuria and briefly diverted its soybean exports to the Soviet Union in return for military aid.[39] But by the mid-1950s, the combined effects of the loss of the Japanese market, a US-led trade embargo of China, and the rise of extensive production in the American Midwest, had displaced Manchuria as the world's supplier of soybeans. At the same time, China's economic planners had switched priorities to value grain more than anything else. The final blow came with the development of chemical fertilisers, which globally supplanted bean cake as domestic and export product.

But China did not give up on soybeans. The country continued to invest in research into new soybean varieties and planting techniques suited to China's many climatic and soil conditions.[40] Collective agriculture did introduce certain efficiencies by replacing household plots with large-scale farms and by making investments in irrigation and transport. Nevertheless, even as new migrants moved to settle the far north of Manchuria in a government scheme to ‹develop the great Northern expanse›, national soybean production fell from a peak of 10.1 million tons at the outset of collectivisation in 1957 to

---

38   Li is discussed in J.C. Fu, *The Other Milk*.

39   J. Zhu (ed.), *Dongbei jiefangqu caizheng jingjishi gao 1945.8-1949.9.* [Draft finance and economic history of the liberated Northeast, 8/1945-9/1949] (Harbin: Heilongjiang renmin chubanshe, 1989), p. 43.

40   These efforts are discussed in contemporary publications spanning the 1950s–1970s, e.g., Tieling diqu nongye kexue yanjiusuo [Tieling regional agricultural science research center], 'Dadou xin pinzhong jieshao' [Introduction to new soybean varieties], *Nongye kexue* [Agricultural science] (1957): 44–47.

7.3 million tons two decades later.[41] How were China's soybeans being used? Significantly, while much of the world had already come to view soybeans as an industrial crop, in China they were officially categorised as a food grain, meaning that they were intended for human consumption.

With the economic liberalisation of the 1980s, China again embraced soybeans in a big way. New policies encouraged domestic production by raising the price the government paid for soybeans, while simultaneously allowing imports, including from the United States. At the same time, domestic producers began competing to create products for the growing consumer market. Tofu making quickly reemerged as a village industry: low-tech tofu workshops were often among the earliest of small-scale businesses known as Township Village Enterprises. But the real sense of opportunity was at the higher end, as provincial governments paired with international investors to capture the added value of processing. Rather than shipping their raw beans to presses in Beijing or Shanghai, soybean-producing provinces like Heilongjiang built their own pressing capacity, as well as laboratories and research institutes. Partnerships with foreign producers such as Hong Kong-based Vitasoy produced new consumer offerings like powdered or Tetra Pak soy milk (often sold mixed with more expensive cow dairy) and found new industrial uses for previously unknown products like soy protein isolate.[42] As the new market value of soybeans grew, so did research into industrial processes, making standard products like bean oil more economical, nutritious, and profitable.

China as a Global Consumer

By the late 1980s, China was exporting over one million tons of soybeans per year, but over the next decade it became a consistent net importer (Figure 1), and in 2004 imports exceeded the country's entire domestic production for

---

41   Chinese official statistics. Quoted in N.R. Lardy, 'Food consumption in the People's Republic of China', in R. Barker, R. Sinha and B. Rose (eds), *The Chinese Agricultural Economy* (Boulder, CO: Westview Press, 1982), pp. 147–62.

42   Vitasoy initiated a joint venture plant in Jiangsu from 1986. This was followed by an explosion of provincial-level enterprises that sought to capitalise from the mass production and branding of what would theretofore have been seen as a homemade commodity. L. Dai, 'Xinxing jiankang yinliao 'weitanai' zai Jiangsu sheng tongguo shengji jianding' [New health beverage 'Vitasoy' undergoes appraisal in Jiangsu]. *Zhongguo rupin gongye* 2 (1986): 36. X. Yang, 'Geiyu chanpin xingxiang fuyu shidai xinyu – cong bizi, doujiangde shichang kaifa tanqi' [Giving products an image to match the times – from the market opening of combs and bean milk] *Jingying yu guanli* 2 (1991): 21–22.

2. *A History of Soy in China*

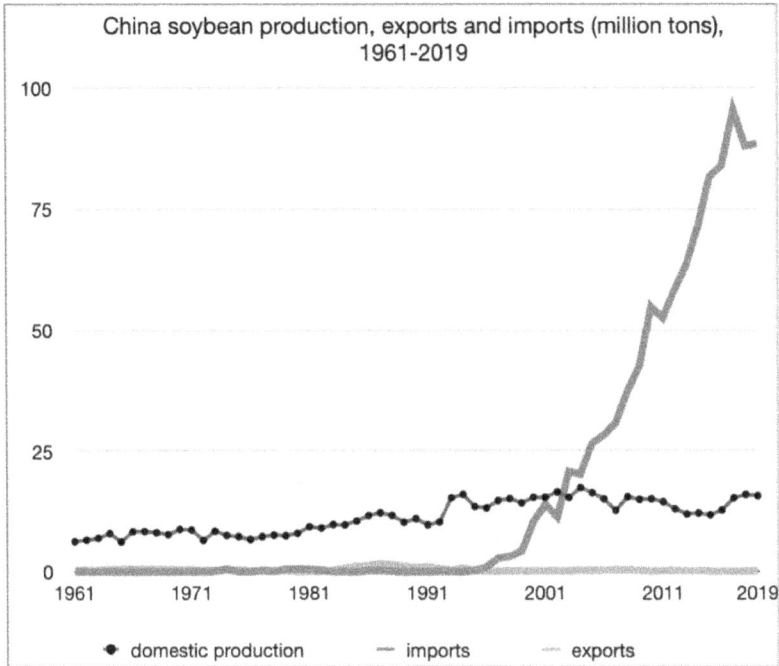

China soybean production, exports and imports (million tons), 1961-2019. Data source: Food and Agriculture Organization of the United Nations. FAOSTAT Statistical Database.

Figure 4.

China soybean production, exports and imports (million tons), 1961–2019. Data source: Food and Agriculture Organization of the United Nations. FAOSTAT Statistical Database.

the first time.[43] This milestone was just the beginning of a spectacular increase in Chinese imports, which by 2008 accounted for more than half the world's soybean trade, and are expected to surpass 100 million tons in 2021 (Figure 2).[44] The foundation for this massive transformation was laid during the 1990s, when China gradually lowered import tariffs first on edible soybean oil, and later on soymeal. While changes to policy and the global markets have caused short-term fluctuations in these two products, the overwhelming trend has been toward higher domestic consumption of both: soybean oil as a kitchen

43 'Woguo dadou jinkouliang shouci chaoguo guonei chanliang' [Chinese soybean imports for the first time exceed domestic production], *Guoji shangbao* 11 Feb. 2004.

44 Food and Agriculture Organization of the United Nations. FAOSTAT Statistical Database, 'China's Soy Imports to Top 100 Million Tons as U.S. Supply Jumps', *Bloomberg News* (22 Dec. 2020).

Brian Lander and Thomas David DuBois

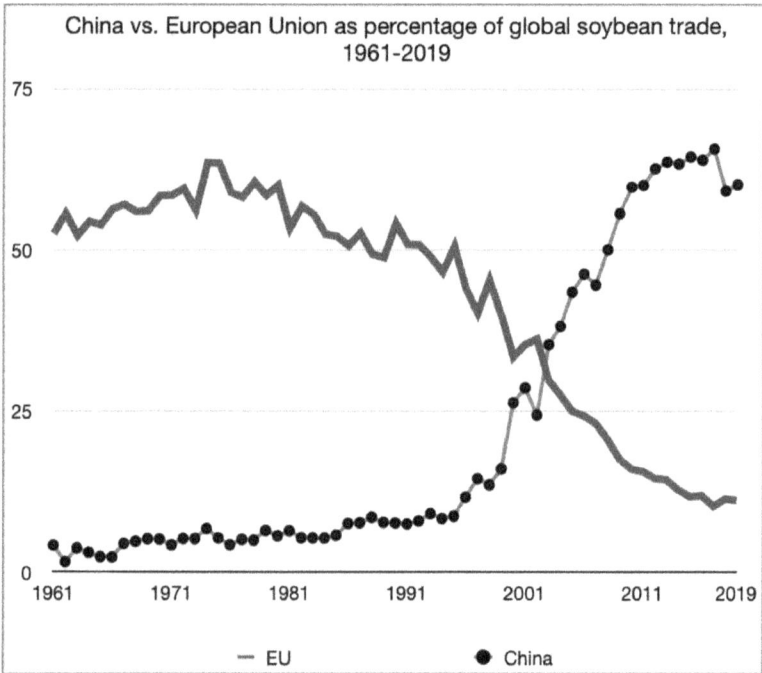

Figure 5.

The comparative position of China and the European Union in the global soybean trade, 1961–2019. As noted in footnote 2, global soybean production increased more than tenfold in this period. Although EU imports did decline slightly (10–15%) from their peak in 2002, the main reason the EU share of the soy trade fell was the enormous increase in overall trade volume driven by Chinese purchases. Data source: Food and Agriculture Organization of the United Nations. FAOSTAT Statistical Database.

staple, and soymeal as a feed component for China's quickly growing livestock industries.[45] Between 1997 and 2014, Chinese per capita consumption of edible oils tripled from 8 to 24 kilograms, almost half of it soybean oil.[46]

Over the past few decades, rising meat consumption has exemplified China's rising standard of living and driven the increased importation of

45   M. Schneider, *Feeding China's Pigs: Implications for the Environment, China's Smallholder Farmers and Food Security.* Institute for Agriculture and Trade Policy, 2011. https://www.iatp.org/sites/default/files/2011_04_25_FeedingChinasPigs_0.pdf

46   Soybean oil surpassed rapeseed in 2003 as the most common oil. J.P. Jamet and J.M. Chaumet, 'Soybean in China: adapting to the liberalization', *Oilseeds and Fats, Crops and Lipids* 23 (6) (2016).

soy, most of which is fed to livestock.[47] Although soy meal only accounts for between ten and twenty per cent of most animal feed formulations, China's livestock numbers have grown so dramatically over the past few decades that the country is now the leading consumer of soy imports worldwide. In 2018 China was home to over 700 million pigs, half the world's total, and soy is also used to feed chickens, ducks, cattle and fish.[48] China's increasing consumption of soy has been one of the main drivers of the expansion of soy acreage globally. In particular, huge tracts of South America's grasslands and forests have been cleared over the past two decades to grow soy to feed livestock in China.[49] It is worth noting that most soybeans planted today are quite different from their ancestors that left China in the twentieth century, having been genetically modified to promote desired characteristics. One innovation fundamental to the recent expansion of soy has been the development of herbicide resistant varieties that allow farmers to kill most of the other plants in fields without harming the soy plants.

The massive influx of cheap soybean imports was bound to affect China's domestic growers, who reduced the area under soy cultivation by 29 per cent between 2010 and 2015. As a result, Chinese leaders began introducing various policy initiatives to increase soy production (notably by replacing corn acreage) across a vast swathe of northern provinces.[50] The disruptions to global trade caused by the outbreak of Covid-19 have further strengthened the chorus of domestic voices calling on China to become less reliant on soybean imports, at least in relative terms.[51] Most recently, this has included new 2021 government guidelines encouraging makers of animal feed to use

47   B. Lander, M. Schneider and K. Brunson, 'A history of pigs in China: From curious omnivores to industrial pork', *Journal of Asian Studies* **79** (4) (2020): 865–89; G de L.T. Oliveira and M. Schneider, 'The politics of flexing soybeans: China, Brazil and Global agroindustrial restructuring', *Journal of Peasant Studies* **43** (1) (2015): 1–28.

48   US Department of Agriculture, Foreign Agricultural Service (USDA), 'Livestock and Poultry: World Markets and Trade; Consolidation and Modernization Continue to Shape China's Livestock Outlook'. 9 April 2018. https://www.fas.usda.gov/data/ livestock-and-poultry-world-markets-and-trade (accessed July 2021).

49   G. de L.T. Oliveira and S.B. Hecht, *Soy, Globalization, and Environmental Politics in South America* (Milton Park: Routledge, 2017).

50   For an overview of these policies in Heilongjiang, see Cheng Yao, 'Gonggeice gaige yu Heilongjiang dadou chanye fazhan yanjiu' [Research on supply side reform and the Heilongjiang soybean industry], *Dadou kexue* [Soybean science] **31** (1) (2018): 126–30.

51   W. Li and G. Zhao. 'Cong xinguan yiqing jan woguo dadou gongying de duiwai yilai wenti' [Viewing the problem of China's dependence on foreign supply of soybeans from the perspective of Coronavirus pandemic], *Shijie zhishi* [World knowledge] **10** (2020): 57–59.

items like rapeseed, cottonseed or peanut to replace soy.[52] Such steps could remove millions of tons of demand from the world market, but would take some time to implement, and would ultimately be constrained by cost as the price of soy drops. More importantly, the basic issue is that China does not have enough land to feed all the animals its people now eat, so any alternative to soy feed will also have to be imported. Whether by necessity or by choice, China is likely to remain a major importer of soybeans for years to come.

Conclusion

Soybeans initially became a widespread food crop in China because they grow so reliably that they could be counted on to provide nutrients if other crops failed. Over millennia, farmers developed varieties that grew in different conditions and had different uses. People also learned to improve soybeans' taste, to extract oil from them and to produce a variety of protein-rich foods. But it was only in twentieth-century Manchuria, a rich frontier of the Japanese empire, that they were first grown as large-scale monocultures for global markets. They have since become one of the world's main agricultural commodities.

Given the enormous impact of soybean cultivation in recent times, it is worth considering their role in China's environmental history. East Asia's lowlands are among the most anthropogenic environments in the world, their natural flora and fauna having largely been replaced by agriculture.[53] By renewing soils and providing people with essential protein, soybeans played a small but vital role in allowing agricultural populations to flourish over many centuries. But soybeans were just one small part of a sophisticated agricultural complex. In contrast to modern soybean monocultures, China's farming systems were among the most productive and sustainable on earth for many centuries because they relied on a wide variety of plants and animals.

Soybeans are often cast as the villains in the recent global expansion of soy monocultures. This perspective is not entirely unreasonable, since the same versatility that has made soybeans valuable to a variety of industries has also

---

52   H. Gu and D. Patton. 'Reshaping grain trade? China moves to change animal feed recipes', Reuters, 21 April 2021. https://www.reuters.com/world/china/reshaping-grain-trade-china-moves-change-animal-feed-recipes-2021-04-21/ . Similar measures were implemented without success in 2018.

53   R.B. Marks, *China: An Environmental History* (Lanham: Rowman and Littlefield, 2017); B. Lander, *The King's Harvest: A Political Ecology of Early China from First Farmers to First Empire* (New Haven: Yale University Press, 2021).

allowed them to displace native crops across five continents. In particular, soy cultivation has led to the destruction of grasslands and forests across large swathes of South America. However, what drives this growth is meat production.[54] Although China has emerged as a massive industrial consumer of soybeans for animal feed, the deeper lesson from China's history is that soybeans are an efficient way to provide vegetal protein for large populations. People in premodern China derived most of their protein from grains and beans, eating far less meat than today. The development of many delicious foods in East Asia from a single species of bean is in fact a model for how billions of humans could live with a much lower impact on the earth.

Acknowledgements

Both authors contributed equally to this paper. We would like to thank Miranda Brown, Madelaine Drohan, David Lord, Dan Ben-Canaan and Linda Grove for comments.

---

54   Note that there is considerable misinformation about the environmental impact of meat production: C. Christen, 'How the Meat Industry is Climate-Washing its Polluting Business Model', https://www.desmog.com/2021/07/18/investigation-meat-industry-greenwash-climatewash/ (Accessed 28 July 2021).

CHAPTER 3.

THE POLISH-BRAZILIAN SOY CONNECTION: CESLAU BIEZANKO AND THE INTRODUCTION OF *SOY UTILITIES* IN RIO GRANDE DO SUL DURING THE 1930S

Rhuan Targino Zaleski Trindade

Introduction

The early 1930s saw the arrival of several Polish intellectuals in Brazil. These people started working with immigrant communities who had settled in the country's southern states of Rio Grande do Sul and Paraná. In this context, they began to write about agricultural experiments with a relatively unknown yet promising product: soy and its derivates

After gaining independence at the end of World War I in 1918, the Second Polish Republic established strong ties with the Poles who had left Europe as emigrants over the previous decades. In the period between 1869 and 1914, Brazil had received around 100,000 of these immigrants, mostly peasants, who had settled in the rural *colonias* of southern Brazil.[1] This phenomenon led to the formulation of a cooperation agenda between Poland and the countries that had welcomed most of its citizens in recent years, mainly Brazil but also other southern American countries.[2]

---

[1]   M. Nalewajko, 'Los inmigrantes polacos en Brasil en sus testimonios', in E. González Martínez and R. González Leandri (eds), Migraciones transatlánticas. Desplazamientos, etnicidad y políticas, pp. 248–67 (Madrid: Catarata, 2015). The term 'colony' (in Portuguese *colônia*), refers to rural nuclei created by public or private entities in Brazil during the period of greatest expression of European immigration, especially between 1824 and 1930. The colonies were configured as allotments of large areas of land for immigrant families, distributed to occupy regions and promote agricultural development in the country, in addition to an immigration policy with broad political and racial purposes. They were occupied by the *colonos*, here translated as settlers.

[2]   K. Smolana, 'Roteiros poloneses na América Latina', in A. Dembicz and K. Smolana (eds), *A presença polonesa na América Latina* (Varsóvia: CESLA, 1996); A. Kicinger. 'Polityka emigracyjna II Rzeczpospolitej', *Central European Forum for Migration Research Working Paper* **4** (2005);

THE AGE OF THE SOYBEAN: 48–61 **doi:** 10.3197/63800040695086.ch03

3. *The Polish-Brazilian Soy Connection*

As part of the new political relations established between Poland and its overseas citizens, dozens of Polish intellectuals – broadly known as 'instructors' – were sent to Brazil as emissaries to immigrant communities, especially between the late 1920s and early 1930s. These people came from different ways of life: from military men to professors, agronomists to teachers and other educational figures. Their most significant actions consisted of organising associations to develop different aspects of sociability (e.g., sportive, rural, cultural, educational). Among these intellectuals, Czesław Marjan Bieżanko, also known as Ceslau Biezanko, rose to prominence for his work with soybean farming in Rio Grande do Sul.

Biezanko was a scientist with a modern standpoint, versed in more than seven scientific areas, ranging from botany to entomology to agronomy and biology. He had received extensive training in Poland before and, primarily, after World War I, as a professor in educational programmes concerning the development of various crops and the study of insects (especially Lepidoptera). In 1930, Biezanko moved to South America, initially to Argentina and then to Brazil, settling in Guarani das Missões, a large Polish colony northwest of Rio Grande do Sul. There, he began various activities with Polish settlers, primarily farmers, introducing different types of crops. Particularly relevant for this chapter are his efforts to introduce twelve different varieties of soy into the region.[3]

Biezanko's most significant work with soy was between 1932 and 1934. In the following decades, when the so-called soy boom would occur in southern Brazil, he was remembered as the introducer of soybeans into the region. Writings celebrating the Polish scientist proliferated in historical accounts on the introduction of soy, and in published biographies, official documents

M. Nalewajko, 'Los polacos hacia América Latina. La política emigratoria del gobierno polaco en el período de entre guerra', in E. Martínez González and A. Fernández (eds), *Migraciones internacionales, actores sociales y Estados. Perspectivas del análisis histórico*, pp. 129–48 (Madrid: Iberoamericana-Vervuert, 2014); A.W. Wychodźcy, 'Emigrants, or Poles? Fears and hopes about emigration in Poland, 1870–1918–1939', *AEMI Journal* 1 (2003): 78–93; M. de Oliveira, 'Origens do Brasil meridional: dimensões da imigração polonesa no Paraná, 1871–1914', *Estudos Históricos* 22 (44) (2009): 218–37; J. Mazurek, *A Polônia e seus emigrados na América Latina (até 1939)* (Goiânia: Espaço Acadêmico, 2016); P. Puchalski, 'Polityka kolonialna międzywojennej Polski w świetle źródeł krajowych i zagranicznych: nowe spojrzenie (1918–1945)', *Res Gestae. Czaopismo Historyczne* 7 (2018): 68–121; R.T. Zaleski Trindade, 'Um Imperialismo Polonês'. Narrativas brasileiras das relações da Polônia com os imigrantes poloneses no período entreguerras (Ph.D. Thesis, Universidade Federal do Paraná, 2020).

3   C. Biezanko, *Relação de plantas exóticas e indígenas cultivadas em Guarani das Missões (1930–1934) e em Pelotas (1934–1949) (Rio Grande Do Sul)* (Curitiba: Gráfica Vicentina, 1964).

and local periodicals.[4] Furthermore, Biezanko's contribution is inscribed in the local memories of the inhabitants of Guarani das Missões and among the whole Polish community in Rio Grande do Sul. He is considered one of the main influences on local Polish identity and received important acknowledgements by Guarani das Missões and Pelotas.[5] Biezanko's contribution is particularly significant, considering that Brazil's early history of soybean farming consisted of several fragmentary experiences. These included state-sponsored attempts to introduce the crop to the country, whether at the federal or state level, and the actions of several agronomists, normally foreigners, who strived to disseminate crop varieties among different rural communities.[6]

The development of soybean farming among Polish communities in Rio Grande do Sul is indissolubly linked to history and memory, touching upon discussions of pioneerism, ethnic and regional identity and present heritage. However, far from indulging in a celebratory tone, this chapter focuses on retracing the soy farming practices and thinking developed by Biezanko in Rio Grande do Sul. It analyses his intellectual production over time, with particular attention to his 1958 text, *Some Notions About Soy and its Cultivation: Soy Utilities*. The manuscripts comprise a series of articles, originally written in Polish for the prestigious Polish-Brazilian periodical, *Lud* [The People], in 1934. By this time, Biezanko was full professor at the Eliseu Maciel College of Agronomy in Pelotas.[7] The text's quasi-propagandistic tone discusses soy's potential for agricultural development. By the time he published the Portuguese translation (1958), the author was interested in disseminating farming experiences and general notions about soy to a broader audience. As a result, the manuscript covers a series of studies and particular views of the author about the new crop that he had introduced more than twenty years earlier among Polish settlers. Aside from remarking on the centrality of soy

4   R.T. Zaleski Trindade, 'Um cientista entre colonos: Ceslau Biezanko, Educação, Associação Rural e o cultivo da soja no Rio Grande do Sul no início da década de 1930' (Master's Thesis, Universidade Federal do Rio Grande do Sul, 2015).

5   Aside from various academic honours, he was honoured with street names in Pelotas and Guarani das Missões and a bust sculpture.

6   Trindade, 'Um cientista entre colonos'. From what has been said so far, we can anticipate that Biezanko was one of these characters and his effort in the early 1930s was part of a specific context of diffusion and experimentation of soybeans in Brazil, as the new product began to be used more systematically. Although some scholars have celebrated Biezanko as the first introducer of soybeans to Rio Grande do Sul, other source discussed in other texts offer a different interpretation.

7   I refer to Polish periodicals published in Brazil.

in opposition to the convictions of the conservative peasantry of the time, the text also includes anecdotes on Biezanko's personal life and serves as evidence for understanding the agricultural entanglements among different communities in Rio Grande do Sul. The economy of Guarani das Missões is to this day strongly linked to soybean farming.[8]

Although Biezanko's efforts were successful, social, cultural and practical difficulties led to the partial abandonment of soybean farming until its international boom in the 1970s.[9] Nonetheless, the Polish scientist's merit can also be measured by his introducing of other plants and the different uses of soybeans that he pioneered in the region among Polish immigrants – from the manufacture of oil to forage in the early 1930s.[10] By the time the north-western region of Rio Grande do Sul (*Alto Uruguai* and *Missões*) entered the great soy production boom of the 1970s, which would progressively expand to other tropical regions of Brazil, local farmers possessed the know-how needed to embark on this experience.[11] The experience of Polish immigrants pioneered by Biezanko during the 1930s offered a privileged cultural reference for these agricultural communities.

A New Product by Biezanko

Biezanko was, above all, a scientist pursuing an agenda of economic development, agrarian modernisation and improvement of his compatriots' living conditions. As a result, his formative and professional trajectory was characterised by the will to combine the study of botany with agricultural production – that is, moving from theory to praxis.

After participating in World War I, Biezanko studied Natural Sciences and Mathematics at the University of Warsaw from 1915 to 1917. Meanwhile,

---

8   According to the Brazilian Institute of Geography and Statistics (IBGE), agriculture and live-stock represents about a third of the municipality's economy. Soy has more than 15,000 planted hectares (having previously reached 18,000), producing almost 36,000 tons today (with a peak record of 60,000 tons in 2017) Source https://cidades.ibge.gov.br/brasil/rs/guarani-das-missoes/pesquisa/14/10193?tipo=grafico&indicador=10370.

9   See P.A. Zarth. 'História agricultura e tecnologia no noroeste do Rio Grande do Sul', in A.I. Andrioli (ed.), *Tecnologia e Agricultura Familiar*, pp. 51–76 (Ijuí: Unijuí, 2009).

10   On the introduction of other crops and plants, see Biezanko, *Relação de plantas exóticas e indígenas cultivadas* and the biography written by Jan Wójcik, 'O Nosso Professor', *Kultura* **6** (224) (1966): 117–26.

11   On the expansion of soy to other tropical regions, see C.M. da Silva and C. de Majo, 'Genealogy of the Soyacene: The tropical bonanza of soyabean farming during the Great Acceleration', *International Review of Environmental History* **7** (2) (2021): 65–96.

between 1913 and 1916, he also studied Chemistry and Natural Sciences at the College of Agriculture's Faculty of Agronomy. Later on, from 1917 to 1920, he studied at Jagiellonian University's Faculty of Philosophy (Natural Sciences) in Krakow. Finally, he studied Chemistry at the University of Poznań from 1920 to 1922 and later took an advanced degree in Natural Sciences, specialising in Zoology and Entomology.[12]

As for his professional experience, Biezanko taught Natural Sciences and Chemistry in several schools in 1916 and between 1924 and 1926. Between 1920 and 1923, he was an assistant in general chemistry at the University of Poznań, and during his vacations, he worked in mills. From 1926 to 1927, he taught at Grudziadz's College of Construction of Machinery and Agricultural Technology, focusing on sugar beet, potatoes, cereals and fruits. In Warsaw, Biezanko taught Agricultural Chemistry and Organic Chemistry at the College of Fruit Culture, directed formation courses for adults and was director and professor of Technological Chemistry. Finally, he taught Natural Sciences at Mickiewicz College in Warsaw.[13] Biezanko's professional experience continued with a 1926 experience at the Sugar Industry Industrial Laboratory in Warsaw, where he also wrote articles relating to beet for sugar production.[14] Such a comprehensive experience with different crops would characterise his activity in Brazil, where he wrote texts on the cultivation methods, uses and diseases of different crops such as soy, onion, coriander, ornamental and medicinal plants and many others.[15]

During his professional experience between 1920 and 1928, Biezanko met with farmers, agriculturalists, technicians and druggists, discussing various topics: 'drinking water, consumer items and their conservation, on milk, on sugar and its importance, or else, harmful insects'.[16] Overall, thanks to his combination of theoretical knowledge and practical experience, Biezanko was an active scientist who directly contacted producers and witnessed the agricultural development of several crops. When he reached Brazil in the

12 See E. Gardolinski, *Ceslau Mario Biezanko: entomólogo de fama mundial* (Curitiba, Gráfica Vicentina, 1965).

13 See João Pedro da Costa's biography in *Diário Popular*, Pelotas, 1981.

14 *Acta de sessão solene: Concessão de Título de cidadão pelotense a Ceslau Bieżanko*, 1971.

15 See, for example, *O uprawie soi. Lud*, 1934; *O pożytkach z soi. Lud*, 1934; 'A cebola', *Cruzeiro do Sul*, Rio Grande, 1935; 'O coentro, seu cultivo, propriedades e utilidades', *Anais do II Congresso de Agronomia*, Porto Alegre, 1940. Other crops include Tanacetum (1952), exotic plants (1964), Nigellas (1957), Ciano (1941). He also tried to raise carps and nutrias, with no great success.

16 João Pedro da Costa's biography published in *Diário Popular*, Pelotas, 1981.

early 1930s, he promptly distributed soybeans among Polish settlers in Guarani das Missões. He aimed to develop a new culture in the region and improve his compatriots' living conditions. In one of his last interviews after the explosion of soy farming, Biezanko confirmed that he brought two kilos of soy of the variety known as *Laredo* to Brazil, which he obtained from his colleague Professor Jan Muszyński.[17] He also discussed the distribution process, facilitated by creating an agricultural association featuring chapters in several locations of Guarani das Missões.[18]

Biezanko's experience in Guarani das Missões involved a complex web of factors. First, the scientist was an *outsider* to the community life of Polish immigrants in the region. Secondly, because he was an intellectual with a broad education, he distinguished himself from the settlers and even from the other teachers present in the colony, who were primarily immigrant farmers. Finally, he proposed the introduction of new products and encouraged different cultivation methods. As such, he was heavily dependent on pre-existing social structures under local parish priest Jan Wróbel.[19] The creation of an *Escola Agrícola* (agrarian school) and a *Centro Agrícola* (agrarian centre), the *União das Sociedades Agrícolas* (in Polish *Centralne Towarzystwo Rolnicze - CTR*), were part of a cooperation process between the agronomist, local parishes and the more distant schools.[20]

Initial farming efforts mostly concerned corn, with soy being a product of secondary interest in the colony.[21] This choice was presumably due to the settlers' need to work for self-subsistence and the provision of surpluses capable of generating some resources in the market. According to Chayanov's theory, soybean farming did not economically compensate for the efforts involved, making social endorsement an essential element for the continuity of small production. While soy would only gain prominence later, Biezanko's efforts also focused on introducing other novelties. These included flax, sorghum,

17   The *Laredo* soybean cultivar is an old forage variety which has its main use as animal feed. Forage soybeans could reach 1.2 m in height. They are dark and not suitable for oil production. Cf. https://hancockseed.com/products/laredo-soybean-seed
https://www.farmprogress.com/livestock/food-plots-and-forage-soybeans

18   See *Cotrifatos*, newspaper of the wheat cooperative of Santa Rosa, Cotrisa.

19   Trindade, 'Um cientista entre colonos'.

20   A.V. Chayanov, *La organización de la unidad económica campesina* (Buenos Aires: Ediciones Nueva Visión, 1974); H. Mendras, *Sociedades Camponesas* (Rio de Janeiro: Zahar, 1978); E. Wolf, *Sociedades Camponesas* (Rio de Janeiro: Zahar, 1976).

21   J. Krawczyk, *Z Polski do Brazylii: wspomnienia z lat 1916–1937* (Warszawa: Muzeum Historii Polskiego Ruchu Ludowego, 2003).

onions, new potato varieties, and pig and silkworm farming.[22] Aside from promoting different products through his writings, Biezanko also imported selected seeds of wheat, onions, Chinese tea and fruit tree seedlings.[23]

In fact, Biezanko did not even introduce soybeans to Guarani in the first place, as demonstrated by several pieces of historical evidence.[24] Instead, his main merit lay in creating socio-scientific networks capable of consolidating the utilisation of new crops such as soy among a conservative peasantry. So what was Biezanko's idea of soybean? Why did he decide to distribute it among the settlers as a product that would benefit them? Analysing his readings, it appears clear that he idealised the crop, defining it as 'one of the plants to which, in recent years, special attention has been paid, especially in Europe and North America, and holds a great number of useful properties'. In this light, it was 'no wonder that, in the experimental stations of many European and North American countries, countless experiments are related to its cultivation, yield, feeding off people and animals, and resistance to diseases and harmful insects, among many others'.[25] In this context, Biezanko attributed to the crop's 'useful properties' the main reason for the growth and development of research and its constant spatial expansion.

The Crop of the Future

According to Biezanko, soy had been known in the Far East since 'remote' times, in Japan, China and India at least 5,000 years ago. What he defined as the crop's widespread 'application and consumption' was motivated by 'the extraordinary nutritional value of its seeds'. At least since the seventeenth century, soy and its derivates had constituted a source of 'basic food in China, Korea, Manchuria and Japan'. In these regions, soy milk had even replaced cow's milk, a clear reflection of the crop's nutritional values and versatility and potential for international dissemination. Another essential characteristic was the crop's commercial value, since soybean seed, oil and flour had been traded from Asia (mainly China, Korea and Japan) to several continents such as Europe, North America and Australia, with England,

22   Ibid., p. 262.
23   Ibid., p. 260.
24   Trindade, 'Um cientista entre colonos'; J.A. de Assis Brasil and B.O. da Silva, 'A soja', *Contribuição do Departamento Estadual de Estatística à 2ª semana ruralista de Ijuí*, 1957.
25   C. Biezanko, *Algumas noções sobre a soja e seu cultivo: utilidades da soja* (Pelotas: Gráfica Artex, 1958).

Germany, Denmark and Holland as the leading importers.[26]

While in western Europe soybeans had been known since the mid-nineteenth century, according to the author, 'in Poland it was also known in the nineteenth century, but only after the Russo-Japanese War did the Poles (mainly officers who served in the Russian army and who returned) bring with them many seeds of various plants of the East'. These included 'several dozen varieties of soybeans', which were later found everywhere in Poland, under the name of 'Japanese beans', and were cultivated in both plantations and small plots and vegetable gardens.[27] Local selection and observation efforts would be coordinated by professors J. Muszyński and Z. Strazewicz at the Garden of Medicinal Plants at Stefan Bathory University, in Vilnius. According to Biezanko, these varieties derived 'the seeds we brought from Europe'.[28] Such information not only confirms Biezanko's relationship with Muszyński but also the historical link between the inception of soybeans in Poland from the mid-1900s onwards, and the Russo-Japanese War (1904–05). This information is particularly relevant considering that several conflict veterans later migrated to Guarani das Missões and helped Biezanko in his projects. In addition, the author confirmed his previous experiences with soybeans during his academic studies and the place from where the seeds came.

Biezanko's writings also detail experiments with soy at the Lavras Agronomic College Experimental Station (Estação Experimental da Escola de Agronomia de Lavras), in Minas Gerais. While travelling to central Brazil, he also found a 'successful' soy farming experience in São Paulo, with seeds presumably 'brought by Japanese emigrants'.[29] Thus, while Biezanko highlighted the existence of soybean farming experiments in Brazil prior to his experience in Guarani das Missões, he also pointed at the steep expansion of soybean farming due to his efforts. Specifically, he asserted that 'the seeds from Poland', which they distributed between north-eastern Argentina and Rio Grande do Sul, quickly spread 'among Polish, German and Italian settlers'. For Biezanko, soy was a plant that deserved to be cultivated on a larger scale; it had extensive utility, it was an easy plant to grow, and it was resistant to insects and diseases.[30]

26   Ibid., p. 5.
27   Ibid., p. 6.
28   Ibid., p. 6.
29   Ibid., p. 6.
30   Ibid., p. 6.

Rhuan Targino Zaleski Trindade

Biezanko believed in the product's potential as a plant capable of bringing agricultural development, income and a better quality of life. In this context, he would fulfil the role of agricultural instructor, creating new bonds between Poland and its emigrants. In order to convince farmers to adopt this product, Biezanko needed to dwell on its qualities both in terms of potential processing products and relative cultivation ease.[31] Thus, Biezanko detailed the crop's characteristics, comparing it to known ones such as beans and peas. He described the shapes of its leaves, stems and pods, the various colouring of its seeds and its adaptation capacities to different climates and soils. In terms of adaptation capacities, he described soy as 'not a very demanding plant regarding climate and soil', generally 'more resistant to drought and perhaps more sensitive to moisture than peas'.[32] The best results would be obtained by planting it in soils rich in humus and clay, even alternating with corn. He also discussed the relationship between soybeans and types and quantities of fertilisers such as manure, artificial fertilisers and lime.[33] Perhaps more importantly, he illustrated the crop's relationship with soil bacteria, the depth needed for effective ploughing and the sowing time (from early spring in September to early December, soy seeds needed to be covered with 4–7 centimetres of soil, depending on the type of seed and the latitude).

Biezanko also produced detailed information on the relation between the required number of seeds and the required space in the planting process; for '24 x 40 cm, we need 20 to 40 kg of seeds per hectare'.[34] Such details were important for local farmers dealing with the crop for the first time, especially those willing to experiment with small amounts. He remarked that under favourable conditions, 'soybeans germinate easily, and the tips of the calluses will appear in a few days. Development is fast. Right after germination, soybeans are very similar to beans. Therefore, we advise planting it in flat fields, which facilitates harvesting.'[35] In this context, different varieties of soybeans could be chosen according to the 'cultivation purpose'. These

31   Trindade, 'Um cientista entre colonos'.

32   Biezanko, *Algumas noções sobre a soja*, p. 7.

33   As for fertiliser, he suggested the use of 336 kg of calcium phosphate and 280 kg of wood ash per hectare, or 28 kg of potassium chloride. These fertilisers must be distributed evenly before planting. Furthermore, the use of lime to treat soil would increase soy production, neutralising acids and making the nitrogen contained in the humus usable, providing the soil with a granular structure.

34   Biezanko, *Algumas noções sobre a soja*, p. 9.

35   Ibid., p. 9.

included a light green variety of tiny seeds consumed for human nutrition, a white and dark red one used for forage, a green one with large seeds and black and light yellow ones to be processed into oil.[36] These could be used in different ways – either as dry fodder (which could take up to ten tons per hectare), green manure and seeding. The average yield per hectare was about 1,200 kilos, but it could be higher depending on the variety.

Predictably, in describing his personal experiences with soybeans, Biezanko primarily focused on the successful impacts of soy in mixed farming regimes with crops such as corn, peas and sorghum, not dwelling on issues and failures. He only discussed the potentially harmful parasites, such as root nematodes, coleopteran larvae, caterpillars, grasshoppers and several Lepidoptera.[37] In this context, it is worth recalling that Biezanko specialised in entomology and was particularly passionate about Lepidoptera, which he collected.[38] While providing relevant information on potential harmful pests, Biezanko stressed that 'none of the enemies and pests' listed could cause significant damage to soy yields.[39] He predicted a rosy future for soybean farming in Brazil, declaring that it would quickly fulfil the country's nutrition needs and soon become a significant export. The planted area would increase each year, becoming the first among oilseeds. The crop's multiple uses and its sturdiness unquestionably made it 'the economic plant of the future'.[40]

Soybeans and Their Impacts, the Utilities of the New Product

After detailing the biological assets of soy, in the second part of his treatise, Biezanko analysed more in-depth its uses. First, he dwelled on the nutritional aspect. Due to its high healthy protein and fat content, the crop could easily replace meat. In addition to eating cooked beans or green peas, soybeans with green seeds could feed humans in multiple ways. Their tender buds could be stewed into a 'tasty soup'. They could be ground into flour, which could make biscuits, cakes, sweets and pasta. Soy flour, combined with wheat, could be even used to treat kidney illnesses and nervous diseases. Soy could also be

36   Ibid., p. 10.
37   Ibid., p. 11.
38   Trindade, 'Um cientista entre colonos'.
39   Biezanko, *Algumas noções sobre a soja*, p. 11.
40   Ibid., p. 11.

turned into milk, butter, cheese and casein.[41] Biezanko's description of the multiple soy derivates for nutrition coincided with their adoption among Polish immigrants.[42] Therefore, the processing of the product was the main target of his observations, as well as its potential for industrialisation, which could potentially bring income to Polish settlers.

Biezanko paid particular attention to 'soy oil', which could replace butter and lard and be used to manufacture soaps, varnishes, lubricants and even lighting. A large part of his work was dedicated to describing the process of oil extraction. He described the modalities of seed squeezing, as seeds were ground in mills that reduced them to fragments and later, boiled and pressed.[43] About 100 kilos of soy could yield fifteen kilos of oil, and the leftovers from the extraction could still be used as flour and fodder. In addition, soy oil could be used to produce glues, paints and celluloid. Soy oil could propel industrial development and entrepreneurship among Polish communities, given its great industrial versatility and potential.[44] The second area of use for soy was animal fodder, perhaps the most relevant to local farmers. According to Biezanko, the crop showed better results than alfalfa and mixed with other foods, such as corn. In addition, the seeds could be used in a cooked (unsalted) form for poultry and fermented. With all these remarkable qualities, soybean was 'a plant of so many uses, easy to grow, with good yield and profit' that would without doubt constitute 'a new source of profit and well-being'.[45]

In order to further demonstrate its utility, Biezanko brought data about the impacts of its expansion, showing an increase in the planted area in the state of Rio Grande do Sul, from 1,050 hectares in 1943 to 6,200 in 1954 and production rates projected to increase from 480 tons in 1942 to 77,100 in 1952.[46] In this context, the state of Rio Grande do Sul was the most prominent national producer, with the municipalities of Santa Rosa and São Luiz Gonzaga (where Guarani das Missões was located) as the main

---

41   Ibid., p. 12.

42   Krawczyk, *Z Polski do Brazylii*.

43   Biezanko, *Algumas noções sobre a soja*, p. 13.

44   Guarani das Missões saw one of the first soy oil processing companies in the northwest region of Rio Grande do Sul. Created by Polish settlers around 1930s, the industry was initially dedicated to flaxseed. With the development of the soybean crop production in the mid-1950s and 1960s, it would mostly reconvert to the production of soy oils and other derivatives.

45   Biezanko, *Algumas noções sobre a soja*, p. 16.

46   Ibid., p. 16.

hubs, producing respectively 59,784 and 15,792 tons.[47] Scholar Líbia Martins Wendling confirmed the data provided by Biezanko. In her research in Guarani das Missões between 1966 and 1967, she identified soy as already the main product, with a total of 88,000 bags produced, over 70,000 litres of oil, principally traded in Santa Rosa and Porto Alegre and two specialised factories.[48] According to Wendling, soybeans would ensure the town's prosperity, considering that 'if the cultivation of soybean were not great, the colony would be poor'.[49] With the increase in prices, local inhabitants expanded soy production on a larger scale, leading to growth in international exports and the expansion of processing facilities. Such trends corresponded to introducing new varieties imported from the United States, and the development of mechanised agricultural techniques presented as almost a magic solution to improve rural livelihoods. Such a transformation also radically transformed local farmer societies, indissolubly linking the fate of second and third-generation European immigrants (by now not only Polish but also Germans and Italians) to the success of soybeans. Social transformations essentially consisted of the radical switch to a capital-oriented production system, stimulated by widespread modernisation. Changes in settlers' production systems and the local environment were significant. The transition from small-scale mixed farming regimes to large-scale monocultures relied on financial support and international market trends.

Thus, although Biezanko's essay seemed to suggest a direct link between the success of local soybean farming businesses and his efforts, the success of soybeans since the 1960s cannot be directly linked to one man, but to large-scale global changes in agricultural techniques and new market pressures. Different reports on the failures of soybean production in the region during the early 1930s seem to confirm this information. These were mainly linked to problems with animal feed (diseases) and difficulties in handling increasing production volumes, as well as endogenous issues of ethnicity, disputes between local leaders and relations with the Polish government, which would eventually lead to Biezanko's abandonment of the colony in 1934, leaving his work unfinished.

47   de Assis Brasil and da Silva, *A soja*. Also see IBGE, *Estatísticas históricas do Brasil: séries econômicas, demográficas e sociais de 1550 a 1988* (Rio de Janeiro: IBGE, 1990). Soybean 2021 production rates in Brazil are estimated at 134 million tons (see https://www.ibge.gov.br/busca.html?searchword=soja).

48   L.M. Martins Wendling, *O imigrante polonês no RS* (São Leopoldo: Universidade do Vale do Rio dos Sinos, 1971).

49   Ibid., p. 26.

Rhuan Targino Zaleski Trindade

Understanding the Polish-Brazilian Connection

During the early 1930s, a highly specialised Polish scientist, with ambitious views on the possibilities of modernisation, industrialisation and social development for his oversea compatriots, campaigned for soybean farming among Polish farmers who emigrated to north-western Rio Grande do Sul. His action was part of a diplomatic relationship between the Polish emigrants and their country of origin, resulting in the creation of socio-agrarian networks and the rise of local associations reiterating traditional religious and cultural values.

The combination of these factors contributed to the spread of soybean farming in the region and the acquisition of new notions on its uses and processing potential. The subsequent expansion of soy production in southern Brazil turned Biezanko into a revered public figure, both among Polish rural communities and the local intelligentsia. Although there were significant discontinuities between Biezanko's experience during the 1930s and the massive expansion of soy production in the following decades, the region certainly bore his distinguishing mark in terms of familiarity with the crop and its multiple utilities. While not eminently successful in outcomes, his experience served as a test for the subsequent soy boom in the region. Ultimately, Biezanko's main merit was to stimulate a breakdown in the rural communities' rather conservative mindset, encouraging the adoption of crops that he considered potentially beneficial for Polish communities and their descendants. Although three full decades were needed before soybeans would become the region's leading crop, many of the practices suggested by Biezanko came into being. His impact figured not simply in the influence of southern Brazilian farmers upon the national expansion of the soybean farming sector from the 1970s, but also in the different uses of soy derivatives which he had envisioned as early as 1934.

Biezanko championed soy and its biological qualities, such as easy growth, resistance to pests and versatility. He also promoted different experiments, encouraging Polish immigrants to adopt it, and described the potential of soy derivate for economic development and the improvement of local livelihoods. He considered soy as the future plant, which would bring economic growth and improve livelihoods. This experience, while little mentioned in many studies about soy, is essential for those willing to understand how this Asian crop rose to prominence in the Latin American Southern Cone within a few decades. Biezanko's capacity to communicate the potential of soy to a broader audience provides an exemplar of the link between global

3. The Polish-Brazilian Soy Connection

history and local experience, radical changes and local memory. Ultimately, the Polish-Brazilian soy connection was a multi-layered process uniquely shaped by specific cultural identities, political agendas and practical needs. Perhaps most importantly, it was shaped by the capacity of scientists such as Biezanko to illustrate the qualities of a new product such as soy and champion its introduction, negotiating the itch for scientific discoveries in tandem with down-to-earth forms of civic commitment.

Part II

Soybean Markets During the Great Acceleration

CHAPTER 4.

GREAT ACCELERATIONS: SOY AND ITS GLOBAL TRADE NETWORK, 1950–2020

Ernst Langthaler

Introduction

Regardless of whether they locate its 'big bang' thirty, 200, or 500 years ago, economic historians have repeatedly identified long-distance trade as a key feature of globalisation.[1] In historical research on global trade and related social and natural issues, three alternative interpretations shape the debate. The Great Specialisation narrative argues that global market integration through long-distance trade from the 1820s onwards connected areas of different factor endowments according to the principle of 'comparative advantage' (primary vs manufactured products), thereby fuelling economic growth, raising public welfare and easing environmental pressures.[2] The Great Divergence narrative concentrates on exploitative transfers of commodities ('unequal exchange') from poor and weak peripheries to rich and powerful core areas in the 'long nineteenth century'.[3] The Great Acceleration narrative downplays the impacts of the above-mentioned trade patterns by emphasising the shift of worldwide resource extraction, exchange, processing, usage and deposition on an unprecedented scale from the mid-twentieth century onwards.[4] Though

1   K.H. O'Rourke and J.G. Williamson, 'When did globalisation begin?' *European Review of Economic History* **6** (1) (2002): 23–50.

2   R. Findlay and K.H. O'Rourke, *Power and Plenty: Trade, War, and the World Economy in the Second Millennium* (New Jersey: Princeton University Press, 2009), pp. 365–428; K.H. O'Rourke and J.G. Williamson, *Globalization and History: The Evolution of a Nineteenth-Century Atlantic Economy* (Cambridge, MA: MIT Press, 1999).

3   K. Pomeranz, *The Great Divergence: China, Europe, and the Making of the Modern World Economy* (Princeton: Princeton University Press, 2009), pp. 211–97; P. Vries, *State, Economy and the Great Divergence: Great Britain and China, 1680s–1850s* (London: Bloomsbury Academic, 2015), pp. 381–407.

4   J.R. McNeill and P. Engelke, *The Great Acceleration: An Environmental History of the Anthropocene Since 1945* (Cambridge, MA: Harvard University Press, 2016), pp. 103–54.

THE AGE OF THE SOYBEAN: 65–90 **doi:** 10.3197/63800040695086.ch04

all these narratives are rich in theoretical arguments, they still need further empirical validation regarding historical and geographical reach.[5]

This chapter adopts a commodity-focused network approach to international trade as a key feature of agro-food globalisation in the era of Great Acceleration. It conceptualises agro-food globalisation as succession of food regimes, i.e. bundles of inter- and transnational power relations connecting food production, distribution and consumption.[6] In addition to statistical analyses of aggregate data on exports and imports, the study includes network analysis for better capturing the (dis-)connective character of (de-)globalisation.[7] Since network analysis requires data on country-to-country commodity flows, the investigation faces some restrictions. While the international trade matrix of agricultural commodities on an annual basis is available from the Food and Agriculture Organisation (FAO) from 1986 only, the network analysis of the previous period builds upon a separate data collection by the US Department of Agriculture (USDA). In addition, FAO figures on exports and imports at the country level complement the dataset.[8]

The international network of agricultural trade is assessed through the lens of soy, which emerged as a commodity of global importance from the mid-twentieth century onwards. Given the socionatural impacts of the worldwide soy expansion, the Anthropocene can even be conceived as the 'Soyacene'.[9] Besides numerous country-case studies, global accounts of soy

5   J. Brolin and A. Kander, 'Global trade in the Anthropocene: A review of trends and direction of environmental factor flows during the Great Acceleration', *The Anthropocene Review* **3** (5) (2020): 1-40; J. Brolin and A. Kander, 'Environmental factors in trade during the Great Transformation: Advancing the geographical coverage before 1950', *Journal of Global History* **15** (2) (2020): 245–67.

6   A. Magnan, 'Food regimes', in J.M. Pilcher (ed.), *The Oxford Handbook of Food History*, pp. 370–88 (Oxford: Oxford University Press, 2012).

7   As a paradigmatic example, see D.A. Smith and D.R. White, 'Structure and dynamics of the global economy: Network analysis of international trade 1965–1980' *Social Forces* **70** (4) (1992): 857–94.

8   US Department of Agriculture, *World Trade in Selected Agricultural Commodities, 1951-65: Vol. 5: Oilseeds, Oil Nuts, and Animal and Vegetable Oils* (Washington, DC: USDA, 1968); Food and Agriculture Organization, *Yearbook of Food and Agricultural Statistics, Part 2: Trade, 1948–1961* (Rome: FAO, 1949–1962); FAO, *Faostat*, http://www.fao.org/faostat (accessed 31 July 2021).

9   C.M. da Silva and C. de Majo, 'Towards the Soyacene: Narratives for an environmental history of soy in Latin America's Southern Cone', *HALAC (Historia Ambiental Latinoamericana y Caribeña)* **11** (1) (2021): 329–56.

as a commodity are still rare.[10] Building on the author's previous work,[11] this chapter goes beyond conventional analyses of country-level export and import figures by adopting a network perspective on international soy trade. It aims at answering the following questions: first, how and why global soy trade unfolded in temporal and spatial terms; second, how and why international links and (sub-)national nodes composed the global soy trade network; third, how and why driving forces and ruling actors at multiple levels shaped soy's trade network and its socionatural outcomes. The results add complexity to simplistic notions of agricultural trade in the Great Acceleration through temporal and spatial differentiation.

Global Contours of Soy Trade

From the mid-twentieth century onwards, soy emerged as the world's most valuable commodity of agricultural trade. The soybean was domesticated from a wild species in China and has been cultivated there as a food crop for millennia.[12] In the twentieth century, the bulk of cultivation shifted from East Asia to North and South America where soy emerged as a cash crop to be sold in domestic and world markets. The harvested bean contains a unique combination of about forty per cent protein and twenty per cent fat, thus enabling flexible uses in its whole form (e.g., as a food ingredient) and, after processing, in separated form as oil (e.g., as industrial raw material) and cake (e.g., as animal feed).[13] The shifting rankings of agricultural commodities according to trade value impressively illustrate soy's post-war rise

---

10   I. Prodöhl, 'Versatile and cheap: A global history of soy in the first half of the twentieth century', *Journal of Global History* **8** (3) (2013): 461–82; C.M. Du Bois, C-B. Tan and S. Mintz (eds), *The World of Soy* (Urbana: University of Illinois Press, 2008); C.M. Du Bois, *The Story of Soy* (London: Reaktion Books, 2018).

11   E. Langthaler, 'Gemüse oder Ölfrucht? Die Weltkarriere der Sojabohne im 20. Jahrhundert', in C. Reiher and S.R. Sippel (eds), *Umkämpftes Essen: Produktion, Handel und Konsum von Lebensmitteln in globalen Kontexten*, pp. 41–66 (Göttingen: Vandenhoeck & Ruprecht, 2015); E. Langthaler, 'The soy paradox: The Western nutrition transition revisited, 1950–2010', *Global Environment* **11** (1) (2018): 79–104; E. Langthaler, 'Ausweitung und Vertiefung: Sojaexpansionen als regionale Schauplätze der Globalisierung', Österreichische Zeitschrift für Geschichtswissenschaften **30** (3) (2019): 115–47; E. Langthaler, 'Broadening and deepening: soy expansions in a world-historical perspective', *HALAC (Historia Ambiental Latinoamericana y Caribeña)* **10** (1) (2020): 244–77.

12   G.-A. Lee et al., 'Archaeological soybean (*Glycine max*) in East Asia: Does size matter?,' *PloS One* **6** (11) (2011): 1–12.

13   C.M. Du Bois and S. Mintz, 'Soy', in S.H. Katz (ed.), *Encyclopedia of Food and Culture*, pp. 322–26 (New York: Scribner, 2003); T. Sorosiak, 'Soybean', in K.F. Kiple and K.C. Ornelas (eds), *The Cambridge World History of Food*, pp. 422–27 (Cambridge: Cambridge University Press, 2000).

to the top: in 1961, aggregate beans, oil and cake ranked twelfth, amounting to only one quarter of the trade value of wheat, which was the leading crop. In 2018, soy products had already gained first place, surpassing wheat more than twofold. A huge stream of soy flooded the highly concentrated world market, controlled by a handful of transnational corporations (TNCs): 153 megatons (Mt) of soybeans, 67 Mt of soycake and 11 Mt of soyoil, adding up to a trade value of 94 billion US dollars.[14] Accordingly, in a 2017 headline the *Financial Times* enthusiastically labelled soy as the 'crop of the century'.[15]

The expansion of global soy trade can be regarded as a feature of the Anthropocene in general and, in particular, of the Great Acceleration: the skyrocketing extraction, circulation, processing, usage and excretion of na-tural resources by (post-)industrial societies based on mass production and consumption from the 1950s onwards.[16] The volume of soy trade, including beans, oil and cake, grew 263-fold from 880,000 t in 1950 to 231 Mt in 2018. Rather than expanding steadily over the period, soy trade experienced two separate boosts, separated by a period of stagnation. This tripartite periodi-sation is indicated by shifting average annual growth rates from decade to decade (1950s: 19.8 %, 1960s: 13.4 %, 1970s: 12.3 %, 1980s: 2.1 %, 1990s: 5.2 %, 2000s: 5.6 %, 2010s: 5.2 %). According to these figures, the Great Acceleration of the societal appropriation of natural resources since the mid-twentieth century has encompassed two accelerations of soy trade: the *First Great Acceleration,* with rapid growth from a low level of trade volume between the 1950s and 1970s, and the *Second Great Acceleration* with more moderate growth from a much higher level from the 1990s onwards. In the 1980s, the growth of soy trade decelerated significantly and stagnated on a medium level (Figure 1).

Global soy trade unfolded unevenly not only in temporal but also in spatial terms. The distribution of net exporting and net importing countries in the post-war decades reveals an international division of labour that involved world regions as diverse as North and South America, Western Europe and East Asia. The First Great Acceleration was more or less limited to the Global North, with the USA as the dominant exporter and Western Europe and Japan as major importers. During the deceleration and Second Great

14   Data source: FAO, *Faostat*, http://www.fao.org/faostat (accessed 31 July 2021).

15   *Financial Times*, 20 June 2017: https://www.ft.com/content/35af007e-49f6-11e7-919a-1e14ce4af89b (accessed 31 July 2021).

16   McNeill and Engelke, *The Great Acceleration*; Brolin and Kander, 'Global trade in the Anthro-pocene'.

4. Great Accelerations

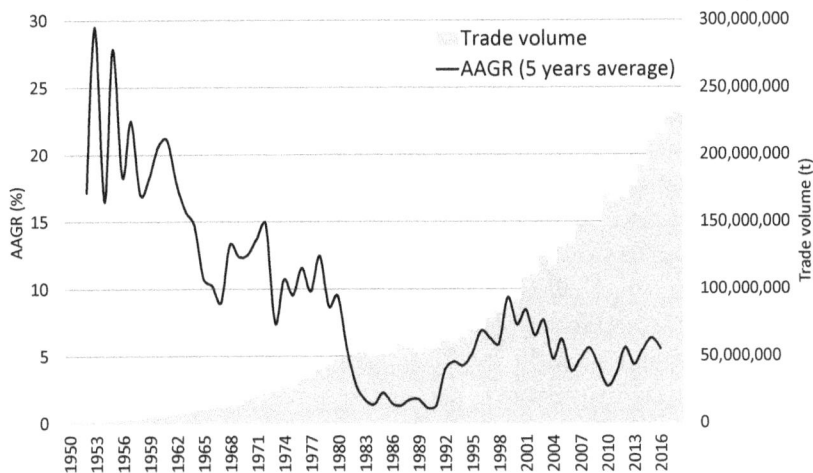

Figure 1.

Global trade in soy products, 1950–2018 (1950–1960: excl. soycake).[17]

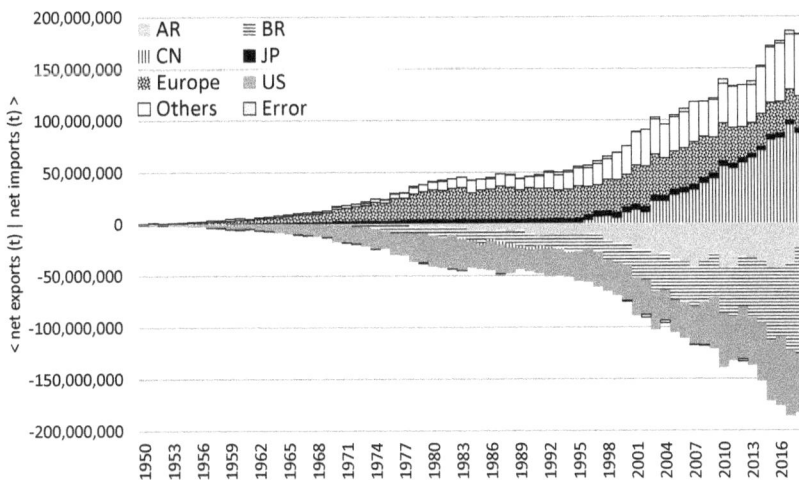

Figure 2.

Physical trade balance of soy products, 1950–2018 (1950–1960: excl. soycake).[18]

17   Data source: Food and Agriculture Organization, *Yearbook of Food and Agricultural Statistics, Part 2: Trade, 1948–1961*; FAO, *Faostat*, http://www.fao.org/faostat (accessed 31 July 2021).
18   Data source: Ibid.

Acceleration, the Global South entered the scene, with Brazil and Argentina as major exporters and China as the dominant importer. The Northern and Southern trade connections did not simply overlap; they amalgamated in complex ways. Though other exporters and importers emerged, the rest of the world played a minor role in the world market for soy (Figure 2).

In a market economy, supply and demand of soy as a commodity are reflected by its price. The series of real (i.e., inflation-adjusted) prices of soy products in the world market reveals different cycles. Real prices tended to fall by about one half from the 1960s to the 1990s, the era of 'cheap soy.' However, expectations of a long-term downward trend in commodity prices were disappointed in the 2000s: real prices almost doubled over a decade, indicating a 'soy boom' as part of a comprehensive commodity boom, before falling again from the early 2010s. These long- and medium-term cycles were interrupted by short-term price spikes, the strongest of which occurred in 1973/74 (Figure 3). Though market prices played an important role in global soy trade, other long-, medium- and short-term impacts were relevant as well. The 1973/74 price spike provides a telling case: it resulted neither from a single fall in supply nor from a single rise in demand, but from multiple human and non-human impacts, including weather conditions, oceanic currents, fish migrations, trade policies, business strategies and public discourses.[19] Consequently, any analysis of global soy trade has to take into account this complex bundle of (ecological, economic, political, cultural, social etc.) relations beyond the simplistic notion of 'market forces.'

Soy's temporally and spatially uneven emergence as a key commodity in the post-war era was embedded in the dynamics of global capitalism in general and food regimes in particular.[20] Global food regimes hierarchically connect different regions of production and consumption through dominant modes of accumulation and regulation. Capitalist accumulation tends to maximise value extraction through expanding the frontiers of a commodity chain to encompass human and non-human resources – labour and nature – that have not yet been incorporated. Capital accumulation through market expansion, including social and natural disruptions, is often contested between the actors involved. In order to proceed, it needs to be regulated

19   M. Roth, *Magic Bean: The Rise of Soy in America* (Lawrence, Kansas: University Press of Kansas, 2018), pp. 196–99; R. Patel, *Stuffed and Starved: Markets, Power and the Hidden Battle for the World Food System* (London: Portobello Books, 2008), pp. 181–87.

20   F. Krausmann and E. Langthaler, 'Food regimes and their trade links: A socio-ecological perspective', *Ecological Economics* 160 (2019): 87–95.

4. *Great Accelerations*

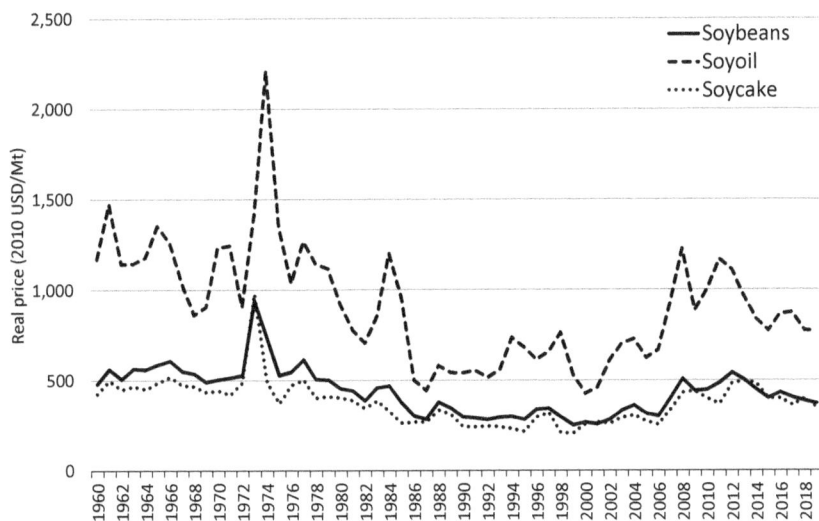

Figure 3.

World real price of soy products, 1960–2019.[21]

according to the powerful interests and values of nation states, capitalist enterprises and social movements. Once a regime can no longer contain its endogenous and exogenous tensions, it falls into crisis and eventually transforms into another regime. Food regime scholars distinguish the 'first', 'British-centred' or 'colonial-diasporic' regime (1870–1929), the 'second', 'US-centred' or 'mercantile-industrial' regime (1947–1973), and the 'third', 'WTO-centred' or 'corporate-environmental' regime (since 1995).[22] Food regime theory provides a framework for investigating the ways in which soy's trade network both *shaped* and *was shaped by* capitalist globalisation.

21   Data source: http://pubdocs.worldbank.org/en/226371486076391711/CMO-Historical-Data-Annual.xlsx (accessed 31 July 2021).

22   P. McMichael, *Food Regimes and Agrarian Questions* (Halifax, Winnipeg: Fernwood Publishing, 2013); H. Friedmann, 'From colonialism to Green capitalism: Social movements and emergence of food regimes', in F.H. Buttel and P. McMichael (eds), *New Directions in the Sociology of Global Development*, pp. 227–64 (Amsterdam: Elsevier, 2005); J.W. Moore, *Capitalism in the Web of Life: Ecology and the Accumulation of Capital* (London, New York: Verso, 2015).

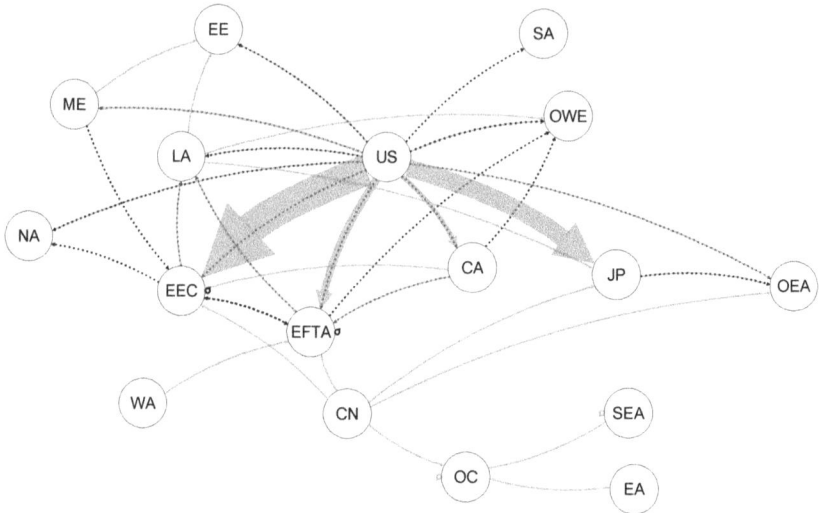

Figure 4.

The global soy trade network in 1962 (top-fifty flows).[23]

Legend: ⇒ soybeans, ⇢ soyoil, CA: Canada, CN: China and other Communist Asia, EA: East Africa, EE: Eastern Europe, EEC: European Economic Community, EFTA: European Free Trade Association, JP: Japan, LA: Latin America, ME: Middle East, NA: North Africa, OC: Oceania, OEA: other East Asia, OWE: other Western Europe, SA: South Asia, SEA: Southeast Asia, US: USA, WA: West Africa.

International Links of Soy's Trade Network

Soy's trade network in the First Great Acceleration evolved in the Western Hemisphere during the Cold War with the USA as political, economic and cultural hegemon. As shown by the network of the top-ifty soy flows (excluding cake) in terms of trade volume in 1962, the USA represented the central node with eleven export links, accounting for 87.9 per cent market share. Major flows of US soybeans (4.4 Mt in total) were shipped to Western Europe (EEC: 1.9 Mt, EFTA: 547,000 t) and Japan (1.1 Mt), while the rest went to Canada, the Middle East and non-Communist East Asia. Minor streams of US soyoil (560,000 t in total) targeted different parts of Europe (including Communist states), Asia, Africa and Latin America. Meanwhile, the People's Republic of China could not maintain the global

23 Data source: US Department of Agriculture, *World Trade in Selected Agricultural Commodities, 1951–65.*

market dominance of Manchurian soy during the interwar period. China's five export links (342,000 t in total, 6.1 per cent market share) went to East Asia and Oceania, particularly Japan, and Western Europe. Latin America (five export and three import links, 2.2 per cent market share) and Canada (three export and one import links, 1.7 per cent market share), both new-comers on the global scene, were playing marginal roles as suppliers of soy products by the 1960s (Figure 4).

The global soy trade network in 1962 was inserted into the US-centred (or mercantile-industrial) food regime. Due to its technology-driven wartime mobilisation of productive resources, the USA as post-war superpower gained dominance in the Western segment of the world market, demarcated by the 'Cold War dam.' State-subsidised surpluses of agricultural commodities, including soy, served not only as sources of revenue for farmers, processors and traders but also as geopolitical weapons of the Western regime of 'cheap food' that fought both poverty and communism with American-style deve-lopment. Based on the 1944 Bretton Woods System and the 1947 General Agreement on Tariffs and Trade (GATT) as key institutions, streams of agricultural commodities flooded overseas markets. US soybeans targeted Western Europe and Japan as food aid (under the 1948–52 European Re-covery Program) or duty-free commodities (after the 1960/61 Dillon Round of the GATT). After crushing, soycake as a main product fed the expanding livestock complex that served the middle-class appetite for prestigious animal products. The remaining soyoil as by-product was either processed by the domestic food industry (cooking oil, margarine, dressings, etc.) or exported to food-deficient 'Third World' countries under the 1954 'Food for Peace' programme. Consequently, 'cheap soy' became a hidden but effective ingredient of industrialised agriculture (i.e., intensification, specialisation and concen-tration) as well as nutrition (i.e., 'meatification' and 'oilification'), thereby enlarging the 'ecological footprint' of affluent societies.[24] At the bottlenecks of state-regulated agro-food value chains, processors and traders – first and foremost the 'ABCD companies' (US-based ADM, Bunge and Cargill and French-based Dreyfus) – emerged as powerful players in the world market.[25]

24 Between 1961 and 1973, the ecological footprint, measured by the number of earths needed to support the average resource use per capita, increased from 2.6 to 4.3 in the USA and from 1.4 to 2.7 in Western Europe. Data source: Global Footprint Network, https://data.footprintnetwork. org (accessed 31 July 2021).

25 McMichael, *Food Regimes and Agrarian Questions*, pp. 21–40; Friedmann, 'From colonialism to green capitalism', 240–45; Bill Winders, *The Politics of Food Supply: U.S. Agricultural Policy in the World Economy* (New Haven, CT: Yale University Press, 2009), pp. 31–158; Langthaler, 'The soy paradox'.

Ernst Langthaler

By the 1980s, the period of deceleration, US dominance in global soy trade became increasingly contested by emerging exporters from South America. As the network of the top-fifty trade links in 1987 reveals, the USA still represented the central node with nineteen export links, representing 57.3 per cent market share. The bulk of US exports were soybeans (17.8 Mt in total) destined for the long-established markets in Western Europe (Netherlands: 3.9 Mt, Spain: 1.6 Mt, others: 2.7 Mt) and Japan (3.8 Mt) as well as for geopolitically sensitive areas such as Taiwan, South Korea, Mexico, Israel and Brazil (4.9 Mt). Small tonnages of US soybeans (837,000 t) went to Communist countries, namely China and Romania. US soycake (3.5 Mt in total) was shipped mainly to Western Europe (2.2 Mt), as well as Canada, Venezuela and the USSR. In the shadow of US dominance, Brazil (13 export links and one import link, 24.2 per cent market share) and Argentina (four export links, 9.9 per cent market share) had emerged as remarkable soy exporters, though not directly competing in US soybean markets. Brazil specialised in soycake (6.4 Mt in total) for overseas markets in Western Europe (Netherlands: 1.4 Mt, France: 1.2 Mt, others: 1.2 Mt), Eastern Europe (1.7 Mt) and the USSR (915,000 t). In addition, Brazilian soybeans (2.0 Mt in total) began to challenge US market dominance in Western Europe (1.7 Mt) and Japan (301,000 t). Considerable exports of Brazilian soyoil (573,000 t in total) flowed to Iran and India. Argentina also specialised in soycake exports (2.6 Mt in total), complemented by exports of soybeans (1.1 Mt in total), supplying first and foremost Western Europe (cake: 1.2 Mt, beans: 437,000 t) and the USSR (cake: 1.0 Mt, beans: 636,000 t). The reach of China (four export and one import links, 4.6 per cent market share) as an exporter of soybeans (1.4 Mt in total) and soycake (319,000 t in total) was limited to neighbouring countries such as the USSR, Hong Kong, Japan and Indonesia. Within Western Europe, Belgium-Luxembourg (three import links and one export link, 1.8 per cent market share) and the Netherlands (one export link and four import links, 1.1 per cent market share) stood out not only as importers of North and South American soy (mainly beans), but also as re-exporters (mainly cake and oil) to neighbouring countries (Figure 5).

The global soy trade network in 1987 was affected by a crisis of the US-centred food regime that involved multiple elements: in the world food market, the shift from abundance to scarcity in the 1972–75 'world food crisis,' triggered by major US grain sales to the USSR and cuts of food aid; in the international monetary system, the dissolution of the Bretton Woods System by the US, EEC and Japanese governments 1971–73; in the

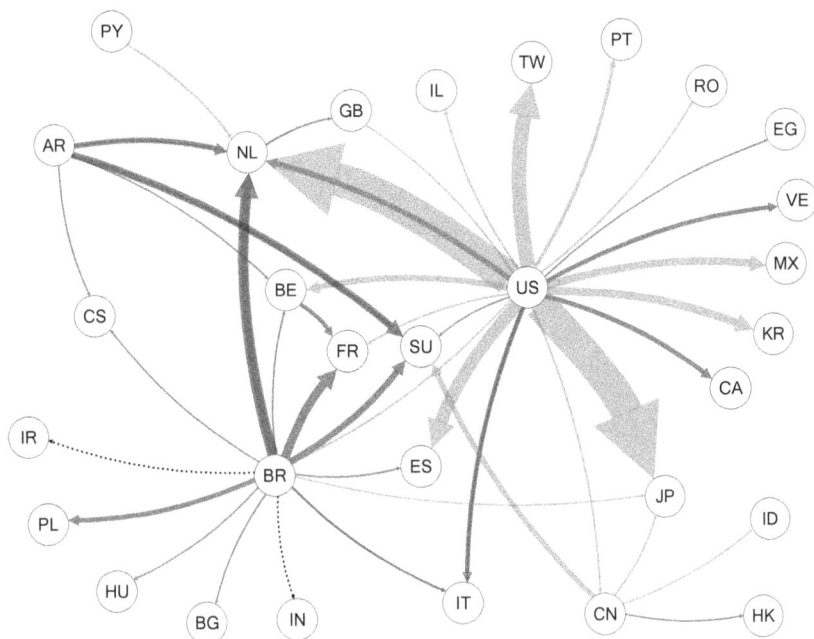

Figure 5.

The global soy trade network in 1987 (top-fifty flows).[26]

Legend: � soybeans, ⤍ soyoil, ➡ soycake, ISO country codes

USA, the shift of power from farmer lobbies to TNCs as agents of market liberalisation; in Western Europe, state-sponsored dumping of agricultural surpluses due to the adoption of the US model of productivist farming, thus provoking international trade disputes; in Eastern Europe, the dissolution of the 'Communist bloc' and its capitalist transformation after 1989; among emerging countries, export-oriented members of the Cairns Group (including South America's 'soylandia'), pressing for liberalisation of agricultural trade in the 1986–94 Uruguay Round of the GATT; among developing countries, over-indebtedness due to deteriorating terms of trade, a problem tackled by World Bank credits in combination with 'structural adjustment

26   Data source: FAO, *Faostat*, http://www.fao.org/faostat (accessed 31 July 2021).

programmes'.[27] In addition to these general tendencies, soy played a special role in the food regime crisis. Rising international demand in reaction to the El Niño-induced shortfall of Peruvian fishmeal, massive feed purchases by the USSR and expanding livestock complexes in Europe and Japan in combination with falling domestic supplies led the US government to impose an export embargo on soybeans in 1973. Though soon relieved, the embargo eroded the trustworthiness of the USA as a trading partner, thus leading European and Japanese customers to search for alternative suppliers.[28]

In the Second Great Acceleration, the centre of gravity of soy's trade network shifted from the Global North to the Global South, as depicted by the network of the top-fifty commodity flows in 2018. According to trade relations, the USA still represented the central node with eighteen export links, but its market share had decreased to 27 per cent. The bulk of US soybeans (77.0 Mt in total) flowed to China (8.2 Mt), Japan (2.3 Mt) and other East and South-east Asian countries (10.7 Mt), followed by Western Europe (Netherlands: 3.5 Mt, others: 2.0 Mt), Mexico (4.8 Mt) and Egypt (3.2 Mt). US soycake (4.6 Mt in total) mostly targeted countries in its vicinity. In terms of trade volume, however, Brazil had already left the USA far behind, commanding 55.2 per cent market share through thirteen export links. Of Brazil's total soybean exports (77.0 Mt), its shipments to China (68.8 Mt) were not only the country's but the world's biggest commodity flow, complemented by sales to European and Asian countries. Apart from the main axis to China, the Brazilian-centred trade network channelled soycake (13.3 Mt in total) to South-east Asia (7.0 Mt) and Europe (Netherlands: 2.7 Mt, others: 3.7 Mt). Besides Brazil, Argentina, with twelve export and two import links, accounted for a 12.3 per cent market share and was another challenger for US dominance in the world market for soy. Bean shipments to China (3.4 Mt) were the exception from the Argentinian rule of specialising in export-oriented soy processing. Accordingly, Argentina marketed soycake (14.8 Mt in total) to South-east Asia (6.6 Mt), Europe (5.0 Mt) and other countries, while selling soyoil (1.9 Mt in total) to India. Other exporters of minor importance included Paraguay (one export link, 2.5 per cent market share) and Canada (one export and one import link, 2.2 per cent market share), both providing soybeans for the Argentinian (4.1 Mt) and Chinese

27   McMichael, *Food Regimes and Agrarian Questions*, pp. 38–39; Friedmann, 'From colonialism to green capitalism', 245–49; Magnan, 'Food regimes', 379–80.

28   Roth, *Magic Bean*, pp. 196–99.

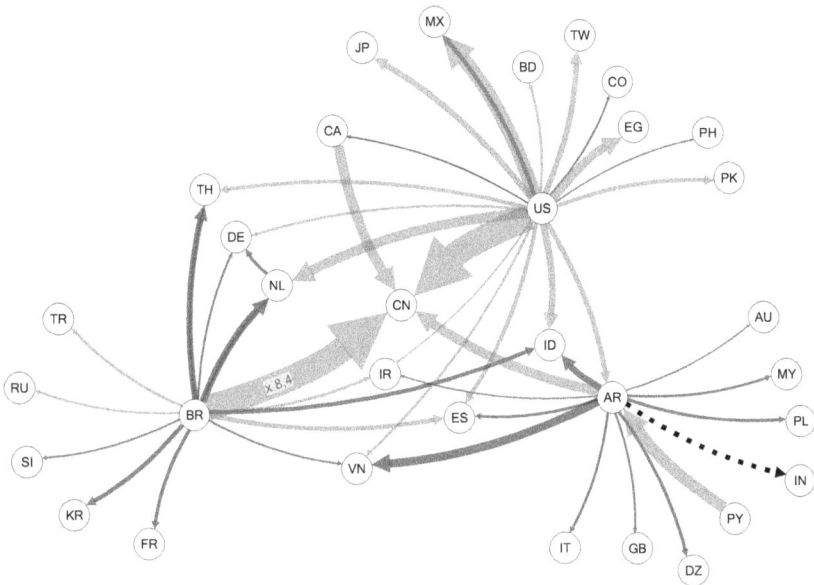

Figure 6.

The global soy trade network in 2018 (top-fifty flows).[29]

Legend: ⇨ soybeans, ⇢ soyoil, ➡ soycake, ISO country codes.

processing industries (3.6 Mt). Within Europe, the Netherlands (one export link and two import links, 0.9 per cent market share) served as a main hub of soy products, sourced mainly in Brazil (cake: 2.7 Mt, beans: 1.3 Mt) and the USA (beans: 3.5 Mt) and directed mainly towards Germany (cake: 1.4 Mt). Last but not least, China (4 import links, zero market share) had fundamentally changed its position in the global soy trade network: from a minor exporter to the major importer of soybeans – the world's 'soy vacuum cleaner' – from South (72.2 Mt) and North America (11.8 Mt), accounting for more than half of total trade volume (Figure 6).

Soy flows from Brazil to China, the trade network's strongest link in 2018, gained ample revenues for agribusiness corporations and the Brazilian treasury, but also caused severe burdens for society and nature. The Brazilian-Chinese soy trade, amounting to 27 billion USD, was dominated by the 'ABCD companies' that controlled 47 per cent export and 37 per cent import trade

78

Ernst Langthaler

value. These top-four transnationals were followed by US-based Gavilon, Chinese-based Cofco, Swiss-based Glencore and Brazilian-based Amaggi and Engelhart, which together covered 25 per cent export and 21 per cent import trade value. The remaining 27 per cent export and 41 per cent import trade value were distributed among other companies, indicating that the Brazilian node of the transpacific trade link was more concentrated than the Chinese node (Figure 7). From the 29.3 mio. hectares (ha) cropland used for soybean production, only nineteen per cent were used for domestic consumption. The remaining area, mainly in the *Cerrado* and *Mata Atlântica* (Atlantic Rainforest) biomes, was appropriated as 'ghost acres' (i.e., land used by a country outside its territory) by China (55 per cent), the European Union (eleven per cent) and other foreign customers. The associated deforestation risk amounted to 61,500 ha, from which 34,600 ha resulted from the Brazilian-Chinese trade. Since forests serve as a sink of greenhouse gases, Brazilian deforestation caused 10.0 Mt emissions of carbon dioxide to the atmosphere, of which the Chinese share was 5.6 Mt.[30] These numbers indicate that the Chinese-driven expansion of the soy frontier in Brazil – alongside impacts from Europe and other world regions – not only reduced domestic socio- and biodiversity in savannah and rainforest biomes but also contributed to global warming as a threat to human life. However, the indirect effects of soy frontier expansion on deforestation and carbon dioxide emissions, through the relocation into rainforests of cattle ranching and other more extensive land uses, were even stronger than its direct effects. The burden of the export-driven expansion of transgenic soy for nature was closely associated with its burden for society, the dislocation of peasant and indigenous communities as well as the health risks of the remaining population due to the excessive application of herbicides.[31] All in all, the Brazilian-Chinese soy trade complemented the '(socio-)ecological teleconnections' typical of the Anthropocene, spanning wide distances with deep impacts on society and nature at both ends.[32]

30   Data source: *Transparency for Sustainable Economies (Trase)*, https://trase.earth (accessed 31 July 2021).

31   A.A.R. Ioris, *Agribusiness and the Neoliberal Food System in Brazil: Frontiers and Fissures of Agro-Neoliberalism* (London, New York: Routledge, 2018), pp. 140–70; C.M. da Silva, 'Between Fenix and Ceres: The Great Acceleration and the agricultural frontier in the Brazilian Cerrado', *Varia Historia* **34** (65) (2018): 409–44; S. Dutra e Silva, 'Challenging the environmental history of the Cerrado: Science, biodiversity and politics on the Brazilian agricultural frontier', *HALAC (Historia Ambiental Latinoamericana y Caribeña)* **10** (1) (2020): 82–116.

32   J.R. McNeill, 'The global environment and the world economy since 1500', in T. Roy and G. Riello (eds), *Global Economic History*, pp. 157–74 (London: Bloomsbury Academic, 2019).

4. Great Accelerations

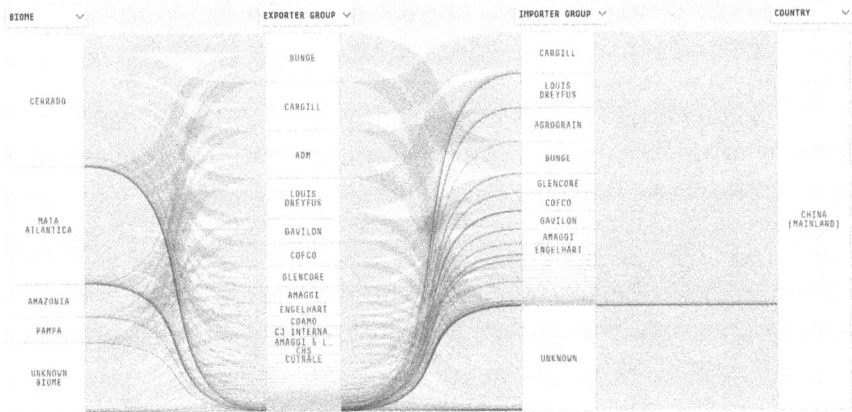

| BIOME ⌄ | EXPORTER GROUP ⌄ | IMPORTER GROUP ⌄ | COUNTRY ⌄ |

(Figure labels — left to right)

BIOME: CERRADO, MATA ATLANTICA, AMAZONIA, PAMPA, UNKNOWN BIOME

EXPORTER GROUP: BUNGE, CARGILL, ADM, LOUIS DREYFUS, GAVILON, COFCO, GLENCORE, AMAGGI, ENGELHART, COAMO, C3 INTERNA, AMAGGI & L, CRS, CUTRALE

IMPORTER GROUP: CARGILL, LOUIS DREYFUS, AGROGRAIN, BUNGE, GLENCORE, COFCO, GAVILON, AMAGGI, ENGELHART, UNKNOWN

COUNTRY: CHINA (MAINLAND)

Figure 7.

Monetary value of soy trade from Brazilian biomes to China by companies, 2018.[33]

In 2018, the global soy trade network entered into a more ambiguous food regime. There is ongoing controversy over whether there was a transition to a solid 'WTO-centred regime,' to a bifurcating 'corporate-environmental regime' or to competing regimes ('food from nowhere' vs. 'food from somewhere'); some even claim that the crisis of the US-centred regime was prolonged.[34] In any case, the USA as former centre found itself in a more polycentric setting, co-determined by the old industrial sub-centres Europe and Japan, the New Agricultural Countries (NACs), as well as TNCs old and new. While the international monetary system was replaced by free-floating currencies, the World Trade Organisation (WTO), founded in 1995 as the successor of the GATT, and bi- and multilateral free-trade agreements emerged as key institutions for the restructuring of agricultural trade according to 'neoliberal' policies. In this both de-regulated and re-regulated framework, the USA, the European Union and Japan managed to sustain state support for agricultural surpluses. They faced fierce competition from the NACs that were challenging US dominance in the world market. South American NACs, most importantly Brazil and Argentina, emerged as major suppliers of soybeans and soycake,

---

33   Data source: *Transparency for Sustainable Economies (Trase)*: https://trase.earth (accessed 31 July 2021).

34   Magnan, 'Food regimes', 381–84; McMichael, *Food Regimes and Agrarian Questions*, pp. 41–47.

driven by skyrocketing Chinese demand for animal feed. Profiting from the dispersion of export hubs across the Americas, the 'ABCD companies' exploited price differentials between harvests in the Global North and the Global South through flexibly sourcing soy as a standardised commodity in semi-annual rhythms. However, the dominating commodity web around 'food from nowhere,' based on cheap soycake and soyoil, faced counter-movements in favour of 'food from somewhere,' hailing whole soybeans as alternative to the highly processed 'neoliberal diet.' Supermarkets as agents of 'green capitalism' took advantage by stretching their marketing strategies to both segments, thereby exploiting the dietary preferences of lower-income as well as higher-income classes.[35]

(Sub-)national Nodes of Soy's Trade Network

Besides international trade links, soy's trade network also consisted of (sub-)national key nodes of production and consumption, including the USA, Brazil and Argentina as major exporters, Japan and China as major importers and the Netherlands as both importing and exporting hub. In the post-war era, the USA replaced China as the major exporter in the world market for soy. The expansion of US soy exports in the First and Second Great Accelerations was interrupted by shrinkage in the 1980s. Throughout this period, the bulk of US exports comprised unprocessed soybeans, while soycake and soyoil played a minor role in overseas markets (Figure 8). This imbalance reveals the US oilseed processing industry to be more oriented towards the domestic market, driven by industrial demand for soycake and soyoil. The USA emerged as a 'soy powerhouse' in the Great Depression and the Second World War, when market regulation by the federal government boosted the domestic production of soybeans as an alternative to price-depressed crops such as wheat, corn and cotton, and as a replacement for scarce tropical oilseeds during the Pacific War. It was not only due to state-led campaigns and favourable market prices that commercial family farms in the Corn Belt and Mississippi Delta adopted soybeans. The originally Asian crop also fitted well into Midwestern and

---

35   McMichael, *Food Regimes and Agrarian Questions*, pp. 41–60; Friedmann, 'From colonialism to green capitalism', 251–57; H. Campbell, 'Breaking new ground in food regime theory: Corporate environmentalism, ecological feedbacks and the "food from somewhere" regime?' *Agriculture and Human Values* **26** (4) (2009): 309–19; B. Vorley, *Food, Inc.: Corporate Concentration from Farm to Consumer* (London: IIED, 2003).

4. Great Accelerations

Southern farming systems (e.g., corn-soy rotation) that were in the midst of a transition to industrialised high-input high-output agriculture. In the post-war decades, rising supplies of US soybeans satisfied rising domestic demand for cake and oil according to the twin-trends of 'meatification' and 'oilification' of middle-class diets. Moreover, US soy served as a weapon in the Cold War that provided the country's military allies and trading partners in Western Europe and East Asia with cheap animal feed for their expanding livestock complexes. Starting with deliveries under the European Recovery Program (ERP), the USA traded soybeans and soycake duty-free to the EEC and Japan in line with the 1960/61 Dillon Round of the GATT. Surpluses of US soyoil were dumped under the 1954 Public Law 480 ('Food for Peace') to food-deficient 'Third World' countries. After international struggles over protectionist trade policies in the 1980s, negotiated in the Uruguay Round of the GATT, the USA expanded soybean exports in the free-trade frameworks of the WTO and NAFTA.

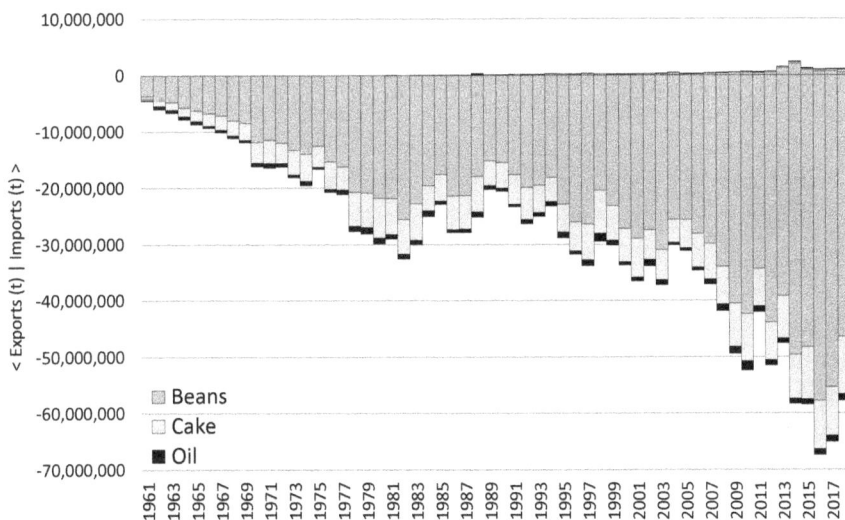

Figure 8.

Soy trade in the USA, 1961–2018.[36]

36   Data source: from FAO, *Faostat*, http://www.fao.org/faostat (accessed 31 July 2021).

With accession to the WTO in 2001, post-reform China emerged as major customer of US soybeans. However, the recent US-Chinese trade war led to a setback that could not be fully compensated for in other foreign markets.[37]

US leadership in the world market for soy soon deteriorated as Brazil and Argentina emerged as soy exporters at the end of the First Great Acceleration and gained market dominance during the Second Great Acceleration. The Brazilian soy expansion resulted from the ambitious project of 'conservative modernisation' implemented by the military dictatorship for geopolitical and fiscal reasons from the mid-1960s onwards. State-led development programmes expanded national agribusinesses and Western technologies northwards from the moderate regions in the very south of the country to the tropical savannahs (*Cerrado*) and rainforests. Indigenous communities and near-natural biomes were disrupted through 'mega-farms' and 'soy deserts,' managed by white settlers. Despite a more contentious relationship between authoritarian and democratic governments and export-oriented agro-elites in Argentina, the situation was similar. Farmers of European descent, with state support, adopted capital-intensive agriculture using Green Revolution technologies in the sparsely populated and extensively used *pampas* from the late 1960s onwards, which promoted soy expansion. The Brazilian and Argentinian soy expansions from the 1990s onwards were driven by both exogenous and endogenous forces. First, 'neoliberal' restructuring of the institutional arrangement in the frameworks of the WTO and Mercosur shifted power from state agencies to (trans-)national companies that adopted a highly flexible agro-export model based on soy as a standardised commodity. Even the 'post-neoliberal' leftist governments from 2003 to 2015/16 stuck to this model in order to gain 'extractive rents,' albeit in different ways. Second, the biotechnological package of transgenic seeds, herbicides and no-till farming, approved in Argentina in 1996 and in Brazil in 2005, simplified and cheapened soy cultivation, thereby serving capitalist interests and the technocratic values of large-scale farmers, agribusiness and investors at the expense of rural communities and their near-natural habitats. Third, rising demand for animal feed from Chinese and other overseas livestock complexes boosted soy prices in the world market from the turn of the century,

37   Roth, *Magic Bean*; Winders, *The Politics of Food Supply*; C.M. Du Bois, 'Social context and diet: Changing soy production and consumption in the United States', in C.M. Du Bois, C-B. Tan and S. Mintz (eds), *The World of Soy*, pp. 208–33 (Urbana: University of Illinois Press, 2008); I. Prodöhl, 'From dinner to dynamite: Fats and oils in wartime America', *Global Food History* **2** (1) (2016): 31–50.

fuelling a commodity boom in general and a 'soy boom' in particular to the advantage of corporate balance sheets and nation-state budgets. After their establishment as suppliers of soycake to European markets in the 1970s and 1980s, the South American 'soy powerhouses' followed different trading strategies from the 1990s onwards. While Brazil specialised in deliveries of unprocessed soybeans to China (Figure 9), Argentina emphasised more diversified exports of soycake and soyoil from the domestic processing industry to South Asia and Europe (Figure 10). This resulted from distinct trade policies: Argentina encouraged domestic crushing of soybeans by imposing heavy taxes on unprocessed exports, reflecting the centralised fiscal federalism and the rather conflictive relationship between agribusiness elites and government. In contrast, Brazil removed such taxes in order to make its soybeans more competitive in international markets, reflecting the autonomy of the subnational states and the involvement of agribusiness elites in political decision-making. Both global players not only competed with the

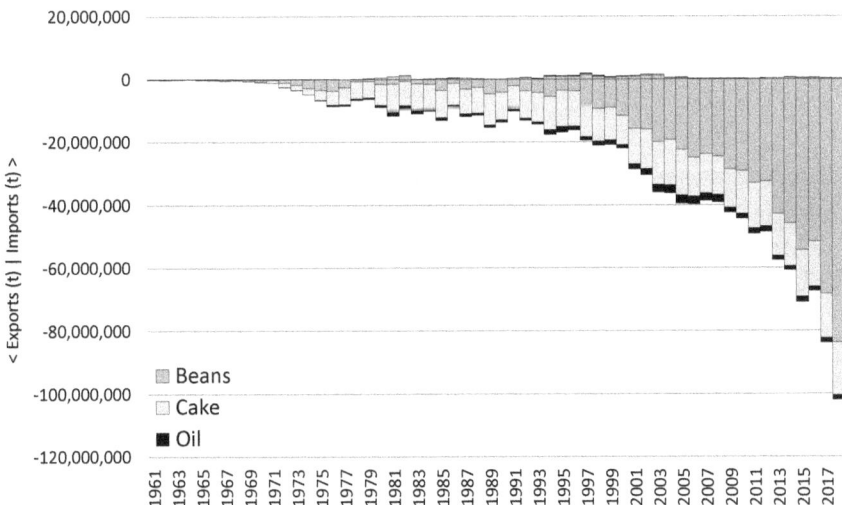

Figure 9.

Soy trade in Brazil, 1961–2018.[38]

38   Data source: FAO, *Faostat*, http://www.fao.org/faostat (accessed 31 July 2021).

Ernst Langthaler

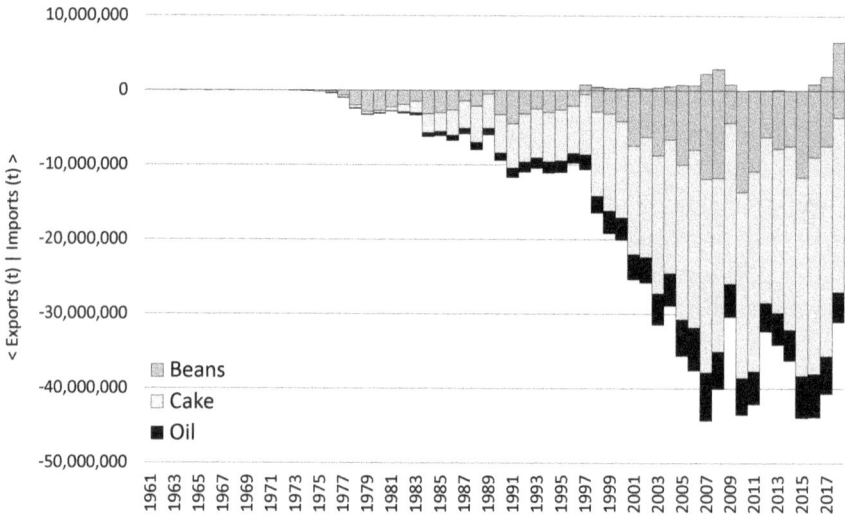

Figure 10.

Soy trade in Argentina, 1961–2018.[39]

USA in European and East Asian markets but also conducted business in countries beyond the US geopolitical sphere (Iran, Russia, Vietnam etc.).[40] After World War Two, Japan became the biggest international market for US soybeans in the First Great Acceleration and, despite shrinking trade

39   Data source: Ibid.

40   G. d. L. T. Oliveira and M. Schneider, 'The politics of flexing soybeans: China, Brazil and global agroindustrial restructuring', *The Journal of Peasant Studies* **43** (1) (2015): 167–94; A. Leguizamón, *Seeds of Power: Environmental Injustice and Genetically Modified Soybeans in Argentina* (Durham: Duke University Press, 2020), pp. 47–58; Ioris, *Agribusiness and the Neoliberal Food System in Brazil*, pp. 49–72; K. Fischer and E. Langthaler, 'Soy expansion and countermovements in the Global South: A Polanyian perspective', in R. Atzmüller et al. (eds), *Capitalism in Transformation*, pp. 212–27 (Cheltenham, Northampton, MA: Edward Elgar Publishing, 2019); M. Baraibar Norberg, *The Political Economy of Agrarian Change in Latin America: Argentina, Paraguay and Uruguay* (Cham: Palgrave Macmillan, 2019), pp. 165–299; M.E. Giraudo, 'Taxing the 'crop of the century': The role of institutions in governing the soy boom in South America', *Globalizations* **65** (1) (2020): 1–17; M. Turzi, *The Political Economy of Agricultural Booms: Managing Soybean Production in Argentina, Brazil, and Paraguay* (Cham: Palgrave Macmillan, 2017); A. Zarrilli, '¿Una Agriculturización Insostenible? La Provincia Del Chaco, Argentina (1980–2008)' *Historia Agraria. Revista de Agricultura e Historia Rural* **51** (2010): 143–76.

4. *Great Accelerations*

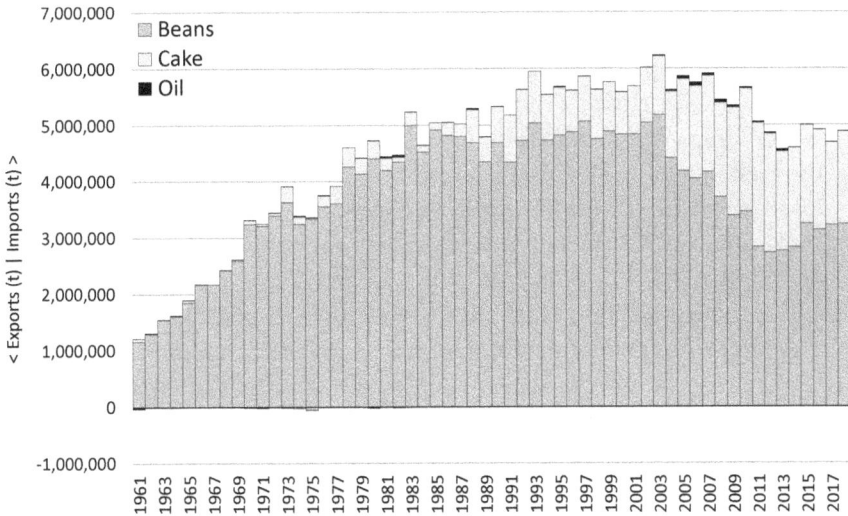

Figure 11.

Soy trade in Japan, 1961–2018.[41]

volumes in the Second Great Acceleration, has served as an important customer ever since. Japan's shift to Western diets rich in meat, dairy products
and wheat accelerated during the post-1945 Allied occupation, e.g. through
introduction of improved school lunches according to US standards, and
the subsequent economic boom with 'American food' symbolising affluent
lifestyles. In addition to Western-style modernisation through oil for food
purposes and cake as high-protein feed for chickens, pigs and other livestock,
soybeans were used for traditional Japanese foods such as tofu, miso and soy
sauce. Since demand for soybeans soon exceeded domestic supply, a business
opportunity for US farmers and traders emerged. US dominance in the
Japanese market resulted from state-supported marketing efforts (product
surveys, public exhibits, demonstration buses etc.) organised by the American
Soybean Association and the USDA's Foreign Agricultural Service, as well
as trade liberalisation through elimination of import tariffs in 1961. Since
the bulk of imports consisted of whole soybeans, the domestic processing
industry was a major player in the US-Japanese commodity chain. From

41   Data source: FAO, *Faostat*, http://www.fao.org/faostat (accessed 31 July 2021).

the turn of the century, Japanese soybean imports were increasingly complemented by soycake, mainly of Chinese origin (Figure 11). Simultaneously,
soyfoods with ingredients of US origin raised public concerns about safety,
especially with regard to genetically modified seeds.[42]

After China had lost its former leadership as a soy exporter to the USA
in the First Great Acceleration under Communist rule, the country regained
global dominance as soy importer in the Second Great Acceleration. In fact,
Chinese demand for soybeans was the key driver of skyrocketing supply
from the Americas in general and the USA and Brazil in particular (Figure
12). This major shift resulted from the post-1978 reforms by the Communist political and economic elites, oriented towards 'modernising' society
through strengthening the urban middle classes with meat-rich diets as an
identity-sustaining lifestyle element. Though tofu and other soyfoods remained common in Chinese diets, urban dwellers of fast-growing mega-cities
strongly increased their intake of meat from soycake-fed pigs, chicken and
other livestock, and of soyoil for cooking. As a crucial step in this direction,
domestic oilseed processing and livestock feeding industries were established through market reforms and government spending. Prior to China's
accession to the WTO in 2001, the authorities liberalised the soy market,
thereby facilitating imports of whole beans from overseas for the growing
feed-livestock complex. The state aimed to foster China's participation in the
world market without losing control over value-adding. During the '2004
Soybean Crisis', however, Chinese processing companies defaulted on their
contracts with transnational traders due to an unexpected price spike in US
soybeans, forcing many of them into high debts and bankruptcy. The 'ABCD
companies' and other foreign agribusinesses acquired large portions of the
Chinese crushing and refining industry, thus controlling the transcontinental
commodity chain from export to import. The government successfully tried to
regain control over the commodity chain through legal and financial support
for the domestic processing industry in general and state-owned companies
in particular. The huge inflow of cheap soybeans from overseas, priced at the
Chicago Board of Trade, outcompeted domestic soybean production, causing
cultivated area to fall and forcing smallholders into rural-urban migration.[43]

42   Du Bois, *The Story of Soy*, pp. 99–101; M.Conlon, 'The history of U.S. soybean exports to Japan'
(JA9502, USDA/FAS, 2009); Katarzyna J. Cwiertka, 'Contemporary issues in Japanese cuisine',
in S.H. Katz (ed.), *Encyclopedia of Food and Culture* (New York: Scribner, 2003), pp. 324–27.

43   Oliveira and Schneider, 'The politics of flexing soybeans'; M. Schneider, 'Feeding China's pigs:
Implications for the environment, China's smallholder farmers and food security' (Minneapolis,

4. *Great Accelerations*

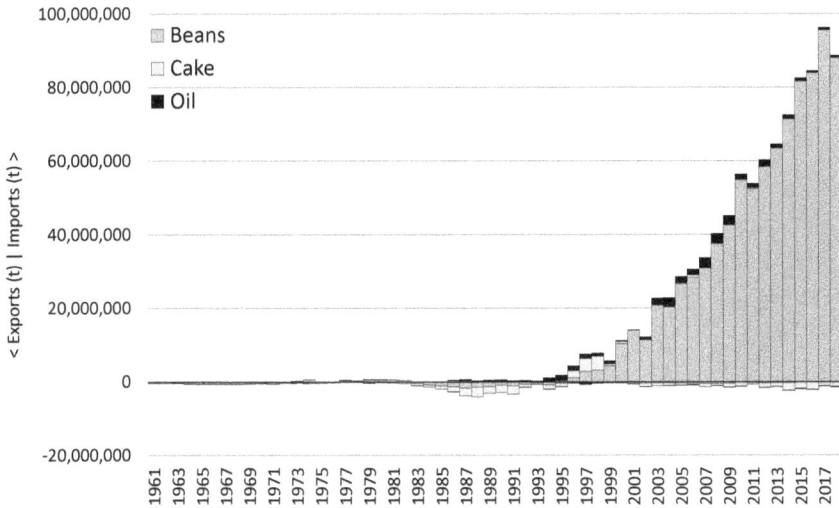

Figure 12.

Soy trade in China, 1961–2018.[44]

In the post-war era, the Netherlands reached a key position in the global soy trade network as both importing and exporting hub within Western Europe. The country strengthened the development path that it had taken under the British-centred food regime in the late nineteenth century: specialising in livestock production in combination with feed imports and meat exports. Government policy towards agriculture as well as the EEC's Common Agricultural Policy aimed at raising productivity, especially production per unit of labour, in order to provide adequate incomes to farming families and cheap food to the rest of the population. These policies contributed to the post-war boost in public welfare that promoted middle-class lifestyles oriented towards animal-based diets. Dutch agro-productivism involved the hegemony of a high-modernist discourse, the application of science-based expert knowledge, the substitution of technology for labour, the expansion of

Washington, D.C., Berlin: Institute for Agriculture and Trade Policy, 2011); M. Schneider, 'Dragon head enterprises and the state of agribusiness in China', *Journal of Agrarian Change* **17** (1) (2017): 3–21.

44   Data source: FAO, *Faostat*, http://www.fao.org/faostat (accessed 31 July 2021).

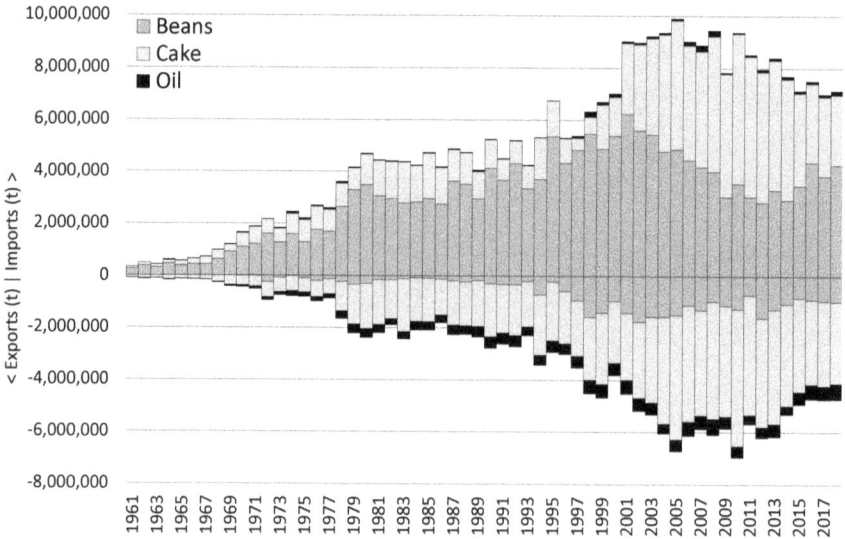

Figure 13.

Soy trade in the Netherlands, 1961–2018.[45]

livestock units per farm and the optimisation of animal performance (e.g., by means of high-value feed). Accordingly, the Netherlands emerged as a major producer and exporter of pork, poultry and dairy products, as well as a major importer of animal feed. Starting with the ERP, the Netherlands imported soybeans (mainly from the USA and Brazil) and, to a lesser extent, soycake (mainly from Brazil and Argentina), at the same time as re-exporting processed co-products to adjacent countries (Belgium, UK, Germany etc.). According to agreements between the USA and the EEC under the GATT, feeding stuffs, including soy products, passed national borders duty-free from 1961 onwards. The global 'protein crisis' of 1972/73, resulting from the collapse of Peruvian fishmeal due to an El Niño in combination with a US trade embargo on soy in light of spiking meat prices, led Dutch factory farms to shift animal feeding from animal-based to plant-based proteins, including soycake as a replacement of fishmeal. The tendency towards soy-based animal feed was reinforced by the BSE crisis, pushing the European Union to ban the use of animal protein for feeding stuffs in 2001. The development of

45   Data source: FAO, *Faostat*, http://www.fao.org/faostat (accessed 31 July 2021).

Dutch soy imports and exports reflected these institutional and technical arrangements: expansion from the 1960s to the 1970s and from the 1990s to the mid-2000s, interrupted by stagnation in the 1980s. Shrinking import volumes from the mid-2010s were accompanied by rising shares of cake in relation to beans (Figure 13). This shift in the composition of soy imports reflects the business strategies of major traders such as the Amaggi Group, which prefers to sell the produce of its own crushing facilities in Brazil to the Dutch market. Domestic crushing is dominated by ADM's and Cargill's facilities close to the ports of Rotterdam and Amsterdam, through which about a quarter of European soy imports enter the continent.[46]

Conclusion

By taking a commodity-focused network perspective on global agricultural trade since 1950, this chapter's findings render simplistic notions of the Great Acceleration more complex. As the case of the soy trade reveals, agro-food globalisation from the 1950s to the 2010s decomposes into three different phases: the First Great Acceleration with rapid growth from low levels of trade volume from the 1950s to the 1970s; the Second Great Acceleration with more moderate growth from much higher levels from the 1990s onwards; and the deceleration on a comparatively medium level in the 1980s. From phase to phase, the global soy trade network and the related positions of its key nodes (Argentina, Brazil, China, Japan, the Netherlands and the USA) shifted due to endogenous and exogenous dynamics. This tripartite periodisation calls for a shift of the term 'Great Acceleration' from singular to plural, emphasising the different trajectories and their respective conditions and consequences. As exemplified by the tremendous flow of soybeans from Brazil to China, the global trade network has involved not only accelerated capital accumulation, but also severe socionatural burdens for affected communities and biomes.

---

46   J. Bieleman, *Five Centuries of Farming: A Short History of Dutch Agriculture, 1500–2000* (Wageningen: Wageningen Academic Publishers, 2010), pp. 239–310; F. Haalboom, 'Oceans and landless farms: Linking Southern and Northern shadow places of industrial livestock (1954–1975)', *Environment and History* **28** (4) (2022): 571–99; A. Schuurman, 'Agricultural policy and the Dutch agricultural institutional matrix during the transition from organized to disorganized capitalism', in P. Moser and T. Varley (eds), *Integration Through Subordination: The Politics of Agricultural Modernisation in Industrial Europe*, pp. 65–84 (Turnhout: Brepols, 2013); J.D. van der Ploeg, *The Virtual Farmer: Past, Present, and Future of the Dutch Peasantry* (Assen: Royal van Gorcum, 2003), pp. 229–73; J.W. van Gelder, B. Kuepper and M. Vrins, 'Soy Barometer 2014: A Research Report for the Dutch Soy Coalition' (Profundo, 2014), pp. 27–29.

Ernst Langthaler

The different trajectories of the soy trade can be explained with regard to the formation, crisis and transition of global food regimes. While the First Great Acceleration was inserted into the US-centred food regime, the deceleration and Second Great Acceleration emerged from the crisis of the old and the transition to a new, more ambiguous food regime. At the global level, the decisive driving forces and actors comprised the USA as Western hegemon, the GATT members and US- and European-based processing and trading companies in the US-centred food regime as well as the Cairns Group, the WTO and agribusiness TNCs in the WTO-centred food regime. The case of the soy trade also highlights the impact of driving forces and actors at (sub-)national levels: first and foremost, national governments that have shaped international trade flows through market regulation (trade contracts, fiscal policies, state support etc.); provincial agencies, pressure groups and social movements that have represented self-serving interests; nationally-based companies that have competed with the transnational 'ABCD companies' in the domestic market (e.g., post-reform China). Consequently, future research should more seriously take into account the impacts of (sub-) national entities in global food regimes. Last but not least, soy's crucial role as a 'more than human' actor should not be underestimated. The bean's unique combination of nutrients enabled commodification to switch flexibly among multiple forms of value extraction, from oil to cake and back. In this sense, soy and its global trade network co-constituted each other.

Acknowledgements

The author would like to thank Jonas Albrecht, Peer Vries and the editors of this volume for their helpful comments. This work has been supported by the Johannes Kepler Open Access Publishing Fund.

CHAPTER 5

SOJIZACIÓN AS A NEW *FIRST MOVEMENT:* A POLANYIAN ANALYSIS OF THE SOUTH AMERICAN SOYBEAN 'BOOM'

Matilda Baraibar Norberg

Introduction

South America – specifically Brazil, Argentina, Paraguay, Bolivia and Uruguay – has become an increasingly specialised world provider of soybeans. Indeed, over the last two decades, more than 33 million additional hectares of land (roughly a surface area equivalent to that of Vietnam, or to all the arable surface of Ukraine) have been incorporated into soybean production.[1] This land-use change, here referred to as *sojización*, has brought multiple consequences, ranging from deforestation, soil degradation and water pollution to agribusiness domination, displacement of family farmers and 'foreignisation' of land.[2] While the consequences differ from one place to another, *sojización* has brought dramatic technological, productive and social transformations throughout the region, leading to increased land concentration and land-use intensification. The consequences of this dramatic change have rightfully received much scholarly attention. Less thoroughly addressed, however, is the preceding history that shaped the preconditions for *sojización* to occur.

The standard (liberal) story about the soybean expansion in the Southern Cone often depicts the boom as a spontaneous (market) response: South America had available (cheap) land, a new technological package had 'emerged', centred on glyphosate tolerant soybeans (which allowed for cutting costs and expanding the frontier by the incorporation of more marginal land) and

---

1   The soybean acreage in these countries has increased from 30 million hectares (Mha) in 2000 to more than 63 Mha in 2021: USDA, *Oilseeds: World Markets and Trade* (Washington: United States Department of Agriculture, Foreign Agricultural Service, 2021).

2   M. Baraibar Norberg, *The Political Economy of Agrarian Change in Latin America: Argentina, Paraguay and Uruguay* (Cham: Springer, Palgrave, 2020), pp. 57–116.

THE AGE OF THE SOYBEAN: 91–114 **doi:** 10.3197/63800040695086.ch05

a sharp rise in international demand for soybeans.[3] In this narrative, *sojización* was simply the natural consequence of 'rational' profit-seeking farmers and traders responding to new price relations (in which soybeans yielded the highest annual returns). However, this naturalised and de-politicised story ignores all the decisions and regulatory shifts that in fact preceded and created the necessary conditions for the boom. A more accurate analysis of the historical and political dynamics behind *sojización* in South America calls for a careful inquiry into the history of gradual and intersecting shifts at different scales (international to national) that paved the way for the boom. For this purpose, I have taken inspiration from Karl Polanyi's seminal 1944 book *The Great Transformation*, in which he exhaustively identified and examined the necessary conditions for a wave of marketisation to emerge. While the empirical focus is on Argentina, Paraguay and Uruguay, these are situated in the wider South American context and put in dialogue with the broader world history of capitalist agri-food relations. This chapter thus contributes to a deeper understanding of *sojización* by a historically informed exploration of the multiple shifts and continuities that made it possible.

Tackling *Sojización* as a New First Movement

Polanyi famously argued that *market society* in nineteenth century England did not emerge spontaneously, organically releasing man's natural 'propensity to barter, truck and exchange', as per the standard liberal narrative; rather, as he demonstrated, the onset of the free market was intentionally staged.[4] He described in detail how a multitude of gradual, but far-reaching, regulative shifts (e.g. the removal of restrictions to mobility, enclosures, the establishment of freedom of contract, the repeal of the Corn Laws, the Gold Standard and the strengthening of private property rights to land) served to disembed humans and nature from their traditional institutions and social ties.[5] This disembeddedness, in turn, allowed for their gradual commodification – i.e. the transformation of nature and of human life time into the interchangeable

---

3   CEPAL, FAO, IICA *Perspectivas de la agricultura y del desarrollo rural en las Américas: Una mirada hacia América Latina y el Caribe 2015–2016* (San José: IICA, 2015) http://repositorio.cepal.org/bitstream/handle/11362/39023/PerspectivasAgricultura2015-16_es.pdf?sequence=1 (accessed 20 Oct. 2021).

4   K. Polanyi, *The Great Transformation: The Political and Economic Origins of Our Time*, Second Beacon Paperback edition (Boston: Beacon Press, 2001), p. 45

5   Ibid., pp. 313–14.

commodities of pieces of land and labour in the self-regulating market.[6] However, humans (labour) and nature (land) are only fictitious commodities; they are not 'produced' to become interchangeable goods (through a market price) in profit-driven markets. However, as this fiction was staged, all other functions of humans and nature, bar those that could produce gain, came to be ignored. This brought stupendous economic achievements, but at the price of great harm and violent destruction.[7]

Famously, Polanyi argued that the unleashing of market forces – the *first movement* – became too destabilising, disruptive and destructive to be able to last for long. The overexploitation of humans and nature thus soon pro-voked societal responses for protection – a *countermovement* – which pushed back markets and ultimately re-embedded them, in what he called a *double movement*. This mechanism explains why England actually experienced 'pure' industrial capitalism only between 1834 and 1870.[8] After that, labour, money and land were still commodified, but their price (wages and rents) ceased to be subject to market mechanisms alone.[9]

When writing *The Great Transformation* in 1944, Polanyi believed that people and societies had learned important lessons from the devastating effects caused by the deliberate expansion of profit-driven 'self-regulating' markets. As a more realistic appraisal of markets had come to the fore, he asserted the death of the unfettered market economy, believing a new era of unprecedented freedom could emerge.[10] Polanyi's predictions about the future may have been overly optimistic. The post-war compromise between states and market, *embedded liberalism*, undeniably put some restrictions on markets, but state intervention proved to be perfectly compatible with an economy based in mass production and consumption and a move towards the universalisation of the commodity form.[11]

Many scholars have taken inspiration from Polanyi to identify a new

6   Ibid., pp. 68–113.
7   Ibid., pp. 110–13; 195–97.
8   Ibid., pp. 87–120; 201–30.
9   Ibid., pp. 240–65.
10   Ibid., p. 322.
11   G. Dale, *Karl Polanyi: A Life on the Left* (New York: Columbia University Press, 2016), p. 6.
     P. McMichael, 'Rethinking globalization: the agrarian question revisited', *Review of International Political Economy* **4** (4) (1997): 630–62.

first movement emerging after the collapse of the Bretton Woods system.[12] Neoliberal globalisation is here understood to represent an increasingly institutionalised *market society* world-wide.[13] Many a scholar has also explored contemporary potential Polanyian *countermovements*.[14] However, such scholars have largely ignored the fact that the most dramatic processes of commodification in recent decades have taken place in natural resource-rich countries in the Global South. More specifically, the most dramatic processes of commodification have happened in peripheral frontier areas, in spheres that until recently were regarded as impossible (or inconvenient) to define, isolate, privatise or price. *Sojización* is arguably a particularly illuminating example of the dynamics of this new *first movement*, responding to an increased global appetite for food, feed and fuel, as well as to the market strategies of transnational agribusiness firms and financial capital. Notwithstanding the fact that *sojización* emerged out of the uneven power-relations of neoliberal agrofood globalisation, it is clear that South America states have also been important actors in (co-)creating the necessary conditions for this 'great agrarian transformation'.[15]

Inspired by Polanyi's holistic and historically informed account of the multiple interacting regulatory shifts necessary for the emergence of marketisation in nineteenth century England, I will now turn to the many-layered history preceding *sojización* in Argentina, Paraguay and Uruguay. This analysis builds on several years of research on various aspects of regulatory shifts in relation to agrarian change, drawing on multiple and diverse sources, inclu-

12   J. Fairhead, M. Leach and I. Scoones, 'Green grabbing: a new appropriation of nature?' *The Journal of Peasant Studies* **39** (2) (2012): 237–61. N. Fraser, 'Can society be commodities all the way down? Post-Polanyian reflections on capitalist crises', *Economy and Society* **43** (4) (2014): 541–58. R. Munck, 'Karl Polanyi for Latin America: markets, society and development', *Canadian Journal of Development Studies / Revue canadienne d'études du développement* **36** (4) (2015): 425–41.

13   B.J. Silver and G. Arrighi, 'Polanyi's "double movement": The belle èpoques of British and U.S. hegemony compared', *Politics & Society* **31** (2) (2003): 337–48. E. Langthaler and E. Schüßler, 'Commodity studies with Polanyi: Disembedding and re-embedding labour and land in contemporary capitalism', Österreich Z Soziol **44** (2019): 209–23. K. Hart and C. Hann, 'Introduction: Learning from Polanyi', in Hann and Hart (eds), *Market and Society: The Great Transformation Today*, pp. 1–16 (Cambridge: Cambridge University Press, 2009).

14   M.E. Warner and J. Clifton, 'Marketisation, public services and the city: the potential for Polanyian counter movements', *Cambridge Journal of Regions, Economy and Society* **7** (1) (2013): 45–61. O. Worth, 'Polanyi's magnum opus? Assessing the application of the counter-movement in international political economy', *The International History Review* **35** (4) (2013): 905–20; W. Carton, 'On the nature of the countermovement: A response to Stuart et al.'s 'Climate change and the Polanyian countermovement: Carbon carkets or degrowth?' *New Political Economy* **25** (1) (2018): 1–6.

15   Baraibar Norberg, *The Political Economy of Agrarian Change*, pp. 68–74.

ding more than fifty interviews with politicians, agribusiness firms, family famers, cooperatives, state officials and NGOs in Argentina, Paraguay and Uruguay. I also use national legislation, white papers and business reports, as well as insights from literature in agrarian history, land-systems science and political economy.

Persistence of Non-market Arrangements Alongside Long-term Commodification (1860s–1970s)

South America has been inserted in capitalist world markets since colonial demand for export crops and wage labour started to transform the region.[16] Argentina, Paraguay and Uruguay soon became specialised in livestock ranching. The long-term trend since then has been that of the constant commodification of new areas (frontier land in agricultural systems) and human activities (wage labour). Notwithstanding this long-term trend, non-market arrangements have also persisted over time.[17] In this section, I explore the history of this dual dynamic from the late nineteenth century until the neoliberal turn in the 1970s. This longer historical contextualisation provides a sense of both the continuities and breaks enabling the process of *sojización* .

The incomplete staging of market society under laissez-faire liberalism (1860s–1920s)

In the new international division of labour that emerged with the first wave of globalisation (1860–1914), South America became an increasingly specialised commodity provider for the great industrial centres.[18] In this era of industrial capitalism in Europe, a spectacular and far-reaching commodification process took off all over the continent. Natural resource-based products became highly 'competitive' due to the region's vast amounts of (cheap) land, combined with rapidly declining international freight costs, a liberal trade regime (pushed particularly by Great Britain) and the Gold

16   A. de la Torre, T. Didier, A. Ize, D. Lederman and S. Schmukler, *Latin America and the Rising South : Changing World, Changing Priorities* (Washington DC, World Bank Group, 2015).

17   The 'incomplete' transition to capitalist farming has been seen to represent backwardness in both liberal and Marxist traditions (e.g. the 'agrarian question').

18   R. Prebisch, *The Economic Development of Latin America and its Principal Problems* (New York, ECLAC, 1950) https://repositorio.cepal.org/bitstream/handle/11362/29973/002_en.pdf?sequence=1&isAllowed=y (Accessed 12 Oct. 2021).

Standard.[19] *Laissez-faire* liberalism was the dominant ideology among both foreign and domestic elites. The big landowners of the *Rio de la Plata* region often acted in alliance with foreign investors, holding down trade and land taxes and defending open migration policies. This led to a rapid extension of the frontier and spurred further rounds of commodification in new areas.[20] The lines dividing the colonial farms from the territories dominated by indigenous groups gradually moved outwards, triggering a series of border conflicts, including military campaigns which sometimes sought to 'free' new lands for agricultural exploitation. Indigenous peoples, and their mestizo allies were displaced, marginalised, forced to settle, or killed – the normal sequence in historical processes involving settler colonialism.[21]

The ports of Montevideo and Buenos Aires shipped wool, hides and meat from a vast 'hinterland' of big cattle and sheep herds, few horses and even fewer men. By the early twentieth century, Argentina managed to diversify exports by adding cereals (wheat and maize) and flax.[22] Under this *belle époque* of primary export-oriented development, Uruguay and Argentina entered a relatively long period of sustained economic growth (while most other export-oriented countries in South America suffered from intense commodity-price volatility and short cycles of booms and busts).[23]

While economic growth rates were high, income distribution worsened sharply during the period.[24] This was linked to the fact that land was the main basis for wealth, and land ownership was extremely concentrated. The unequal, bimodal land tenure structure of *latifundio* (big estates controlling vast amounts of land – typically engaged in ranching activities) and *minifundio* (small farmers with very small parcels of land – typically engaged in

19   L. Bértola and J.A. Ocampo, *The Economic Development of Latin America since Independence* (Oxford: Oxford University Press, 2012), pp. 35–70.

20   P. Winn, 'British informal Empire in Uruguay in the nineteenth century', *Past & Present* **73** (1976): 100–23.

21   J.C. Chasteen, *Heroes on Horseback: A Life and Times of the Last Gaucho Caudillos* (Albuquerque: UNM Press, 1995), pp. 141–70. While similar campaigns existed in many countries, a seminal example was the so-called 'Conquest of the Desert', in the 1870s in Argentina, which defeated, killed and displaced indigenous communities: J.R Scobie, *Revolution of the Pampas: A Social History of Argentine Wheat, 1860–1910*, (Austin: University of Texas Press, 1964), pp. 117–21.

22   V. Bulmer-Thomas, *The Economic History of Latin America since Independence*, Second edition (Cambridge: Cambridge University Press, 2003), p. 154; Scobie, *Revolution of the Pampas*.

23   H. Williebald and L. Bértola, 'Uneven development paths among settler societies, 1870–2000', in C. Lloyd, J. Metzer and R. Sutch (eds), *Settler Economies in World History* (Leiden/Boston: Brill, 2013), p. 106; Bulmer-Thomas, *The Economic History*, p. 154.

24   Bulmer-Thomas, *The Economic History*, p. 117.

5. Sojización *as a New* First Movement

plant agriculture) had been established already during colonial times, but the dominance of the *latifundio* was reinforced after independence.[25] In Uruguay the agrarian frontier was already exhausted by the mid-nineteenth century with most of the land in the hands of large ranchers.[26] In Argentina, the lion's share of the 'new' lands ended up in the hands of the big landowners. The majority of the many European migrants that arrived in both countries never became land property owners, but actually ended up in the big cities. The rest became sharecroppers or (seasonal) labourers on large estates.[27] Meanwhile, the area controlled by the *latifundio* expanded from the 1850s to the 1930s.[28]

Paraguay at first followed a different developmental path from its neighbours. After overthrowing the local Spanish administration in 1811, it promoted autarchy, self-subsistence and family agriculture. Land was nationalised, expropriated from the church, foreigners and big landowners. The state also monopolised the commercialisation of lucrative trade branches, such as *yerba mate*.[29] This path was nevertheless terminated by the defeat in the *War of the Triple Alliance* (1864–1870) against the joint forces of Brazil, Argentina and Uruguay. The war nearly wiped Paraguay off the map; it lost an impressive share of its male population, territory, economy and autonomy. Large tracts of public land were sold off to pay off its war debt; around 26 million hectares of land were privatised between 1870 and 1914 and a sizable amount ended up in foreign hands.[30] Consequently, *latifundio* (many of which now came under foreign ownership), achieved a dominant position

25   Private property rights to land were strengthened legally and in practice through fencing, which facilitated enclosure and frontier expansion. At some point in the turbulent times between the eighteenth and nineteenth centuries, big ranchers occupied state land, and were able to keep it. N. A Biangardi, *Expansión territorial, producción ganadera y relaciones de poder en la región Río de la Plata* (Ph.D. Thesis, Universidad Nacional de La Plata, 2015). pp. 125–40. E. Von Bennewitz, 'Land tenure in Latin America: from land reforms to counter-movement to neoliberalism', *Acta Universitatis Agriculturae et Silviculturae Mendelianae Brunensis* **65** (5) (2017): 1793–98; Baraibar Norberg, *The Political Economy of Agrarian Change*, pp. 66–97.

26   J. Álvarez, 'Institutions, the land market and income distribution in New Zealand and Uruguay, 1870-1940', paper presented at the XIV International Economic History Congress in Helsinki, Finland, 2006.

27   Scobie, *Revolution of the Pampas*, pp. 114–18. Bulmer-Thomas, *The Economic History*, p. 145.

28   Von Bennewitz, 'Land tenure in Latin America', 1796.

29   Baraibar Norberg, *The Political Economy of Agrarian Change*, pp. 69–70; G. Raidán Martínez, 'Medio ambiente y agricultura en el Paraguay', *Población y Desarrollo* (35) (2008): 107–19.

30   D. Abente, 'Foreign capital, economic elites and the state in Paraguay during the liberal republic (1870–1936)', *Journal of Latin American Studies* **21** (1– 2)(1989): 61–88; F. Masi and D. Borda, *Estado y economía en Paraguay, 1870–2010*, (Centro de Análisis y Difusión de la Economía Paraguaya (CADEP), 2011) p. 24.

98

Matilda Baraibar Norberg

in Paraguay's agrarian structure.[31] Besides selling off land, Paraguay adopted similar export-oriented policies to those of Argentina and Uruguay, leading to the expansion of ranching activities and new export products (particularly *quebracho*, a hardwood that yields an extract used in tanning).[32]

While markets expanded during this period (with land and labour becoming increasingly commodified in areas close to urban centres, transport nodes and ports), a high degree of self-subsistence, use of family labour and non-market access to land continued to prevail among peasants and family farmers, the groups who still constituted the vast majority of the rural population. People's time was not sold on a 'free' labour market, and people consistently engaged in different transactions for a plethora of social motives, rather than exclusively in the search for profit. Particularly in rural areas, nature and people remained largely non-commodified, embedded in traditional institutions and social ties.[33]

One partial explanation for this persistence of non-market arrangements was the fragility and weakness of the relatively newly independent South American states. State intervention had freed trade from the confines of the colonial legacies of mercantilism but, as Polanyi underlined, an efficient free market system requires much more than deregulation. It needs constant maintenance by a strong 'market-police' state, imposing strict competitive rules.[34] In contrast to late nineteenth-century England, the 'new' South American states – exposed to external as well as internal threats – were too weak to guarantee the capitalist 'rules of the game'. Moreover, chronic budget constraints often curbed infrastructural projects that would have allowed for the integration of more humans and land into commodity circuits.[35] Most of the *de facto* investments in this respect were instead made by foreign trading firms (mainly British) and exclusively related to an extractive model of

31   Masi and Borda, *Estado y economía en Paraguay*, p. 25.

32   Ibid., pp. 26–31; Raidán Martínez, 'Medio ambiente y agricultura en el Paraguay',112.

33   According to Polanyi, the institution of the market was fairly common from the later Stone Age, but its role was no more than incidental to economic life, and it was always been embedded in social relations. Polanyi, *The Great Transformation*, pp. 45–56, 60.

34   Ibid., pp. 60, 110–13.

35   G. Flichman, 'The state and capital accumulation in Argentina', in C. Anglade and C. Fortín (eds), *The State and Capital Accumulation in Latin America* (London: Macmillan, 1990), pp. 11–23. H. Finch, *A Political Economy of Uruguay since 1870* (New York: St. Martin's Press, 1981).

agrofood exports.[36] Notwithstanding some country-specific variation,[37] all states lacked the financial and organisational muscles needed to secure private property rights, 'free' competition and the necessary infrastructure (railways, roads, rural police, public services, etc.) required for a more thorough staging of market society. In geographically and socio-economically remote areas, far away from the sites of high concentration of capital and power, the state was almost completely absent, with no capacity to provide basic services. In the absence of state-backed markets, other social arrangements prevailed.

The states not only lacked resources to enforce market society, but also acted in contradictory ways, playing a dual role in relation to marketisation. The landowning elite had a huge influence over the state and thus often shaped agrarian institutions in their own interest. The result was intensely ambivalent. The rural associations – e.g. the *Argentine Rural Association* (SRA), the *Uruguayan Rural Association* (ARU) and the *Rural Association of Paraguay* (ARP) – emerged as powerful lobby organisations in the late nineteenth century. Together with their allies the (British) investors and the Catholic Church, they pushed for the removal of export and land taxes.[38] On the other hand, this kind of captive state's ambition to establish bourgeois property rights on the land (i.e. commodify land) was often compromised by its alliances with traditional landed classes, essential to securing social control and political order.[39] The great landowners had little interest in a transition to a pure market system, as they wanted to maintain their vast lands and paternalistic, dependent and pseudo-feudal labour relations with their workers (*peones*).[40] Different forms of unfree labour (peonage), such

36   Foreign trading firms expanded into finance (loans and insurance), infrastructure (railways and ports), processing (slaughterhouses). G. Jones, *Merchants to Multinationals: British Trading Companies in the Nineteenth and Twentieth Centuries,* (Oxford: Oxford University Press, 2000) pp. 43–48.

37   Argentina and Uruguay developed a more independent state apparatus and domestic markets than Paraguay, but were still largely in a dependent position. Baraibar Norberg, *The Political Economy of Agrarian Change,* pp. 69–70.

38   Due to capital constraints, no country completely eliminated trade duties to raise revenues from the tariff system. Only Uruguay managed to impose a land tax (despite ranchers' opposition). Bulmer-Thomas, *The Economic History,* p. 170. See M. Baraibar, *Green Deserts or New Opportunities? Competing and Complementary Views on the Soybean Expansion in Uruguay, 2002–2013* (Stockholm: Stockholm Studies in Economic History, 2014).

39   J.P. Barrán and B. Nahum, *Un dialogo difícil, 1903–1910* (Montevideo: Ediciones de la Banda Oriental, 1981).

40   Since a competitive labour market did not exist, the seemingly contradictory elements of deteriorating labour conditions and labour shortages co-existed.

Matilda Baraibar Norberg

as debt-labour, were often used to keep labour costs down.[41] The states reinforced the power of landlords over 'their' workers by making vagrancy, conscription and passport requirements stricter, restricting labour mobility.[42] This effectively prevented the establishment of a 'free' labour market, standing in the way of the spread of capitalist relations.

Another way in which the *latifundio* curbed the spread of capitalism, was the fact that the large estates generally incorporated very little technology and labour, as they could compensate low productivity per hectare with vast holdings of land. Capital accumulation originated more from land rents (rentier capitalism) than from the surplus extracted from labour.[43] The South American landlords could live well and achieve high status from the 'richness of nature' alone, without having to subject themselves to the market 'laws' of constant productivity increases. The small farms, the *minifundio*, on their hand, lacked both the capital and the technology to invest in the land, and remained to a large degree engaged in subsistence farming.

Economic behaviour is always contingent on historically-produced social relations. These relations had to be rewritten before economic behaviour could be recast, but the export-oriented model reinforced the wealth and power of the great landowners and it did not create a national bourgeoisie.[44]

An incomplete 'agrarian modernisation' under state intervertionism (1930s-1970s)

After the Great Depression, faith in self-regulating markets and export-led growth plummeted. *Laissez-faire* policies were declared to have failed throughout the world, but they were understood to have been particularly detrimental for South America, since the price relations between primary commodities and manufactured goods in the period from 1870–1930, had moved against the first, creating a long-term deterioration in the terms

41   Bulmer-Thomas, *The Economic History*, p. 145.

42   R. Slatta, 'Rural criminality and social conflict in nineteenth-century Buenos Aires province', *The Hispanic American Historical Review* **60** (3) (1980): 453–67. Parlamento del Uruguay, *Código Rural de la República Oriental del Uruguay*, (Montevideo: Códigos y leyes del Uruguay, 1875) https://lawcat.berkeley.edu/record/188708?ln=en (Accessed 10 Sept. 2021).

43   Baraibar Norberg, *The Political Economy of Agrarian Change*, p. 66

44   J. Martinez-Alier, 'Ecology and the poor: A neglected dimension of Latin American history', *Journal of Latin American Studies* **23** (3) (1991): 18; Williebald and Bértola, 'Uneven development paths among settler societies', 108.

5. Sojización *as a New* First Movement

of trade.[45] Development was no longer sought through specialisation in comparative-advantage (i.e. natural resources), but the new leitmotif was state-led inward-oriented development and industrialisation, which meant changing the productive structure away from high reliance on a few commodities.[46] From the 1930s to the early 1970s, pervasive policy intervention was adopted, for example Import Substitution Industrialisation, which ultimately promoted the industrial sector at the expense of a heavily taxed agricultural sector, using both direct and indirect policies (e.g. export taxes, export quotas and licences on major agricultural commodities). In addition, exchange rate controls resulted in a highly overvalued currency, adding further disincentives to agricultural production. Paraguay differed from the rest of the region in so far that it continued with a free-trade regime, while also increasing state interventionism.[47]

While 'true' development was understood to require industrialisation, agrarian reform and modernisation were also on the policy agenda. The *latifundio* – the power base of the old rural oligarchy, with its feudal-like social relations and low use of technology and labour – was seen to stand in the way of total factor productivity growth, the formation of domestic markets and emancipation of the rural poor.[48] At the same time, smallholders (*minifundio*) and landless peasants were trapped in poverty or subsistence farming.[49] The need for agrarian reform in South America was soon expressed by a broad range of actors – from struggling peasants (*campesinos*), to scholars and politicians, and the president of the USA.[50]

45   Prebisch, *The Economic Development of Latin America*, pp. 7–10.

46   C. Kay, *Latin American Theories of Development and Underdevelopment* (London: Routledge Library Editions, 1989); Williebald and Bértola, 'Uneven development paths among settler societies', 118.

47   Masi and Borda, *Estado y economía en Paraguay*, pp. 29–33

48   C. Carlson, 'Agrarian structure and underdevelopment in Latin America: Bringing the latifundio "back in"', *Latin American Research Review* **54** (3) (2019): 678–93; A. García, 'Proceso y frustración de las reformas agrarias en América Latina', *Estudios Internacionales* **1** (3–4) (1967): 353–410; Kay, '¿El fin de la reforma agraria en América Latina?', 66–69.

49   Ibid.; Kay, *Latin American Theories of Development and Underdevelopment*; Von Bennewitz, 'Land tenure in Latin America', 1795.

50   E. Feder, 'The Campesino's perspectives in Latin America', *The Developing Economies* **7** (2) (1970): 233–46. The US-led *Alliance of Progress* expressed its aim to: 'encourage, in accordance with the characteristics of each country, programs of comprehensive agrarian reform leading to the effective transformation, where required, of unjust structures and systems of land tenure and use, with a view to replacing latifundia and dwarf holdings by an equitable system of land tenure'. Inter-American Economic and Social Council, *Charter of Punta del Este* (Punta del Este: OEA, 1961) https://avalon.law.yale.edu/20th_century/intam16.asp (Accessed 11 Nov. 2021)

Consequently, the majority of these states launched new public bodies for state-led redistribution of land to the rural poor.[51] *The social function of land* – the doctrine that land ownership is not an inviolable right of the individual against the state, but entails societal responsibility beyond private utility – became popular. By the middle of the twentieth century, this doctrine had been incorporated into most of the constitutions in the region, opening up for public confiscation of large private landholding that failed to fulfil its social purpose.[52] Land reform was often combined with tax incentives, extension services, credits, price supports and price regulation. The aim was to transform big and petty producers alike into modern high-yielding, market-oriented, farmers. The states particularly promoted the *green revolution* technologies (high yielding seeds, fertilisers, machines, irrigation and pesticides).[53] In this vein, the states also invested in public agricultural research and development, as well as in infrastructure (silos, elevators, roads and ports) to reduce transport costs and to incorporate new areas in market circuits.[54]

It was within this context that soybeans started to become more important. In Argentina, with its well-developed seed science, locally adapted soybean-seed varieties with improved yields started to emerge in the 1960s, after decades of public and private investments in research and development on soybean seeds.[55] Soybeans were also promoted in extension services as an ideal wheat rotation crop.[56] Gradually, soybeans started to become adopted by some commercially-oriented medium size farmers. Soybeans proved ideal

51   Bulmer-Thomas, *The Economic History*; Kay, '¿El fin de la reforma agraria en América Latina?'.

52   T. Ankersen and T. Ruppert, 'Tierra y Libertad: the social function doctrine and land reform in Latin America', *Tulane Environmental Law Journal* **19** (69) (2006): 70–122; Feder, 'The Campesino's perspectives', 237.

53   D.M. Jones, 'The Green Revolution in Latin America: Success or failure?' *Publication Series (Conference of Latin Americanist Geographers)* **6** (1977): 55–63; H. Bernstein, 'Is there an agrarian question in the 21st Century?' *Canadian Journal of Development Studies/Revue canadienne d'études du développement* **27** (4) (2006): 449–60; Baraibar Norberg, *The Political Economy of Agrarian Change*, pp. 77–89.

54   M. Teubal, 'Soja y agronegocios en la Argentina: la crisis del modelo', *Lavboratorio, Cambio Estructural y Desigualdad Social* **10** (22) (2008): 1–33; Raidán Martinez, 'Medio ambiente y agricultura en el Paraguay', 112–13.

55   W. Shurtleff and A. Aoyagi, *History of Soybeans and Soyfoods in South America (1882–2009): Extensively Annotated Bibliography and Sourcebook* (Lafayette: SoyInfo Center, 2009), pp. 88, 173–75, 242, 318, 449.

56   D. Martínez 'Historia de la soja en la Argentina: Introducción y adopción del cultivo', in H. Baigorri and L. Salado (eds), *El cultivo de soja en Argentina* (Buenos Aires: Agroeditorial, 2012), pp. 17–23.

for Argentina, since they were not consumed nationally and thus faced no price or quota restrictions, which from time to time affected most other crops.[57] Even so, it would take many more decades until soybeans began to gain prominence as an export commodity in Argentina.

Public research on soybeans also had a long history in Uruguay. During the 1960s and 1970s soybeans were hailed as an ideal rotation crop in mixed farming systems, capable of boosting the productivity of pastures due to their nitrogen-fixing capacity. Yields in Uruguay remained relatively low, however, and only a small segment of farmers adopted the crop.[58]

In Paraguay, soybean production started to expand within the realm of the governmental efforts to expand wheat production under the 'National Wheat Program', in which soybeans were promoted as the ideal complementary rotation crop.[59] Soybeans became particularly important in the Atlantic Forest areas; Paraguay aimed to turn its vast 'unproductive' forests, into agricultural areas. Infrastructure projects (e.g. the construction of the *Itaipú* mega hydroelectric dam) and economic incentives for farmers to clear the forest, spurred settlement in the area.[60] Many of the settlers were Brazilian farmers (so-called *Brasiguaios*), with a strong specialisation in cash crops such as cotton and soybeans.[61] Leapfrogging into the technological advances from decades of Brazilian publicly funded research in adapting temperate-zone soybean varieties to the geographic conditions of the Atlantic forest,[62] the economic margins of soybean production increased. Paraguay achieved an exportable surplus of soybeans in 1970, but the production was still negligible compared with the expansion of the coming decades.[63]

Notwithstanding the dynamic development of soybeans and some other crops during this period, the bulk of agricultural production was stagnating, and South America's share of international agricultural trade was in steady

57   Klein and F. Luna, 'The growth of the soybean frontier in South America: The case of Brazil and Argentina', *Journal of Iberian and Latin American Economic History* **39** (2020): 1–42.

58   E. Errea, J. Peyrou, J. Secco and G. Souto, *Transformaciones en el agro uruguayo - Nuevas instituciones y modelos de organización empresarial* (Montevideo: Ucudal, 2011), p. 12.

59   Baraibar Norberg, *The Political Economy of Agrarian Change*, pp. 90–93, 167.

60   Ibid,. pp. 93–96, 104–12.

61   Raidán Martinez, 'Medio ambiente y agricultura en el Paraguay', 114.

62   C.M. da Silva and C. de Majo, 'Genealogy of the soyacene: The tropical bonanza of soya bean farming during the great acceleration', *International Review of Environmental History* **7** (2) (2021): 79–80.

63   USDA, *Oilseeds*.

decline.[64] One reason was the relatively protectionist agricultural trade policies of the advanced economies. The USA and Europe heavily subsidised their agriculture, which soon led to considerable surpluses driving international food prices downward.[65] In addition, the *General Agreement on Tariffs and Trade* (GATT) – well aware of the unique political status that agriculture enjoyed in some major economies – included articles that allowed protective tariffs on key agricultural products.[66] Another reason was the already-mentioned domestic policies favouring industry at the expense of agriculture, particularly in Argentina and Uruguay, through restrictions and levies on agricultural trade and regulations on land market.[67] These and other regulations put breaks on market forces, slowing down the pace of commodification.

Another obstacle to the development of capitalist farming in the region was the extreme resilience of the landowning structure. After decades of 'agrarian reform', South America remined the region with the most unequal land structure in the world, with large estates dominating the landscape.[68] Powerful landlords successfully managed to curb all initiatives that would seriously change the land-owning structure.[69] On a country-by-country basis, agrarian reform was least implemented in Argentina and Uruguay. In Paraguay, Stroessner had, under half a century of 'agrarian reform', expanded the agrarian frontier, with 12.23 MHa of public land, mainly in the Atlantic forest region – most of it ending up in the hands of big landowners.[70] The

64   Between 1948–1952 and 1965, the share of Latin America in global agricultural trade declined. The declining share of the region continued between 1963 and 2000; R. Serrano and V. Pinilla, 'The declining role of Latin America in the global agricultural trade, 1963–2000', *Journal of Latin American Studies* **48** (1) (2016): 115–46; C. Kay, '¿El fin de la reforma agraria en América Latina?'.

65   J. Clapp, *Food*, 2nd ed. (Cambridge: Polity Press, 2016), pp. 97–99.

66   R. Sharma, *Agriculture in the GATT: A Historical Account* (Rome: FAO, 2000) https://www.fao.org/3/x7352e/X7352E04.htm (Accessed 11 Nov. 2021)

67   Baraibar Norberg, *The Political Economy of Agrarian Change*.

68   Coordinadora de Derechos Humanos del Paraguay, *Informe Chokokue* (Asunción: Codehupy, 2007); Kay, '¿El fin de la reforma agraria en América Latina?'; Carlson, 'Agrarian structure and underdevelopment'; Ankersen and Ruppert, 'Tierra y Libertad', 99–106.

69   C. Kay, 'Why East Asia overtook Latin America: Agrarian reform, industrialisation and development', *Third World Quarterly* **23** (6) (2002):1073–102; Von Bennewitz, 'Land tenure in Latin America', 1796; Feder, 'The Campesino's perspectives', 239.

70   By 1960, latifundio represented roughly 5% of farm units and about four-fifths of the land, while minifundio represented four-fifths of farm units but had only 5% of the land: E. Botella-Rodríguez, 'La cuestión agraria en América Latina: desafíos recurrentes y nuevas preguntas para la historia rural', in D. Soto and J.M. Lana (eds), *Del pasado al futuro como problema: la historia agraria contemporánea española en el siglo XXI* (Prensas universitarias de Zaragoza, 2018), pp. 285–311.

5. Sojización *as a New* First Movement

survival of the *latifundio* during the twentieth century played the function of retarding the mobilisation of the land similarly to how Polanyi argued feudal forms of life did in Central and Eastern Europe in the nineteenth century.[71] Since the landlords were more interested in social prestige and political power than in maximising economic profit, they did not reinvest surpluses into production to incessantly rise productivity. Most of them did not adopt green-revolution technologies, but continued to produce in extensive ways, making little use of capital and labour.[72]

In a parallel way, the great majority of peasants continued with very small parcels of land, and did not become competitive capitalist farmers.[73] Petty producers that had non-market access to land continued to live, produce and reproduce without having to turn themselves into wage laborers. They continued deploying diverse production strategies (including self-subsistence) and family labour on their small parcels of land, instead of specialising land-use and exerting themselves to find the most advantageous employment for whatever capital or labour they could command. They upheld production and consumption patterns in accordance with the social values of tradition, reciprocity, exchange or kinship (not only profit).[74]

The small producers who had to pay rents to access land, however, had no alternative but to try to maximise profit. While some of them became successful capitalist farmers – specialising in cash crops, hiring wage-labour, and reinvesting in the land – most did not. The green revolution technologies spurred the displacement of peasants since the so-called high-yielding seeds required optimum conditions (irrigation, intensive use of fertilisers and chemical pesticides, rich soils and monoculture) to provide good harvests. Since poor peasants did not have these conditions, many ended up outcompeted. Displaced peasants from areas of fertile land and proximity to market outlets often moved to more marginal lands or forest areas. Geographical and socio-economic remoteness from the sites of high concentration of capital and power could thus represent a refuge from the pressures of market society.

71   Polanyi, *The Great Transformation*, pp. 192–93

72   Baraibar Norberg, *The Political Economy of Agrarian Change*, p. 80; García, 'Proceso y frustración de las reformas agrarias', 353–410; Jones, 'The Green Revolution', 57; Botella-Rodríguez, 'La cuestión agraria', 289.

73   Baraibar Norberg, *The Political Economy of Agrarian Change*; Bulmer-Thomas, *The Economic History*.

74   J. Balsa, 'Consolidación y desvanecimiento del mundo chacarero: Transformaciones de la estructura agraria, las formas sociales de producción y los modos de vida en la agricultura bonaerense, 1937–1988'. (Ph.D. Thesis, Universidad Nacional de La Plata, 2004).

While the governments wished to incorporate all land under production, budget constraints often curbed necessary investments in basic infrastructure to make exploitation profitable in many remote places.[75] Many forest areas, swamps and savannahs were thus continuously considered too difficult, too costly and too risky to exploit by capitalised farmers and firms. Many peasants and native communities could still access food knowledge, prestige, farmland and labour (partially or completely) in these areas, outside of market arrangements. Land in these places remained interwoven with multiple human needs, such as source of food (often combining hunting, fishing, gathering and farming), home, identity, sacred place, etc.[76] Overall, and notwithstanding the long-term trend of commodification, significant amounts of production and consumption remained, to a varying degree, socially embedded in ways that Polanyi depicted as characterising pre-market societies.[77] However, with the breakdown of the Bretton-Woods system in 1973, the introduction of floating exchange rates and the oil crisis, the idea of the self-regulating market came again to dominate with strengthened force around the world.

The Neoliberal Turn: Regulative Shifts Paving the Way for *Sojización*

A renewed faith in liberalisation and de-regulation, was already expressed by the military dictatorships in the 1970s, but it was not until after the 1982 debt crisis and the wave of re-democratisation in the same decade (1983 in Argentina, 1985 in Uruguay and 1989 in Paraguay) that neoliberal reforms would be fully implemented. The international lending organisations spurred and deepened the policy shift through their free-market reform prescriptions, and their enhanced power over debt-ridden developing countries, under the

75   Botella-Rodríguez, 'La cuestión agraria', 291; S. Mansourian, L. Aquino, T. Erdmann and F. Pereira, 'A comparison of governance challenges in forest restoration in Paraguay's privately-owned forests and Madagascar's co-managed state forests', *Forests* 5 (4) (2014): 763–83; C. Kay and L. Vergara-Camus, *La cuestión agraria y los gobiernos de izquierda en América Latina* (Buenos Aires: CLASCO, 2018); Ankersen and Ruppert, 'Tierra y Libertad'.

76   Author's interviews with small producers in Argentina, Paraguay and Uruguay, February–March 2017; A. Nunes, R. Dettogni, B. Almeida and E. Fischer, 'Wild meat sharing among non-indigenous people in the southwestern Amazon', *Behavioral Ecology and Sociobiology* 73 (2) (2019): 26; M. Graziano, 'Jevons paradox and the loss of natural habitat in the Argentinean Chaco: The impact of the indigenous communities' land titling and the forest law in the Province of Salta', *Land Use Policy* 69 (31) (2017): 608–17.

77   Polanyi, *The Great Transformation*, pp. 79–88.

5. Sojización *as a New* First Movement

so-called *Washington Consensus*.[78] The powerful rural lobby organisations from the late nineteenth century already mentioned above (SRA, ARP and ARU in Argentina, Paraguay and Uruguay respectively), took an active role in restructuring policies in all three countries, pushing for even lower taxes and a further down-sizing of the state.[79]

While the restructuring reforms varied in intensity, timing and forms,[80] all countries took significant steps towards trade liberalisation (dismantling tariff and nontariff barriers such as taxes and quotas), promotion of foreign direct investments, fiscal austerity (e.g. cutting public research and investments on agriculture, closing down the public boards for price regulation and support programmess) and decentralisation (significantly reducing the role of the state for planning and development of the economy).[81] Moreover, land markets were liberalised. Restrictions on foreign or anonymous shareholders owning land were removed, private property rights to land were strengthened through land-titling programmes, public land was privatised and the *social* function of land was toned down (excluding the traditional usufruct right to land). Furthermore, rules for land renting were relaxed and many other 'obstacles' to investments in land markets (foreign or domestic) were taken away.[82] Soon foreign investments started to flow back into the region, particularly into land deals and agriculture.[83]

78 J. Katz, 'La macro- y la microeconomía del crecimiento basado en los recursos naturales', in A. Bárcena Ibarra and A. Prado (eds), *Neoestructuralismo y corrientes heterodoxas en América Latina y el Caribe a inicios del siglo XXI* (Santiago: Cepal, 2015), pp. 243–59.

79 Ezquerro-Cañete, 'Poisoned, dispossessed and excluded'; Baraibar Norberg, *The Political Economy of Agrarian Change*.

80 Argentina became the shop window for neoliberal reforms under Carlos Menem (1989–1999). Uruguay, with its strong tradition of state interventionism, liberalised less than Argentina. The neoliberal shift was more gradual in Paraguay, since it had always remained open to trade. F. de Castro, B. Hogenboom and M. Baud, *Environmental Governance in Latin America* (Basingstoke: Palgrave Macmillan, 2016); Bulmer-Thomas, *The Economic History*, pp. 393–95; Baraibar, *Green Deserts or New Opportunities*.

81 A. Ezquerro-Cañete, 'Poisoned, dispossessed and excluded: A critique of the neoliberal soy regime in Paraguay', *Journal of Agrarian Change* 16 (4) (2016): 702–10; M. Margulis and T. Porter, 'Governing the global land grab: multipolarity, ideas, and complexity in transnational governance', *Globalizations* 10 (1) (2013): 65–86; Teubal 'Soja y agronegocios', 6.

82 Ezquerro-Cañete 'Poisoned, dispossessed and excluded', 704–07; Baraibar Norberg, *The Political Economy of Agrarian Change*, pp. 230–41; R. Lapitz, G. Evia and E. Gudynas, *Soja y carne en el Mercosur: comercio, ambiente y desarrollo agropecuario* (Montevideo: Coscoroba Ediciones, 2004), p. 11. Author's interview with the Senior consultant for various international organisations and former high official of INTA and SAGPyA, in Buenos Aires, 28 March 2017.

83 M. Turzi, *The Political Economy of Agricultural Booms: Managing Soybean Production in Argentina, Brazil, and Paraguay* (Cham: Palgrave Macmillan, 2017), p. 12; Margulis and Porter, Governing the global land grab, 68, 70.

Matilda Baraibar Norberg

Free-market rule was becoming increasingly institutionalised in the whole world, and regulatory shifts at different levels reinforced each other. For example, the breaks previously imposed on the international agricultural market under GATT were to a large extent removed by the 1995 creation of the *World Trade Organization* (WTO), with its *Agreement on Agriculture*.[84] Agriculture has also become increasingly financialised in line with the whole economy, and as regulations against speculation on agricultural futures markets began to be relaxed in the 1980s and 1990s.[85] Liberalisation and financialisation allowed for agribusiness enterprises to expand exponentially, and become increasingly transnational.[86] These shifts of the neoliberal turn coincided with population increase, economic growth (particularly in China) and dietary shifts towards augmented meat consumption. The rising global demand for animal products in turn increased demand for animal feed, such as soybeans.

Moreover, the WTO institutionalised a strong intellectual property right regime (IPR) through the *Agreement on Trade-Related Aspects of Intellectual Property Right*, obliging countries to provide for the protection of plant varieties (most often via patents). Biotechnology and strong Intellectual Property Rights made it possible to isolate, privatise and apply economic value to new elements of nature. The genetic information of plants also became inserted in the capitalist economy through strong patents on biotech/GM crops.[87] This commodification has played a crucial role for *sojización*, since soybeans are the most widely used GM crops in the world, accounting for almost half of the global biotech crop area.[88] The most common trait is herbicide tolerance, i.e. soybeans designed to be combined with a specific herbicide for cheap weed control and no-tillage farming. The first patented biotech soybeans were Monsanto's (now Bayer) 1996 *Roundup Ready* soybean, modified to tolerate glyphosate (Roundup). Argentina, Paraguay and Uruguay were fast approving the new technology, which marked a new phase of expansion in soybean production, since the new technological package allowed for good yields in less fertile soils, and reduced operating costs.

84 P. McMichael, *Food Regimes and Agrarian Questions* (Winnipeg, Canada: Fernwood Publishing, 2013); Baraibar Norberg, *The Political Economy of Agrarian Change*.

85 J. Clapp 'Financialization, distance and global food politics', *The Journal of Peasant Studies* 41 (5) (2014): 797–814.

86 Clapp, *Food*, pp. 97–99.

87 Clapp, *Food*, pp. 64–67, McMichael, *Food Regimes and Agrarian Questions*, p. 45.

88 See ISAAA, Brief 55: Global Status of Commercialized Biotech/GM Crops: 2019 https://www.isaaa.org/resources/publications/briefs/55/executivesummary/default.asp (Accessed 27 Nov. 2021).

5. Sojización as a New First Movement

The transforming regulatory shifts of the neoliberal turn changed the playing field so that agribusiness could expand and rapidly specialise in land-use that rendered the highest margins in the international market. This turned out to be soy. While the soybean had been known in South America among experts and in experimental fields since the late nineteenth century, and while it had become more widely known and often promoted in different ways by public agencies by the mid-twentieth century, it was not until the early twenty-first century that the soybean became the leading crop *par excellence*.[89] The soybean chain is also the most integrated to world trade of all agricultural commodities, with the lion's share of harvest grown for export, thus forming part of the wider political restructuring that emerged out of the collapse of the Bretton Woods system, in which regulations and restrictions put on markets have been gradually liberalised or removed. These policy shifts, combined with the new biotechnology that allowed for the incorporation of more marginal lands, made possible the dramatic expansion of soybean-area driven by agribusiness firms and financialised big farmers – a new *land rush* had emerged, *sojización*.[90]

The Consequences of *Sojizacón*

Along with the expansion of corporate farming in Latin America, decisions over land-use and productive orientation are increasingly taken in response to shareholder value maximisation (rather than to the experience and expectations of the farmer). Soybean production under the technological packages of genetically modified traits proved to involve significant economies of scale, increasing land concentration in the world's already most unequal land region. Petty producers had either to become more 'competitive' or to abandon the peasant way of life.[91] Consequently, during the last few decades people have been pushed away from the land they worked in Latin America at a pace

89   Shurtleff and Aoyagi, *History of Soybeans*. The percentage of 'soy-complex' exports from the Southern Cone countries of total soy-complex trade was 10 % in 1970, 39% in 1980, 54% in 1990 and 63% in 2000: USDA, *Oilseeds*.

90   L. Gonzaga Belluzo 'La reciente internacionalización del régimen del capital', in A. Bárcena and A. Prado (eds), *Neoestructuralismo y corrientes heterodoxas en América Latina y el Caribe a inicios del siglo XXI* (Santiago de Chile: CEPAL, 2015), pp. 112–25; J. Garcia-Arias, A. Cibils, A. Costantino, V. Fernandes and E. Fernández-Huerga, 'When land meets finance in Latin America: Some intersections between financialization and land grabbing in Argentina and Brazil', *Sustainability* **13** (14) (2021): 8084; Clapp, *Food*, pp. 133–35.

91   Baraibar Norberg, *The Political Economy of Agrarian Change*, pp. 189–96.

comparable to the most contentious wave of land enclosures in England between 1750 and 1850.[92] In contrast to the case of industrialising Europe, however, the recent wave of Latin American peasants migrating to the cities is not absorbed by an expanding industry; rather they become 'surplus', impoverished in the city slum. The petty farmers that manage to remain in activity need to adopt more commodified forms of doing agriculture (buy inputs and services, specialise land-use in the highest economic returns, etc).

Agribusiness firms have not only expanded and displaced peasants and other less productive farming units, but have increasingly entered forest areas and savannah to cultivate soybeans or pastures for cattle. Accordingly, a new commodification process of land emerged, in which an ever-increasing chunk of nature (land) and human lifetime (labour) is transformed into an increasingly socially disembedded commodity. In this process, claims to land based on family, tribe, vicinity, and/or on customary rules and traditions are weakened.[93]

The increased commodification of land has not only caused social exclusion, but also increased environmental exploitation. The most significant land-use conversion has been the recent sizeable expansion of crop (feed) and pastureland over natural ecosystems.[94] As forests, savannahs and wetlands are transformed into agricultural land, the diversity of ecosystems and species declines, and carbon is released into the atmosphere. At the same time, risks of outbreaks of zoonotic diseases rise, since wildlife, livestock and people come into closer contact.[95] Moreover, the increased competition for land not only crowds out peasants and converts forest areas to agricultural land, but also increases the incentives for land-use intensification, further increasing the pressures on the soil, the water and other natural resources This is already causing water pollution, soil-degradation and loss of other ecosystem functions.

---

92   The liberalisation following the AoA alone is estimated to have caused between 20–30 million people in the global south to leave agriculture: McMichael, *Food Regimes and Agrarian Questions*, p. 54.

93   Author's interviews with small producers in Argentina, Paraguay and Uruguay

94   F. Pendrill, M. Persson, J. Godar, T. Kastner, D. Moran, S. Schmidt and R. Wood. 'Agricultural and forestry trade drives large share of tropical deforestation emissions', *Global Environmental Change* 56 (2019): 1–10; J. Graesser, T.M. Aide, R. Grau and N. Ramankutty, 'Cropland/pastureland dynamics and the slowdown of deforestation in Latin America', *Environmental Research Letters* 10 (3) (2015); da Silva and de Majo, 'Genealogy of the soyacene', 83–88.

95   IPBES. Workshop Report on Biodiversity and Pandemics (Bonn, Germany: Intergovernmental Science-Policy Platform on Biodiversity and Ecosystem Services, 2020).

5. Sojización as a New First Movement

Conclusion

The increased commodification of land in South America during the past decades has thus brought about a transformation that in many respect mirrors Polanyi's *first movement*. This chapter has shown that *sojización*, this new 'great transformation', was preceded by extensive and intentional intervention in much the same way as Polanyi described of *market society* expansion in late nineteenth century England. Notwithstanding the long history of uneven commodification, the chapter has showed that many people in rural areas in South America remained embedded in social-economic arrangements along the 'Polanyian' lines of reciprocity, redistribution and householding/self-subsistence up until the neoliberal turn, when market expansion allowed new chunks of nature and human activity to become increasingly commodified.

The chapter has also showed how the recent wave of commodification, illustrated by *sojización*, just like the previous market expansion, brought high 'costs' for humans and nature. Polanyi understood that one core causal explanation for exploitation of humans and nature in market society is the fact that land and labour are actually not like any other goods; they are not created to be sold on the market.

> Labour is only another name for a human activity which goes with life itself, which in its turn is not produced for sale but for entirely different reasons, nor can that activity be detached from the rest of life, be stored or mobilised; land is only another name for nature, which is not produced by man; actual money, finally, is merely a token of purchasing power which, as a rule, is not produced at all. None of them is produced for sale.[96]

They are thus *fictitious* commodities. But when this fiction is staged, when man and nature are no longer allowed to be multidimensional and socially embedded, but commodified (privatised, isolated and rationalised down to tradeable goods subjected to the law of the market), all other functions bar those that can produce profit come to be ignored. Since the profit-creating functions cannot be isolated from the person or whole ecosystem carrying out the particular task, man and nature suffer overexploitation.[97] This mechanism explains the alarming exploitation of humans and nature that emerged with *sojización*.

Besides providing analytical insights into the causes and effects of *sojización*, Polanyi's seminal book can also inspire us to look for protective responses. Famously Polanyi argued that the unleashing of market forces – the *first*

96   Polanyi, *The Great Transformation*, pp. 110–13; 233–60.
97   Ibid., pp. 68–76

movement – became too destabilising, disruptive and destructive to be able to last for long. The overexploitation of humans and nature thus soon provoked societal responses for protection – a *countermovement* – which pushed back markets and ultimately re-embedded them in society in a *double movement*.[98] According to Polanyi, this mechanism explains why England actually experienced 'pure' industrial capitalism only between 1830 and 1870. After that, labour, money and land were still commodified, but their price (wages and rents) ceased to be subject to market mechanisms alone.[99]

Undeniably, there are many social movement reactions in South America that mobilise against *sojización*, suggesting 're-embedding' land and agriculture in other social values than profit, and these could be seen as potential *countermovements*.[100] To date, I nevertheless find them either too small and fragmented, or too short-lived and uncoordinated, to pose any challenge to the system. There has also been discussion of whether the 'post-neoliberal' governments that swept the region in the 'Pink Tide' (*Kirchnerismo* in Argentina, *Frente Amplio* in Uruguay, and the very short period of Fernando Lugo as President in Paraguay) could be seen as some type of new re-embedding. The more interventionist environmental legislation taken by the states in this period, such as the ban on logging in many forest areas (in Paraguay and Argentina) and obligatory rotation plans against soil erosion (in Uruguay), can be seen as putting some breaks on *sojización*. All three countries have also adopted policies in support of family farming.[101] However, while environmental legislation and programmes for social protection undeniably increased between 2005 and 2015, soybeans under agribusiness continued to expand and family farmers are continuously and increasingly out-competed and displaced.[102]

One reason for the lack of effective *countermovements* to protect humans and nature may lie in the severely eroded ability of individual states to push for alternative economic politics in today's globalised agro-food system, in which the investment and trade liberalisation regimes are institutionalised

---

98   Ibid., pp. 230–38.

99   Ibid., pp. 119, 259–60.

100   M. Altieri and C. Nicholls 'Agroecology: a brief account of its origins and currents of thought in Latin America', *Agroecology and Sustainable Food Systems* **41** (3–4) (2017): 231–37; McMichael, *Food Regimes and Agrarian Questions*; Worth, 'Polanyi's magnum opus?'; Carton, 'On the nature of the countermovement'.

101   Baraibar Norberg, *The Political Economy of Agrarian Change*, pp. 209–99.

102   Ibid., p. 298.

5. Sojización *as a New* First Movement

world-wide.[103] The distance between the actual places of exploitation, where social-ecological costs are felt, and the places of decisions and power is abysmal. While there already existed a contradiction between market expansion at the international level and the protectionist movement and policies at the national level under Britain's unilateral adherence to free trade,[104] today's incongruity between the decisions taken by transnational corporate actors and the local people most affected by the decisions taken is vast. This allows for an unequal socio-ecological exchange in which the most extractive and damaging activities are dislocated far away from the core economies. The commodification of international capital has further enlarged the distance and abstraction involved in the agro-food system, obscuring and constraining the feedbacks of the socio-ecological 'costs' involved. Investors can thus make profits whilst their risks and costs can be transferred to geographically disparate locations. Consequently, the social-ecological costs of this new first movement fall disproportionally on nature and poor people in the South – i.e. the depletion of natural resources and the displacement of small farmers and indigenous groups caused by *sojización* take place in specific places in rural South America. Ultimately, while the world economy is increasingly integrated, and the strong economic actors increasingly transnational, humans and nature are still placed-based..

This mismatch of scales between transnational capital and local people and places thus poses new challenges for the emergence of a new *double movement*. However, as shown in this chapter, South America has not been a power-less site, or a mere victim of the strategies of the transnational agribusiness firms, but domestic regulative shifts have played an important role co-shaping *sojización*. The lack of strong interventionist policies (stricter environmental and social legislation, limits to the freedom of enterprise, and other regulation for the rolling back of markets and compensatory protection of humans and nature) is in this way also responding to domestic factors, such as powerful pressure groups in favour of *sojización*. The traditional landed organisations together with the 'new' transnational agribusiness have formed a new powerful alliance. The power of these domestic pressure groups has been particularly clear in the (sometimes-violent) discussion (and shifting practices) on export taxes on soybeans (Argentina and Paraguay) and land

103   Silver and Arrighi, 'Polanyi's "double movement"', 338–40.
104   Dale, *Karl Polanyi*, p. 416.

Matilda Baraibar Norberg

taxes (Uruguay).[105] *Market society* is in this way self-reproducing. However, in line with Polanyi it seems clear that the devastating effects of *sojización* cannot continue indefinitely: 'business as usual' is not an option. In the end, climate change, biodiversity loss and staggering inequality eventually put whole societies at risk, pressing for the use of more vigorous tools to end overexploitation and re-embed the economy.

105   Baraibar Norberg, *The Political Economy of Agrarian Change*, pp. 329–55.

CHAPTER 6.

THE IMPACT OF SOYBEANS ON BRAZIL'S LAST AGRICULTURAL FRONTIER: SCIENTIFIC FOOD PRODUCTION, LANDSCAPE CHANGES AND THE CERRADO'S COMMODITY FRONTIER

Cassiano de Brito Rocha, Kárita de Jesus Boaventura, Giovanni de Araujo Boggione and Sandro Dutra e Silva

Introduction

The expansion of the Brazilian agricultural frontier was marked by drastic environmental transformations, profoundly affecting the landscapes of entire biogeographic regions. Brazilian environmental historiography has traditionally focused on deforestation, especially in the Atlantic Coast Forest.[1] Another recurring theme concerns the advance of the agricultural frontier on forest formations such as those of the Amazon.[2] However, recent debates on Brazilian biogeographic formations have displayed a growing interest in non-forested bioregions and phytophysiognomies, such as grasslands and savannas, including the Pampa, the Caatinga and the Cerrado. Among these, the Cerrado has particularly suffered from the advancement of the agricultural frontier in recent decades, becoming the host environment of the local green revolution and the subsequent agronomic development for the scientific production of grains and commodities.[3]

---

1 W. Dean, *With Broadax and Firebrand: The Destruction of the Brazilian Atlantic Forest* (Berkeley, Los Angeles, London: University of California Press, 1995); J.A. Drummond, *Devastação e preservação ambiental no Rio de Janeiro* (Niterói, EDUFF. 1995); J.A., Pádua, *Um sopro de destruição: pensamento político e crítica ambiental no Brasil escravista 1786–1888* (Rio de Janeiro: Jorge Zahar Editor, 2004).

2 R. Rajão et al., 'The rotten apples of Brazil agribusines', Science **369** (6501) (2020): 246–48.

3 Dutra e Silva, 'Challenging the environmental history of the Cerrado: science, biodiversity and politics on the Brazilian agricultural frontier', *Historia Ambiental Latinoamericana Y Caribeña HALAC* **10** (1) (2020): 82–116; R. Nehring, 'Yield of dreams: marching west and the politics of scientific knowledge in the Brazilian Agricultural Research Corporation – Embrapa', *Geoforum* **77** (2016): 206–17.

THE AGE OF THE SOYBEAN: 115–139 **doi:** 10.3197/63800040695086.ch06

Cassiano de Brito Rocha et al.

The effects of human presence in the Cerrado can be perceived since the historical records that precede the arrival of European colonisers when the region suffered the effects of excessive use of fire and other actions linked to the expansion of agriculture.[4] After the arrival and settlement of European colonisers in the central *chapadões* of Brazil during the eighteenth century, expansion fronts were linked to different natural resources such as mining and the native pastures from the Brazilian Cerrado grasslands.[5]

As a biogeographic formation, the Cerrado is characterised by complex landscapes and significant plant and animal diversity, with contiguous and identifiable phyto-physiognomies on a regional scale and similar geoclimatic conditions.[6] The history of colonising occupation in the Cerrado is recent but with marked circumscriptions of landscape transformation. These include mining, extensive cattle farming, the expansion of pastures with exotic grasses and the green revolution responsible for expanding the agricultural frontier in the Brazilian tropical savannah from the second half of the twentieth century.[7] Historical studies indicate that, between the 1930s and 1950s, there was an important demographic migration to Central Brazil, especially to Goiás, due to the first agricultural colonisation projects. These projects aimed to improve the utilisation of Central Brazilian tropical forests for grain production and went hand in hand with plans to expand railway networks to states such as Goiás and Mato Grosso and the recently created federal capital Brasilia.[8] The patterns of agricultural frontier expansion adopted during this period were based on clearing and burning tropical forests.

4   C. Bachelet, 'Pré-história no Cerrado: Análises antracologicas dos abrigos de Santa Elina e da Cidade de Pedra – Mato Grosso', *Fronteiras: Journal of Social, Technological and Environmental Science* **3** (2) (2014): 96–110; A.S. Barbosa, *Andarilhos da Claridade: os primeiros habitantes do Cerrado* (Goiânia: Universidade Católica de Goiás. Instituto do Trópico Submúmido, 2002).

5   S. Dutra e Silva, *No Oeste, a terra e o céu: a expansão da fronteira agrícola no Brasil Central* (Rio de Janeiro: Mauad X, 2017); R.W. Wilcox, *Cattle in the Backlands: Mato Grosso and the Evolution of Ranching in the Brazilian Tropics* (Austin: University of Texas Press, 2017); D. McCreery, *Frontier Goias, 1822–1889* (Palo Alto: Stanford University Press, 2006).

6   P.S. Oliveira and R.J. Marquis, 'Introduction: development of research in the Cerrados', in P.S. Oliveira and R.J. Marquis (eds), *The Cerrados of Brazil: Ecology and Natural History of a Neotropical Savanna*, pp. 1–10 (New York: Columbia University Press, 2002).

7   Dutra e Silva, *No oeste*; Wilcox, *Cattle in the Backlands*; M.C. Karasch, *Before Brasília: Frontier Life in Central Brazil* (Albuquerque: University of New Mexico Press, 2016); McCrerry, *Frontier Goias*.

8   Dutra e Silva, *No oeste*; Wilcox, *Cattle in the Backlands*; P.E. James, 'Trends in Brazilian agricultural Development', *Geographical Review* **43** (3) (1953): 301–28; L. Waibel, 'Vegetation and land use in the Plateau Central of Brazil', *Geographical Review* **38** (554) (1948): 529; L. Waibel, 'Uma viagem de reconhecimento ao sul de Goiás', *Revista Brasileira de Geografia* **9** (3) (1947): 313–42.

6. The Impact of Soybean Farming in Brazil's Last Commodity Frontier

As argued by Gerd Kohlhepp, between the 1950s and 1960s, this process consisted of the irrational use of tropical forests.[9] In Central Brazil, such patterns would be active until the 1950s, especially in the forested areas of Mato Grosso de Goiás.[10]

In the case of the Brazilian Cerrado, tropical forest occupies small territories in the bioregion's entirety. Conversely, most characteristic grassland domains were considered – at least until the first half of the twentieth century – unsuitable for agricultural production, in the main due to the acidity and low soil fertility.[11] However, from the second half of the twentieth century, the expansion of the agricultural frontier in the Cerrado intensified, driven by the modernisation of agriculture and national developmental policies.[12] In addition to the areas of Goiás, other Cerrado regions in Central Brazil began to receive a significant migration of southern Brazilian farmers during the 1970s[13].

A decisive historical factor for the Cerrado's agronomic development was the creation of regional and national research institutes that sought to investigate its soils' potential for agriculture. A defining point was the creation of the Brazilian Agricultural Research Company (Embrapa), instituted by Law No. 5,851 of 7 December 1972, in which President Emílio Garrastazu Medici (1969–1974) authorised the Executive Branch to establish this public company

9   G. Kohlhepp et. al., *Colonização agrária no Norte do Paraná: processos geoeconômicos e sociogeográficos de desenvolvimento de uma zona subtropical do Brasil sob a influência da plantação de café* (Maringá: Editora da Universidade Estadual de Maringá-EDUEM, 2014); G. Kohlhepp, 'A importância de Leo Waibel para a geografia brasileira e o início das relações científicas entre o Brasil e a Alemanha no campo da geografia', *Revista Brasileira de Desenvolvimento Regional* **1** (2) (2013): 29–75.

10   Dutra e Silva, *No oeste*; S. Faissol, *O Mato Grosso de Goiás* (Rio de Janeiro: Instituto Brasileiro de Geografia e Estatística-IBGE; Conselho Nacional de Geografia, 1952); James, 'Trends'.

11   C.M da Silva, 'Resistência da tradição devastadora à inovação racionalizadora: a atuação do IRI Research Institute (IRI) e os dilemas do avanço da fronteira cafeeira no estado de São Paulo 1950–1960', in J.L. Andrade e Franco, S. Dutra e Silva, J.A. Drummond and V. Silva Braz (eds), *História ambiental. Natureza, sociedade, frontera*, pp. 395–418 (Rio de Janeiro: Garamond Universitária, 2020); C.M. da Silva, *De agricultor à farmer. Nelson Rockefeller e a modernização da agricultura no Brasil* (Curitiba/Guarapuava, Universidade Federal do Paraná – UFPR, Universidade Estadual do Centro-Oeste – Unicentro, 2015); C.M. da Silva, 'De um Dust Bowl paulista à busca de fertilidade no Cerrado: a trajetória do IRI Research Institute e as pesquisas em ciências do solo no Brasil 1951–1963', *Revista Brasileira de História da Ciência – Manguinhos* **5** (1) (2012): 146–55.

12   S. Dutra e Silva, K. de Jesus Boaventura, E.D. Porfírio Júnior and C.M. Silva Neto, 'A última fronteira agrícola do Brasil: o Matopiba e os desafios de proteção ambiental no Cerrado', *Estudios Rurales* **8** (15) (2018): 145–78; Dutra e Silva, *No oeste*.

13   G. Kohlhepp and S. Dutra e Silva, 'Colonização No Brasil Central: a fronteira agrícola em Mato Grosso entre as décadas de 1950 a 1970' *Fronteiras: Revista Catarinense de História* **39** (2022): 50–81.

118

Cassiano de Brito Rocha et al.

linked to the Ministry of Agriculture.[14] Embrapa's agronomic policies were aligned with the II National Development Plan (1975–1979) inaugurated by President Ernesto Geisel (1974–1979), which particularly favoured the states of Goiás, Mato Grosso and Minas Gerais, at the time about 73 per cent of the Cerrado's total surface. In addition to the mentioned states, a report by Embrapa's Cerrado Agricultural Research Center (CPAC) in 1976 also defined as 'cerrados' (at the time they used the term in plural form) portions of the states of Maranhão, Piauí and Bahia.[15] According to Embrapa's principles and modernisation guidelines, the cerrados should fulfil an essential role in Brazilian economic development. In this context, one should note that, between the late nineteenth and early twentieth centuries, the rational use of the cerrado grasslands was a fundamental goal for local agricultural development and, at the same time, a way to protect forested areas.[16]

The Brazilian development effort attributed an essential role to agriculture due to its contribution to the national food supply, the supply of industrial raw materials and valuable exports. The Second National Development Plan established the goal of agricultural modernisation and the conversion of new strategic areas. In this context, the occupation of the cerrados was thus considered a valuable alternative, not only because it incorporated unexploited agricultural soils but also due to its geo-economic centrality.[17] Analysing Embrapa's priority areas for agronomic research and territorial occupation, the region known as Matopiba (including Cerrado portions of Maranhão, Tocantins, Piauí and Bahia) did not initially figure as central for agronomic developmental projects. According to Embrapa's initial reports, this region would await future investments, while investment and research priority would be given to areas of significant demographic expansion such as Goiás, Mato Grosso and Minas Gerais. Yet, from the 1980s, Matopiba began to experience an initial process of agricultural development. During the 1990s, it received significant investment and the migratory influx of southern Brazilian agricultural producers who saw cheap land in the region as a promising business for

14   J.F. Rogério, *Avental Subalterno à Gravata: a mercadorização da ciência e a proletarização do cientista nas pesquisas em nanotecnologia da Embrapa e da Unicamp* (São Paulo: Seven System Internacional Ltda, 2011).

15   Empresa Brasileira de Pesquisa Agropecuária (EMBRAPA), *Centro de Pesquisa Agropecuária dos Cerrados, Planaltina, DF. Relatório Técnico Anual 1975–1976* (Brasília: EMBRAPA, 1976) pp. 1–150.

16   J.A. Drummond and J.L. Andrade Franco, *Proteção à natureza e identidade nacional no brasil, anos 1920–1940* (Rio de Janeiro: ed. Fiocruz 2009), p. 272.

17   EMBRAPA, *Centro de Pesquisa*, p. 13.

agriculture. However, in the first decades of the twenty-first century, soybean farming gained importance and visibility, becoming renowned among different institutes as Brazil's 'last agricultural frontier'.[18]

Matopiba's recognition and territorial delimitation were officialised in a technical cooperation agreement concluded between the National Institute of Colonization and Agrarian Reform (INCRA) and Embrapa's Strategic Territorial Intelligence Group (GITE), established in 2013.[19] The ensuing reports seek to highlight the agricultural potential of this region, where the Cerrado covers 91 per cent of the entire territory (portions of the Amazon and Caatinga cover the region on the eastern and northwest limits, respectively). Another political-administrative delimitation of Matopiba is known as the Legal Amazon, stretching over 62 per cent of the region's total surface.[20] According to the Brazilian Forest Code, rural properties inserted within this legal framework need to be partly protected – eighty per cent of the Amazon's total forested areas and 35 per cent of the Cerrado are located in the planning territory described as the Legal Amazon. For Cerrado properties outside the limits of the Legal Amazon, the protected percentage drops from 35 to twenty per cent.[21]

According to a 2014 survey conducted by Embrapa's GITE, based on data obtained from the Ministry of the Environment (MMA) and the Chico Mendes Institute for Biodiversity Conservation (ICMBIO), Matopiba possesses 42 Conservation Units, covering a total area of 8,838,764 hectares (ha), approximately 12 per cent of the territory.[22] The studies related to the characterisation of the agricultural framework – compiled by GITE, based on data from the National Indian Foundation (FUNAI), the Secretariat

---

18   United States Department of Agriculture (USDA), *Foreign Agriculture Service, Commodities Intelligence Report. Brazil's Latest Agriculture Frontier in Western Bahia and MATOPIBA* (USDA, 26 July 26) https://ipad.fas.usda.gov/highlights/2012/07/Brazil_MATOPIBA/ (accessed 11 Aug. 2021).

19   The Strategic Territorial Intelligence Group (GITE) was created by Embrapa's Board of Directors on 12 May 2013, through Ordinance No. 1801, published in Embrapa's Administrative Communications Bulletin (BCA 53/2013). On the proposal to delimit the territory of the Matopiba, see E.E. Miranda et al., *Proposta de delimitação territorial do MATOPIBA* (Campinas SP: Embrapa. Technical Note 1, 2014).

20   Brazilian Republic Presidency, Law n. 5,173 (L5173) (27 Oct. 1966). Available at http://www.planalto.gov.br/ccivil_03/leis/l5173.html (Accessed 13 Aug. 2021).

21   'L12651'. 25 May 2012, http://www.planalto.gov.br/ccivil_03/_ato2011-2014/2012/lei/l12651.html (Accessed 13 Aug. 2021).

22   L.A. Magalhães and E.E. De Miranda, 'MATOPIBA: Quadro Natural. Embrapa Territorial-Outras publicações técnicas', *INFOTECA-E* (2014).

Cassiano de Brito Rocha et al.

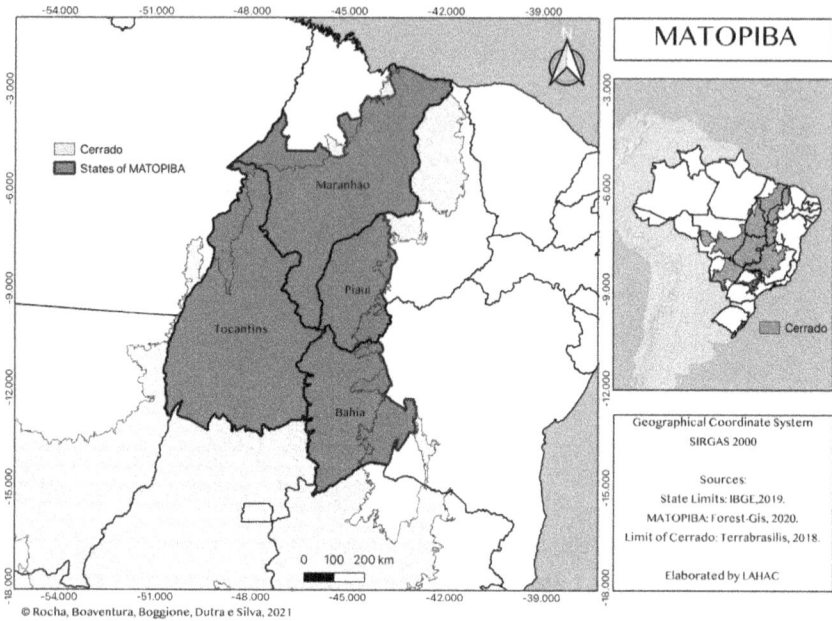

Figure 1.

Geographic location of Matopiba in the Cerrado biome.

for Policies for the Promotion of Racial Equality (SEPPIR) and INCRA, demonstrated the existence of 28 indigenous territories, covering a total area of 4,157,189 ha (5.7 per cent of the territory). In addition to these territories, studies indicate the presence of 865 agrarian reform settlements, which occupy an area of 3,706,699 (5.1 per cent of the territory) and 34 territories occupied by traditional communities, stretching across 249,918 ha (0.3 per cent of the territory).[23]

Over the last decades, Matopiba has experienced pressures from agri-business companies that have appropriated low-value territories, improving their logistics and industrial production chains. The progressive speculative pressure and agrarian valorisation of Matopiba have provoked conflicts with indigenous populations and traditional communities. The increase in land prices promoted by the advancement of soybean farming has pushed the

---

23   M.F. Fonseca and E.E. De Miranda, 'MATOPIBA: Caracterização do Quadro Agrário. Embrapa Territorial-Outras publicações técnicas', *INFOTECA-E* (2014).

cattle frontier into the forest areas of the Amazon, generating new outbreaks of burning and deforestation. Such historical processes help understand the complex socioenvironmental entanglements of this new agricultural frontier. This chapter analyses the historical role of soybean in the agronomic development of this new agricultural frontier, exploring the local expansion of this *Commodity Frontier* and its influence on the Cerrado's natural landscapes.

Materials and Methods

Study Area

The study area corresponds to the territory of Matopiba, with an approximate extent of 73 million ha (Figure 1). This territory covers about 33 per cent of the state of Maranhão (23,982,346 ha), 38 per cent of the state of Tocantins (27,772,052 ha), 11 per cent of the state of Piauí (8,204,588 ha) and 18 per cent of Bahia (13,214,499 ha).[24]

Collection of agricultural data

The surveys of data concerning harvested area (in hectares), and production rates (in tons), were extracted from two databases produced by the Brazilian Institute of Agricultural Research and Statistics (IBGE): Municipal Agricultural Production (PAM) between 1974 and 2019; and Systematic Survey of Agricultural Production (LSPA) from 2020.[25] WFP provides statistical information on 64 agricultural products, 31 temporary and 33 permanent crops. The data are collected through questionnaires devised and applied by State Agricultural Research Supervisors in collaboration with IBGE. The consolidation of WFP data also incorporates other technical information from agricultural research committees belonging to public or private institutions at the regional, state and municipal levels. These include Coordinating Groups of Agricultural Statistics (GCEA), Regional Commissions of Agricultural Statistics (COREA) and Municipal Agricultural Statistics Commissions (COMEA). Drawing from these databases, this research uses

---

24   Ibid.

25   Brazilian Institute of Geography and Statistics (IBGE), *Municipal Agricultural Production (PAM)*: https://www.ibge.gov.br/estatisticas/economicas/agricultura-e-pecuaria/9117-producao-agricola--municipal-culturas-temporarias-e-permanentes.html (Accessed 13 Aug. 2021); Brazilian Institute of Geography and Statistics (IBGE), *Systematic Survey of Agricultural Production (LSPA)*: https://www.ibge.gov.br/estatisticas/economicas/agricultura-e-pecuaria/9201-levantamento-sistematico--da-producao-agricola.html (Accessed 13 Aug. 2021).

data about soybean crops between 1974 and 2019, referring to municipal units with a minimum extension of one hectare and harvest indexes of at least one ton per year. The data were consolidated by Federative Unit, i.e., Maranhão, Tocantins, Piauí and Bahia. For 2020, data from the Systematic Survey of Agricultural Production (LSPA) were utilised since PAM had not incorporated LSPA's data into its database. However, 2020 data were manually incorporated into the database through *Google Sheets*.

Data related to Matopiba's soybean complex exports were obtained from the *Comex Stat* platform of the Brazilian Ministry of Industry, Foreign Trade and Services.[26] This platform provides information on foreign trade statistics allowing the creation and customisation of queries. The historical series available in *Comex Stat* aggregates data from 1997. In this sense, this research consolidated data on products related to fresh soybeans and their processed derivatives, according to the international method of the Harmonized System of Designation and Codification of Goods (HS4): soybeans in grains, soybean meal and soybean oil.

Methodology for detecting changes in satellite images

Monitoring changes on the Earth's surface is a difficult task commonly performed using multispectral remote sensing images. The limiting factors for the complete exploration of this data involve the absence of observations, mainly due to clouds, cloud shadow, low temporal resolution, or even low spatial resolution, especially when used in applications requiring continuous and frequently updated Earth Observation time-series data. The methodological proposal of this research stage follows the flowchart below presented in the form of a transformation diagram (Figure 2).

Google Earth Engine platform

Google Earth Engine (GEE) is considered a *Geographic Information System* (GIS) tool used worldwide for urban and environmental data analysis.[27] Due to its cloud image storage, the platform has more than forty years of global satellite imagery, i.e., historical and current images of the entire planet Earth. There are several applications of use through the platform; some examples

---

26 Ministry of Industry, Foreign Trade and Services (MDIC), *COMEX STAT*: http://comexstat. mdic.gov.br/pt/home (Accessed 13 Aug. 2021).

27 N. Gorelick et. al., 'Google earth engine: Planetary-scale geospatial analysis for everyone', *Remote Sensing of Environment* **202** (2017): 18–27.

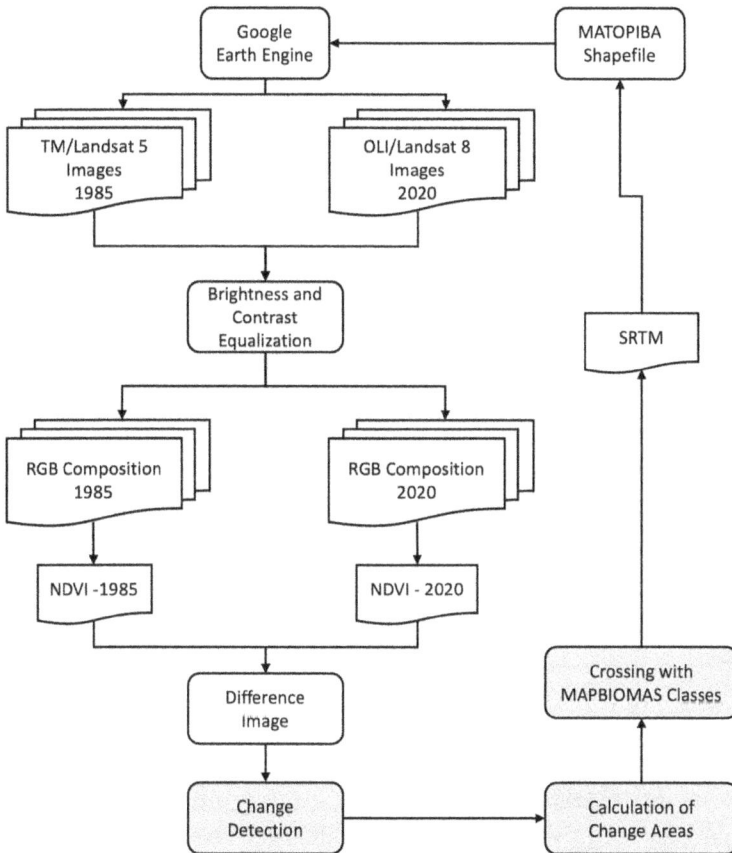

Figure 2.

Data transformation diagram

are deforestation detection, land cover classification and changes in land cover, estimation of forest biomass and carbon. GEE works through scripts written in its environment from navigation programmes. For this research, images of the Landsat Program launched in the mid-1960s were acquired, eight satellites producing images of earth with a frequency of sixteen days. Currently, the operational satellite is Landsat 8, which has two sensors, the *Operational Land Imager* (OLI) and the *Thermal Infrared Sensor* (TIRS). For the analysis of changes, images from the TM/Landsat 5 *Thematic Mapper* of 1985 were also acquired. Both have similar spectral and spatial characteristics.

124

Cassiano de Brito Rocha et al.

For the actual clipping of the area, the vector data of the Matopiba perimeter made available by the Forest–GIS initiative was used and incorporated into the script to obtain the Landsat mosaic of scenes. Other data was acquired from the *Shuttle Radar Topography Mission* (SRTM), a space mission that obtained a digital model of the terrain to generate a base of digital terrestrial topographic charts with a thirty-metre resolution. This product will be used in detection analyses regarding the topography of the Matopiba region.

Vegetation Index

Normalised Difference Vegetation Indexes (NDVI) are images that allow the evaluation of the state of vegetation and indicate the primary production – chlorophyll production – and local humidity using a numerical indicator obtained through satellite images. NDVI consists of the normalised ratio between the reflectivity of the bands in near-infrared and red by the sum of this same reflectivity. The higher the NDVI value, the higher the density of the vegetation cover. For this purpose, histograms were adjusted by equalisation of μ brightness and contrast σ^2. Visually, the images show obvious differences, as seen in Figure 3. The differences are shown in light tones.

Change detection

The change detection technique is performed using the image difference between NDVI images. The difference is made by pixel and shows the pixels that have changed given a certain threshold in grayscale. For this research, the established threshold was fifteen per cent, which means that any change in vegetation that has been altered (suppressed or modified) at a rate of fifteen per cent is shown in the result. The previous stages of histogram equalisation aim mainly to avoid false positives and/or negatives. Detection results were therefore vectorised and transformed into polygons for area calculation. Finally, the landscape change data (change detection) were cross set with Mapbiomas data containing the land use classes for the Matopiba area.

The Soy Frontier in Matopiba and the Scientific Production of Food

Based on the estimates of the MAP and the LSPA of IBGE presented in Figure 4, in 1974, soybean plantations stretched for 238 ha in Matopiba. Although territorial studies were carried out only from 2014, already by the

6. The Impact of Soybean Farming in Brazil's Last Commodity Frontier

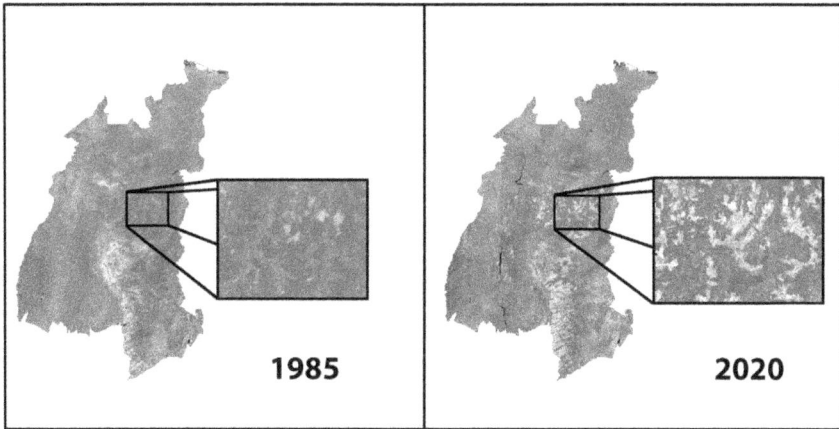

Figure 3.

Visual inspection of landscape change in images of the TM/Landsat5 and OLI/Landsat8 systems 35 years apart. Source: US Geological Survey (USGS).

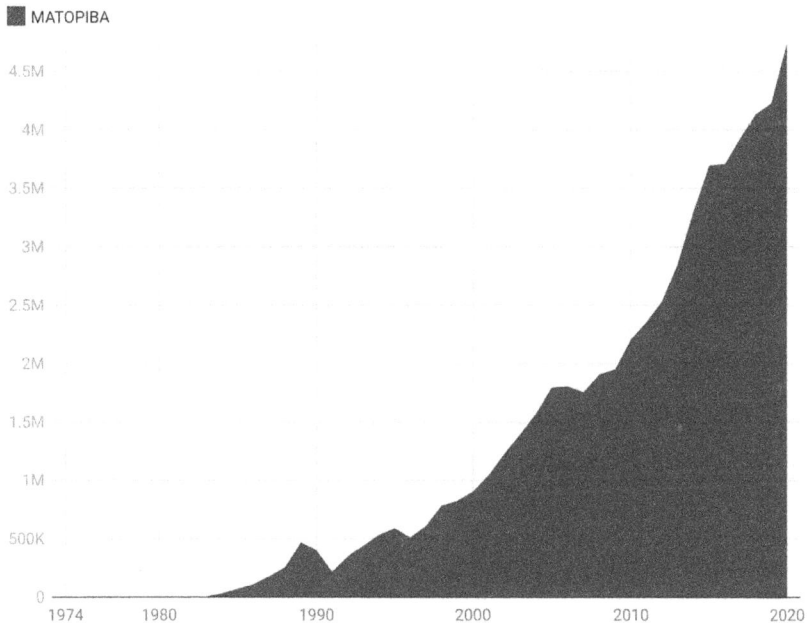

Figure 4.

Matopiba. Area of soybean cultivation 1974–2020

Cassiano de Brito Rocha et al.

1990s soybean harvests had expanded in the region (403,085 ha).[28] Ten years later, in 2000, the harvested area had more than doubled, reaching 904,995 ha. In the following two decades, the expansion of the harvested area continued to record significant values, 2,207,453 ha in 2010 and 4,743,943 in 2020. In the last decade alone, 2,536,490 ha of soybean cultivation were incorporated in Matopiba.

Unlike other Cerrado territories, the expansion of the soybean frontier in Matopiba occurred mainly at the expense of native vegetation.[29] According to surveys conducted by PAM/IBGE, the states of Bahia and Tocantins recorded the largest incorporations of new areas in the reference period, with 1,412,585 and 1,052,930 new hectares. The data estimated in 2020 reveal that the state of Bahia led soybean frontier expansion in Matopiba (1,772,600 ha), followed by Tocantins (1,079,210 ha), Maranhão (1,057.30 ha) and Piauí (834.824 ha). Such data illustrates the rapid expansion of the soybean frontier in the region, by around 92 per cent between 1990 and 2020.

Despite the potential of other large-scale crops in Matopiba, such as cotton and corn, soybeans constitute the region's leading crop with 61 per cent of the total harvested area of cereals, legumes and oilseeds.[30] All the states that form the region have more than 50 per cent of their total harvested area in the referenced crop class. In the state of Tocantins, this percentage reaches a surprising 73 per cent. Following these expansion rates, Brazil's soybean monocultures will likely expand to 12.4 million ha between 2021 and 2050. In this context, Matopiba could potentially accommodate about 86 per cent of total production (9.3 million ha).[31]

Figure 5 shows the growth of soybean production in Matopiba. In 1974 the territory produced 295 tons of soybeans and remained relatively incipient until the 1990s when soybeans began to consolidate as the dominant crop in the region. In 1990, 260,638 tons were harvested and, ten years later, production increased to 2,208,221 tons, reaching 6,295,111 tons in 2010. By

28   Miranda et al., 'Proposta de delimitação'.

29   A. Ram and K. Costa, *The Expansion of Soybean Production in the Cerrado: Paths to Sustainable Territorial Occupation, Land Use and Production* (São Paulo, SP: AGROICONE, 2016).

30   E.E. Sano, 'Land use expansion in the Brazilian Cerrado', in A. Hosono et al. (eds), *Innovation with Spatial Impact: Sustainable Development of the Brazilian Cerrado*, pp. 137–62 (Singapore: Springer, 2019).

31   A.C. Soterroni et al., 'Expanding the soy moratorium to Brazil's Cerrado', *Science Advances* **5** (7) (2019): eaav7336.

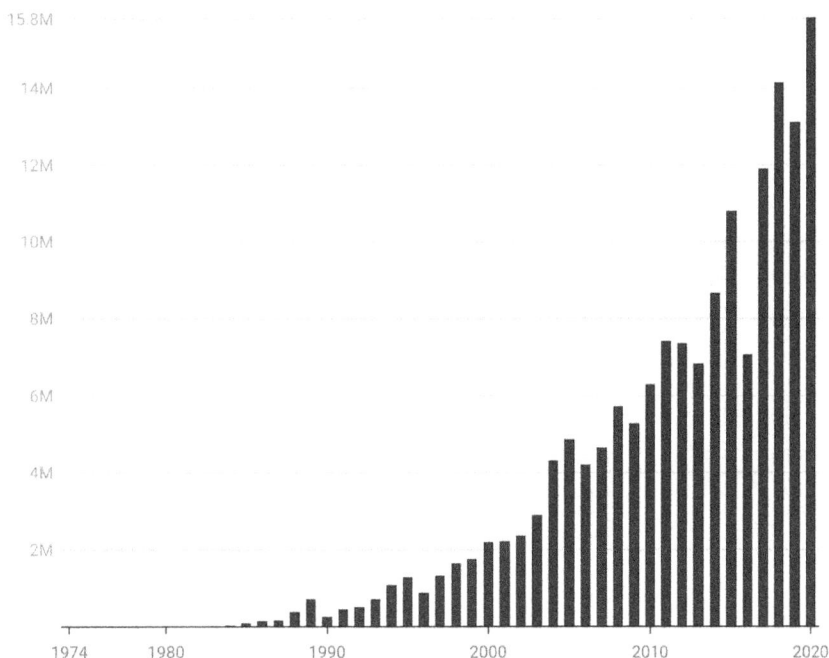

Chart: Rocha C.B.; Boggione, G.A.; Boaventura, K.J.; Dutra e Silva, S., 2021. Source: Municipal Agricultural Research – IBGE / Systematic Survey of Agricultural Production – LSPA/IBGE, 2021. Create with Datawrapper

Figure 5.

Soybean production in Matopiba 1974–2020

2010 *The Economist* published an article, with a rhetorical and adjectival tone, about the Cerrado's agricultural 'miracle', with the transformation of once arid landscapes into highly productive farms, with large soybean plantations and 'oceans of white cotton'.[32] However, the article also reported that what might seem like a green miracle resulted from a long journey of scientific research and agronomic development.

The combination of the region's soil composition and its geographical location between the Tropics of Cancer and Capricorn creates several obstacles

32   'The miracle of the Cerrado', *The Economist*, 28 Aug. 2010: https://www.economist.com/briefing/2010/08/26/the-miracle-of-the-cerrado (Accessed 13 Aug. 2021).

to large-scale soybean farming. One of the factors is photoperiodism, which makes cultivation difficult in tropical/low latitude zones. High soil acidity (with a PH between 4 and 5), nutrient deficiency and high aluminium content create farming issues. These are two main obstacles that have been overcome with the employment of long-term actions that can be consolidated into three framework-groups: i) international cooperation, financing and structuring of sound research institutions engaged in the dissemination of technologies for large-scale agriculture, through a process of continuous innovation; ii) a broad soil improvement programme in the Cerrado biome; iii) adaptation and improvement of cultivars and other biotechnological innovations.[33]

Concerning international cooperation and the development of scientific research institutions, it is essential to highlight the role of international co-operation between 1930 and 1960, such as the US Agency for International Development (USAID) and branches of the Rockefeller Foundation, such as the IRI Research Institute.[34] Another essential international cooperation took place in a second phase from the 1970s, through the Japanese-Brazilian Program for the Development of the Cerrados (PRODECER), resulting from the global soybean supply crisis in the 1970s.[35] Due to an internal production crisis, the United States, the largest oilseed supplier, implemented a blockade on exports. Faced with export reductions in a time of increased world demand, Japan was significantly impacted by the supply crisis due to its dependence on foreign grains (sixty per cent of the national food supply). A 1974 cooperation agreement signed by Prime Minister Kakuei Tanaka and President Geisel committed technical and financial collaboration. This process resulted in the Agricultural Promotion Company (CAMPO), which directly developed 345,000 ha through a US$684 million investment lasting until March 2001. In this initial phase, joint research efforts – first USAID and then Japan International Cooperation Agency (JICA) – were thus fundamental in structuring solid research institutions for agronomic development. These included CPAC/Embrapa Cerrados and the Goias Agricultural Research Company (Emgopa), currently incorporated by the

---

33   Hosono et al., 'Innovation with spatial impact: Sustainable development of the Brazilian Cerrado', in Hosono et al. (eds), *Innovation with Spatial Impact.*

34   J. Foweraker, *The Struggle for Land: A Political Economy of the Pioneer Frontier in Brazil from 1930 to the Present Day* (Cambridge: Cambridge University Press, 2002); Nehring, 'Yield of dreams'; C.M. da Silva, 'Nelson Rockefeller, a Associação Americana Internacional (AIA) e a ideologia da modernização em busca de novas fronteiras 1946–1961', *Tempos Históricos* **17** (1) (2013): 171–84.

35   A. Hosono, 'Economic and social impacts of Cerrado agriculture: Transformation for inclusive growth through clusters and value chains', in Hosono et al. (eds), *Innovation with Spatial Impact*, pp. 19–68.

129

Goias Agency for Technical Assistance, Rural Extension and Agricultural Research (Emater), both created in 1975.[36]
The Japanese-Brazilian cooperation is generally understood as one of the leading agents responsible for transforming 'a barren land into a barn of the world'.[37] More specifically, the first phase of PRODECER was more focused on the state of Goiás. However, the investments of the second and third phases were oriented to agronomic development in the whole Matopiba region.[38] The Japanese-Brazilian cooperation resulted in a Pro-Savana technology package, which aimed to expand this model of agricultural production to Mozambique.[39] Other projects are still under development, seeking to expand production and productivity in the Cerrado, such as digital precision agriculture.[40] The transformations of the Cerrado's barren soils emerged from a long soil improvement programme, with pioneering studies from scientists such as Andrew Colin McClung and other researchers linked to the IRI Research Institute during the 1950s.[41] Studies related to liming and other nutrients to correct soil acidity were fundamental for further research on agronomic soybean development undertaken by Embrapa researchers since the 1970s, especially those on symbiotic nitrogen fixation performed by Johanna Döbereiner.[42]

36 A. Hosono et al. (eds), *Development for Sustainable Agriculture: The Brazilian Cerrado* (London: Palgrave Macmillan, 2016).

37 Japan International Cooperation Agency (JICA), *50 Anos de Cooperação Brasil-Japão 1959–2009* (Brasília DF: JICA, 2009): https://www.jica.go.jp/brazil/portuguese/office/publications/pdf/50anos.pdf (Accessed 24 Aug. 2021).

38 D. Calmon, 'Shifting frontiers: the making of MATOPIBA in Brazil and global redirected land use and control change', *The Journal of Peasant Studies* (2020): 1–25.

39 T.Y. Kobashikawa, 'Brazilian agribusiness in Mozambique: The Prosavana Programme case study', *Revista Nera* **51** (2020): 345–65.

40 Governo Brasileiro, 'Brasil e Japão anunciam colaboração': https://www.gov.br/pt-br/noticias/agricultura-e-pecuaria/2021/04/brasil-e-japao-anunciam-colaboracao-para-agricultura-de-precisao-e-digital (Accessed 26 Aug. 2021).

41 A.C. McClung et. al. 'Alguns estudos preliminares sobre possíveis problemas de fertilidade, em solos de diferentes campos cerrados de São Paulo e Goiás', *Bragantia* **17** (1958): 29–44; L.M.M. Freitas et al., 'Fertilizing experiments in two soils of Campo Cerrado', *IBEC Research Institute* **21** (1959); D.S. Mikkelsen et al., 'Efeitos da calagem e adubação na produção de algodão, milho e soja em três solos de campo cerrado', *Boletim do Instituto-de Pesquisas IRI* **29** (1963); A.C. McClung and L.M.M. Freitas, 'Sulfur deficiency in soils from Brazilian campos', *Ecology* **40** (2) (1959): 315–17.

42 J. Döbereiner, et al., 'Problems in the inoculation of soybeans in acid soils', *9ᵗʰ International Grasslands, Congress* (São Paulo, January 1965) p. 7; J. Döbereiner, 'Manganese toxicity effects on nodulation and nitrogen fixation of beans (*Phaseolus vulgaris L.*), in acid soils', *Plant and Soil* **24** (1966): 153–66; S.M. Souto and J. Döbereiner, 'Fixação de nitrogênio e estabelecimento de duas variedades de soja perene (*Glycine javanica L.*) Com três níveis de fósforo e de cálcio, em solo com toxidez de manganês', *Pesquisa Agropecuária Brasileira* **4** (1) (1969): 59–66.

Concerning crop adaptation and improvement, the technology that represented a balance in the agronomic development of soybean in the Cerrado is related to cultivars with a long juvenile period. These biotechnologies solved issues of photoperiodism, especially for cultivation in tropical areas with reliefs characterised by low latitude.[43] A significant contribution was that of researcher Plinio Mello de Souza, who collected about 3,000 soybean varieties from the South of the United States, Philippines, Japan and other parts of the world, selecting the cultivars that best adapted to the low latitudes of the Cerrado.[44] The most successful breed was lo75-2760, nicknamed *Doko*, initially tested in Londrina, state of Paraná, at the Embrapa Soja Center and later sent for improved tests at CPAC.[45] Overall, the collaboration between Emgopa, Embrapa Soja, CPAC and PRODECER resulted in a continuous programme of innovation which created the most widespread non-transgenic crops planted in the Cerrado, such as BRSGO 7360, BRSGO Luziânia, BRSGO 8660, Emgopa 313 and BRSGO *Chapadões*.[46]

This seventy-year journey of agricultural development was responsible for the extraordinary results in Matopiba (Figure 5). Over the last twenty years, soy production in Matopiba has increased more than seven times, reaching 15,828,817 tons in 2020. This process was not exclusively due to the incorporation of new areas, as productivity gains increased significantly with the introduction of new breeds and agricultural technologies. According to surveys conducted in the reference databases for this research, initial soybean yields in Matopiba reached an average of just over one ton per hectare (1,239 tons per ha in 1974 and 1,183.50 in 1980). By the 2000s, average productivity rates had more than doubled, reaching 2,489.75 tons per ha. In 2020, the average productivity in Matopiba reached 3,280.60 tons per ha.

43   A. Dall'agnol, *A Embrapa Soja no contexto do desenvolvimento da soja no Brasil: histórico e contribuições* (Brasília, DF: Embrapa, 2016).

44   A. Hosono and Y. Hongo, 'Technological innovations that made Cerrado agriculture possible', in Hosono et al. (eds), *Development for Sustainable Agriculture*, pp. 11–34.

45   A. Hosono et al., 'The spatial economics of agricultural development and the formation of agro-industrial value chains: The Brazilian Cerrado', in Hosono et al. (eds), *Innovation with Spatial Impact*, pp. 1–17.

46   R.K. Zito et. al., *Cultivares de soja: macrorregiões 3, 4 e 5 Goiás e Região Central do Brasil* (Embrapa Soja-Fôlder/Folheto/Cartilha - INFOTECA-E, 2012).

The Impact of the Soy Frontier on the Changing Landscapes in Matopiba

In addition to the implications obtained from the Municipal Agricultural Production and the Systematic Survey of Agricultural Production from the Brazilian Institute of Geography and Statistics (LSPA/IBGE), this work also used the methodology of detecting changes by satellite images, which allowed an understanding of the historical transformations of Matopiba's landscapes. These data point to other variables of soybean frontier expansion, focusing mainly on landscape changes. The first significant implication refers to the degree of vegetation change in this region. These data allow us to compare the vegetation existing in the 1980s – a period in which the process of agricultural occupation was in an embryonic phase – with the anthropised landscapes in the early twenty-first century, a time of significant soybean frontier expansion. Thus, when comparing the data of temporal landscape change between 1985 and 2020, the area of total change (anthropised area) amounts to approximately 17 million ha, equivalent to 23 per cent of the total area of Matopiba (Figure 6). In this context, the primary agents of

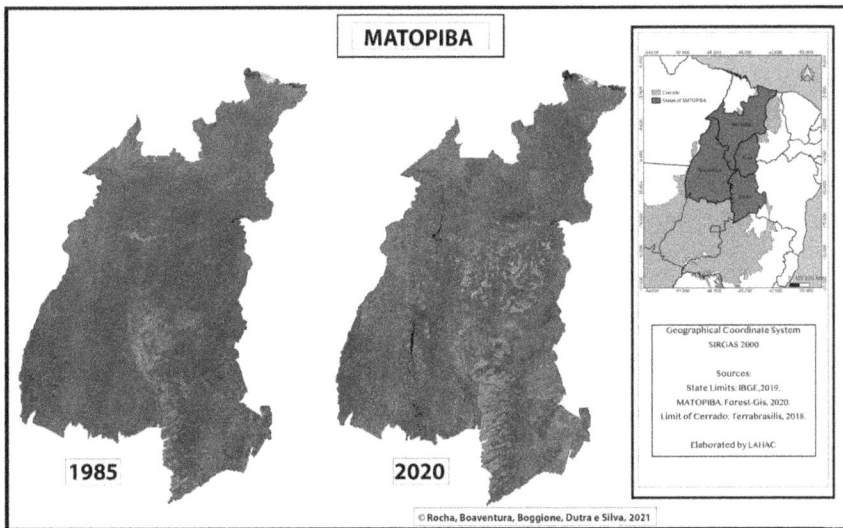

Figure 6.

Comparative image of change detection in Matopiba's landscapes (1985–2020)

132

Cassiano de Brito Rocha et al.

change were livestock and soybean, respectively 16.17 and 4.79 per cent of the total forests transformed into farmed areas.

Considering Matopiba's total area, approximately 73 million ha, soybeans constitute one of the most impactful anthropogenic elements, covering a total area of 3,499,239 ha, while cattle use 11,806,288. Such numbers appear particularly significant if compared to the distribution of the main natural landscape: savanna (25,920,783 ha), forests (9,912,097 ha), grasslands (6,801,201 ha), *apicum* (6,604,079 ha), rocky outcrop (9,663,204 ha).[47] Soybean is, therefore, the second anthropogenic element determining landscape changes in Matopiba, impacting 4.8 per cent of the territory's native vegetation (20.9 per cent of the total anthropised area and 75.8 per cent of the total farmed area), 6.7 per cent when taken together together with other crops.

The soybean frontier expansion in Matopiba is particularly pronounced in the plateaus of western Bahia (*chapadões*), the eastern Tocantins, southern Maranhão and southwest of Piaui. Matopiba's highest altitude areas present the maximum value of 1,148 metres. The greatest soybean concentration thus occurs in the plateau areas, which play a fundamental role in regulating the hydrological balances of the Midwest, North and Northeast of Brazil.[48] Conversely, deforestation phenomena in the region's low-lying areas are associated with the expansion of cattle farming.

As of 2020, Matopiba's soybean production indexes per state saw Bahia in first position (1,076,303 ha), followed by Tocantins (895,419), Maranhão (610,971) and Piauí (543,197). In percentage terms, soybean distribution and its impact on the transformation of natural landscapes show Bahia with 8.14 per cent, followed by Piauí (6.62), Tocantins (3.22) and Maranhão (2.54). The territory of Bahia thus experienced the most significant transformative impact. The state's territory within Matopiba amounts to approximately

47 The transition region between the mangrove and the mainland is commonly described as apicum. On this topic, see A.J. Schmidt et al., 'Sobre a definição da zona de apicum e sua importância ecológica para populações de caranguejo-uçá Ucides Cordatus (linnaeus, 1763)', *Boletim Técico Científico CEPENE* **19** (1) (2013): 9–25.

48 In these *chapadões* are located the springs of the Parnaíba River, which form the basin of the same name. In this plateau region there are also other springs that follow the water course towards the rivers that supply the Tocantins River basin. In water terms, the springs of the Rio Preto, tributary of the São Francisco River, also stand out. In geological terms, *chapada das Mangabeiras* is characterised by the high concentration of sandstone Urucuia, of the Triassic Age, which further north, in the state of Piauí, merges with another even older old sandstone, Poty. Thus, the Parnaíba River, carries in its course a considerable amount of sediment, which is deposited in the Atlantic Ocean, forming the Parnaíba Delta, the second largest delta in South America (see Barbosa, *Andarilhos da Claridade*).

13.2 million ha, 2.8 million ha of which are anthropised (about 20.97 per cent of Matopiba's total area). By covering 8.14 per cent of the total area of Bahia within Matopiba, soybean farming areas amount to 38.9 per cent of the anthropised area and 69.5 per cent of the state's total farmed area.

While to lesser extents, soybean plays a significant role in other states forming Matopiba. The area of Tocantins within Matopiba is equivalent to approximately 28 million ha. The total anthropised area is approximately 9.9 million ha, which is equivalent to 35.71 per cent of the state's total area included in Matopiba. In this context, soybean cover about 3 per cent of the territory (9 per cent of the anthropised area and 77.8 per cent of the farmed area). Together, the other crops correspond to 2.5 per cent of the anthropised area. Similarly, the area of Maranhão within Matopiba is equivalent to approximately 24 million ha. The total anthropised area is about 4.5 million ha (18.75 per cent of the state's total area within Matopiba). In this context, soybean monocultures cover 2.5 per cent of the total area (13.6 per cent of the anthropic area and 73.5 per cent of the farmed area). Other crops correspond to 4.88 per cent of the anthropised area. Finally, the portion of Piauí in Matopiba amounts to approximately 8 million ha. The state's total anthropised area is approximately 3.4 million ha (41.8 per cent of the total area), of which soybean monocultures cover 15.81 per cent. Overall, soybean monocultures cover 83 per cent of Piauí's farmed areas in Matopiba, while the other crops correspond to 3.2 per cent of the anthropised area.

Commodity Frontier in the Global Soybean Market

The theoretical roots of *the commodity frontier* are linked to Adam Smith's theory on the international division of labour, considering that the improvement of productive forces reaches its 'maximum perfection' when integrated into the global market scale.[49] American sociologists Terence Hopkins and Immanuel Wallerstein appropriated the classic concept of the international division of labour to propose the category of *commodity chains*.[50] These sociologists presuppose the existence of a complex business network with wealth production processes integrated beyond political boundaries. Drawing from this concept, environmental historian Jason Moore developed the concept of *commodity frontier*, considering that traditional frontier approaches attribute a

49   A. Smith, *The Wealth of Nations* (New York: Bantam Classics, 2003).

50   T.K. Hopkins and I. Wallerstein, 'Commodity chains in the world-economy prior to 1800', *Review Fernand Braudel Center* **10** (1) (1986): 157–70.

primary role to the state and disregard the power and influence of commodity chains.[51] This concept is based on the historical analysis of the Atlantic trade of commodities such as sugarcane, understood in the typology of capitalist geographic expansion.[52] In market systems, the relationship between society and nature forms part of capital accumulation processes, whose inherent expansive characteristics lead to the opening of new commodity frontiers. Therefore, environmental transformations emerge from the appropriation of productive forces, as new productive frontiers are incorporated due to the elasticity of a market system organised by *commodity chain agents.*

In this context, scientific agricultural production can be understood as part of improving productive forces in the international division of commodities, such as soybeans, which incorporate new frontiers for wealth production. Thus, Matopiba's soybean frontier analysis stems from the conjuncture of global capitalist dynamics. In this sense, the soybean frontier can be analysed in terms of the following variables: i) the territorialisation of soybean in Matopiba from a perspective of transnational interests; ii) soybean production in Matopiba in the relationship between domestic and foreign markets; iii) commodity chains of the *process of incorporating* new frontiers and expanding soybean production in Matopiba.

Concerning soybean territorialisation in Matopiba, it is interesting to retrace the scenario of how soybeans became the world's leading agricultural commodity and the reasons that allowed this crop to play a vital role in the Cerrado's agricultural expansion.[53] To understand this phenomenon, one should consider the process of 'creative destruction' and convergence dynamics between opportunity and innovation that culminated in incorporating new soybean production frontiers.[54] Its incipit was the natural disaster caused by predatory anchovy fishing off the coast of Peru in the late 1960s, which resulted in a significant reduction in the supply of fishmeal, used mainly in

---

51   J.W. Moore, 'Sugar and the expansion of the early modern world-economy: Commodity frontiers, ecological transformation, and industrialization', *Review Fernand Braudel Center* **23** (3) (2000): 409–33.

52   Ibid.

53   S.B. Hecht and C. Mann, 'How Brazil outfarmed the American farmer', *Fortune* **157** (2008): 92–105; G. Oliveira et al. 'Soy production in South America: globalization and new agro-industrial landscapes', *Journal of Peasant Studies* **43** (2) (2016): 251–610; 'Se consolida la "Patria Grande" de los transgénicos', *El País Tarija*, 14 July 2019: https://elpais.bo/economia/20190714_se-consolida-la-patria-grande-de-los-transgenicos.html (Accessed 30 June 2021).

54   J.A. Schumpeter, *Capitalism, Socialism and Democracy* (New York: Harper & Brothers, 1942).

6. The Impact of Soybean Farming in Brazil's Last Commodity Frontier

Europe for the fattening of pigs, poultry and cattle.[55] Fishmeal's scarcity turned soy into a substitute protein supplement. However, after the United States, the largest soybean producer, experienced a severe drought, soybean commodity chains quickly mobilised to incorporate new territories into production, especially in South America.[56]

This complex conjuncture has turned soybeans into South America's most dynamic agricultural segment and the world's fastest-growing commodity, with three countries of the Southern Cone (Brazil, Argentina and Paraguay) among the world's five largest producers.[57] This scenario promoted a new territorialisation by incorporating new frontiers to production, creating the so-called *Soylandia*.[58] The territorialisation of soybeans, in turn, aroused the interests of transnational farmers, especially the United States, expanding the international borders of production to vast areas of Matopiba.[59] Thus, this territory included new actors, attracting transnational pioneers, who began to be inserted into the global chains of soybean commodities.[60]

55   M.M. Schaefer, 'Men, birds and anchovies in the Peru current – dynamic interactions', *Transactions of the American Fisheries Society* **99** (3) (1970): 461–67; C.P. Idyll, 'The anchovy crisis', *Scientific American* **228** (6) (1973): 22–29; S. Tveteras et al., 'Individual vessel quotas in Peru: Stopping the race for anchovies', *Marine Resource Economics* **26** (3) (2011): 225–32.

56   P.M. Fearnside, 'Soybean cultivation as a threat to the environment in Brazil', *Environmental Conservation* **28** (1) (2001): 23–38.

57   OEC – Observatory of Economic Complexity 'Soybeans (HS: 1201) Product Trade, Exporters and Importers | OEC': https://oec.world/en/profile/hs92/soybeans (Accessed 13 Sept. 2021).

58   M.M. Schaefer, 'Men, birds and anchovies in the Peru current – dynamic interactions', *Transactions of the American Fisheries Society* **99** (3) (1970): 461–67; C.P. Idyll, 'The anchovy crisis', *Scientific American* **228** (6) (1973): 22–29; S. Tveteras et al., 'Individual vessel quotas in Peru: Stopping the race for anchovies', *Marine Resource Economics* **26** (3) (2011): 225–32.

59   A. Ofstehage, 'Farming is easy, becoming Brazilian is hard: North American soy farmers' social values of production, work and land in Soylandia', *The Journal of Peasant Studies* **43** (2) (2016): 442–60; A. Ofstehage, 'Encounters with the Brazilian soybean boom: transnational farmers and the Cerrado', in S. Sherwood et al. (eds), *Food, Agriculture and Social Change*, pp. 60–72 (London: Routledge, 2017); A. Ofstehage, 'Farming out of place: Transnational family farmers, flexible farming, and the rupture of rural life in Bahia, Brazil', *American Ethnologist* **45** (3) (2018): 317–29; J.K. Hansen, 'Foreword' in J.W. Gibson and S.E. Alexander (eds), *In Defense of Farmers: The Future of Agriculture in the Shadow of Corporate Power*, pp. xi–xvi (Lincoln: University of Nebraska Press, 2019); A. Ofstehage and R. Nehring, 'No-till agriculture and the deception of sustainability in Brazil', *International Journal of Agricultural Sustainability* **19** (3–4) (2021): 335–48.

60   J. Ye al., 'The incursions of extractivism: moving from dispersed places to global capitalism', *The Journal of Peasant Studies* **47** (1) (2020): 155–83; C.M. da Silva and C. de Majo, 'Towards the Soyacene: Narratives for an environmental history of soy in Latin America's Southern Cone', *Historia Ambiental Latinoamericana y Caribeña (HALAC)* **11** (1) (2021): 329–56; G. de Oliveira and S. Hecht (eds), *Soy, Globalization, and Environmental Politics in South America* (London: Routledge, 2017).

These interests are more clearly manifested not only by the presence of transnational farmers but also in the relationship between production and internal and external markets. In order to better understand this relationship, this research has cross-referenced information, using data from municipal agricultural production (PAM/IBGE) with export volumes using the international method of the Harmonized System of Goods Designation and Codification (HS4).[61] The result of this crossing of information on soybean production in Matopiba demonstrates the influence of the international commodity market in the expansion of the soybean frontier. In Matopiba, it was not domestic market demands that led the soybean production process. On the contrary, the global soybean market has driven production, while its surplus has been directed to domestic demand. In particular, from 2015, there has been a strong trend towards the internationalisation of soybean production in Matopiba, with over seventy per cent of the grains and commodities destined for international markets.

Finally, data from the survey show that Matopiba's primary international soybean exports demand comes from China (about 62 per cent of the total demand). In addition to China, other global markets participate in foreign demand for soybeans, especially Spain (6 per cent of production), France, Thailand and Germany (4 per cent of production each). These countries, therefore, represent approximately eighty per cent of Matopiba's soybean exports, and at the same time, the growing international demand is generating pressure to advance the soybean frontier in this region of the Brazilian Cerrado. Matopiba is a clear example of the soybean market's bilateral interdependence between China and Brazil, a determinant process in establishing prices.[62] Despite the apparent risks of this interdependence, the surge in demand guarantees China's investment in Brazilian soybean farming.[63]

The socio-environmental impacts of this relationship are already evident in the transformation of the Cerrado and other biogeographic formations in Brazil. In Matopiba, both cooperatives of small producers and traditional communities have seen increasingly restricted access to the Cerrado's natural

---

61   HS4 considers the following soybean yield variables: i) soybean in grains; ii) soybean meal and iii) soybean oil.

62   Brazilian soy is the cheapest soybean commodity imported by China (see N. Yevchenko et al., 'Ensuring sustainable imports of soybeans to China: a comparative study of bilateral foreign trade with Brazil', *E3S Web of Conferences* **273** (2021): 08014).

63   E.K. Sorrow, 'Chinese investment in the Brazilian soybean sector: Navigating relations of private governance', *Journal of Agrarian Change* **21** (1) (2021): 71–89.

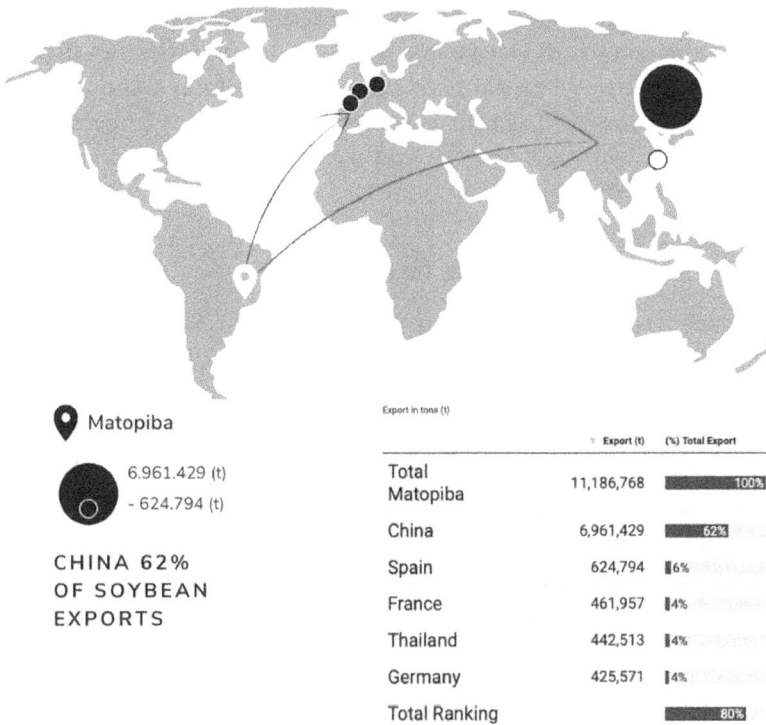

	Export (t)	(%) Total Export
Total Matopiba	11,186,768	100%
China	6,961,429	62%
Spain	624,794	6%
France	461,957	4%
Thailand	442,513	4%
Germany	425,571	4%
Total Ranking		80%

Matopiba

6.961.429 (t)
- 624.794 (t)

CHINA 62% OF SOYBEAN EXPORTS

Export in tons (t)

C. B. Rocha; K. J. Boaventura; G. A. Boggione; S. Dutra e Silva, 2021. Source: Comex Stat - MDIC (Brazilian Government) - 2021

Figure 7.

Matopiba's international soybean market (2020).

resources, impacting their commercial and subsistence activities.[64] These social segregation dynamics propelled by the soybean frontier's advancement have also been observed in other regions of the Southern Cone, such as Argentina and Paraguay.[65] The rise of the soybean frontier in the Cerrado and the consequent loss of native vegetation have significantly impacted the

---

64   G.R. Lopes et al., 'Maldevelopment revisited: Inclusiveness and social impacts of soy expansion over Brazil's Cerrado in Matopiba', *World Development* **139** (2021): 105316; M.G.B. Lima and U.M. Persson, 'Commodity-centric landscape governance as a double-edged sword: The case of soy and the Cerrado Working Group in Brazil', *Frontiers in Forests and Global Change* **3** (2020): 27.

65   J. Henderson et al., 'The Paraguayan Chaco at a crossroads: drivers of an emerging soybean frontier', *Regional Environmental Change* **21** (3) (2021): 1–14; C.M. Graziano and E. Zepharovich, 'Jevons paradox and the loss of natural habitat in the Argentinean Chaco: The impact of the indigenous communities' land titling and the Forest Law in the province of Salta', *Land Use Policy* **69** (2017): 608–17.

regional climate, and studies demonstrate that climate change and rainfall alterations may constitute a scenario of potential risk for soybean cultivation itself.[66] In Matopiba, soybeans have changed patterns of access and allocation of essential resources such as land and water, besides having impacted local feeding systems, with social results that point to a contradictory scenario within narratives of progress.[67]

Final Remarks

This research has demonstrated that the expansion of Matopiba's soybean frontier was driven by internal and external factors, characterised by scientific agricultural production to meet the demands of global commodity chains. In this context, soybeans acted as an essential historical actor in local processes of landscape change, demographic occupation and agricultural expansion. Similarly, soybean frontier expansion acted toward the displacement of other frontiers, such as pushing cattle to the Amazon's cheaper lands, resulting in the advancement of burning and deforestation. The soybean frontier was also associated with socio-environmental problems in the region, with threats to local communities who drew their subsistence from the natural resources of Matopiba's *chapadões*.

Considering these transformations, the expansion of soybean farming in Matopiba is part of a perfect commodity frontier case. In an evident process of territorialisation by the international division of soybean production, agricultural outputs in the region are projected to significantly expand between 2021 and 2050. This estimate reckons with the incorporation of about 9.3 per cent of new hectares to the soybean frontier, a potentially devastating impact on the Cerrado's ecological balance.

As a result, when looking at soybean's trajectory in the Cerrado's biogeographic context, one should consider this crop as embodying a history of accumulated agronomic development knowledge to overcome various environmental challenges. This set of challenges, such as photoperiodism and low soil fertility, was overcome through international cooperation, promotion of genetic research and improvement and other biotechnological innovations.

---

66   R. Flach, et al., 'Conserving the Cerrado and Amazon biomes of Brazil protects the soy economy from damaging warming', *World Development* **146** (2021): 105582.

67   J. Henderson et al., 'The Paraguayan Chaco at a crossroads: drivers of an emerging soybean frontier', *Regional Environmental Change* **21** (3) (2021): 1–14; Graziano and Zepharovich, 'Jevon's pardox and the loss of natural habitat', 608–17.

6. The Impact of Soybean Farming in Brazil's Last Commodity Frontier

The soybean's agronomic development history in the Cerrado still displays great research potential, especially in analysing historical sources of research institutes such as Embrapa and Emgopa.

Moreover, the relationship between the soybean frontier and the global commodity market highlights the historical peculiarities of this crop's trajectory in the Brazilian Cerrado. The most curious aspect is the recent and rapid incorporation of this frontier region within international markets, with more than seventy per cent of grain production destined for export. Finally, a more in-depth study of China's role in expanding the soybean might help understand the current market pressures affecting Matopiba and other territories like Mato Grosso and Goiás.

In Goiás, for example, the soybean frontier has already evolved to create an agro-industrial complex, with demand for grains, bran and oils that serve the domestic and external market of soybeans and meat (poultry and pigs).[68] Despite the obvious risks of this interdependence, China and Brazil continue to strengthen these commercial partnerships, allowing the stability of Brazilian agribusiness regardless of the recent economic decay of services and industrial production. While these complex relationships still lack further historical analyses, their socio-environmental consequences are already visible: if, on the one hand, soybean farming generates dividends for the agribusiness sector, on the other hand, the socio-environmental shortcomings might be unprecedented for Cerrado regions such as Matopiba.[69]

Acknowledgements

This article is the result of collaborative research developed at the Laboratory of Environmental History of the Cerrados of the Evangelical University of Goiás, in partnership with the State University of Goiás, coordinated by Sandro Dutra e Silva. Sandro Dutra e Silva appreciates the support of the National Council for Scientific and Technological Development (CNPq) with a research productivity grant. Giovanni de Araujo Boggione thanks the support of the Foundation for Research Support of the State of Goiás (FAPEG) through a postdoctoral fellowship under the supervision of Sandro Dutra e Silva at the Evangelical University of Goiás.

---

68   D.A.L. Lunas, 'Constituição do complexo agroindustrial da soja no sudoeste de Goiás' (Master's Thesis, Universidade Federal de Uberlândia, 2019).

69   B.B.N. Strasbourg et al., 'Moment of truth for the Cerrado hotspot', *Nature Ecology & Evolution* **1** (4) (2017): 1–3; Dutra e Silva, 'Challenging'.

Part III

The Elements of Soy and Environmental Degradation

CHAPTER 7.

A SOYACENE OF FIRE: KNOWLEDGE, SCIENCE AND THE IGNEOUS EXPANSION OF SOYBEAN IN TROPICAL/SUBTROPICAL SOUTH AMERICA

Eduardo Relly and Claudio de Majo

Over the last decades, soybean farming has become a hallmark of technological modernisation, contributing to worldwide economic growth. In particular, its expansion in tropical and subtropical bio-regions of the Global South has enhanced the financial and geopolitical relevance of several developing countries. On the other hand, no other global crop conveys a sense of global ecological degradation more than the soybean. Among its many 'evil' impacts, the conversion of forested land to soybean monocultures through illegal arson has aroused global concerns and political rejection. As soybean farming reached the tropics and the subtropics, it joined ecosystems whose natural regeneration cycles heavily relied on fire – either through self-igneous or anthropogenic drives, emerging from millennia of coevolutionary interactions.[1]

However, the pathways for soy to become a fire-induced crop in the most biodiverse landscapes of the planet followed a non-linear path. After landing in the United States in the first half of the twentieth century, soybean cultivation gained momentum, surpassing production rates in Manchuria.[2] This transfer epitomised an early globalisation process allowed by the progress of the agricultural sciences, as soybean farming expanded in the North-Atlantic Region and Japan. Since its global inception, science, capitalism, and the need to overcome production limits have contributed, as soybean farming channelled hopes and dreams of progress in international nutrition regimes and bioenergy production.[3] Unlike many other old key crops such

---

1   M. Flitner, 'Gibt es einen "deutschen Tropenwald"? Anleitungen zur Spurensuche', in Michael Flitner (ed.), *Der deutsche Tropenwald. Bilder, Mythen, Politik*, pp. 9-22 (Frankfurt am Main: Campus Verlag, 2000), p. 13.

2   I. Prodöhl, 'Versatile and cheap: a global history of soy in the first half of the twentieth century', *Journal of Global History* **8** (3) (2013): 463.

3   See, for example, M.R. Finlay, 'Old efforts at new uses: A brief history of chemurgy and the American search for biobased materials', *Journal of Industrial Ecology* **7** (3–4) (2003): 33–46; and

THE AGE OF THE SOYBEAN: 143–163 **doi:** 10.3197/63800040695086.ch07
© Eduardo Relly and Claudio de Majo

144

Eduardo Relly and Claudio de Majo

as wheat, rice and corn, soy emerged in an age of growing optimism about the power of science in agriculture. While intensive soybean farming was still confined to Manchuria, German scientists such as Justus Liebig and Hermann Hellriegel were already applying chemical methods in agricultural fields by the mid-nineteenth century. These experiments demonstrated that leguminous plants such as soybean could absorb and fix nitrogen out of the atmosphere.[4] Although the early transfer of soybeans from eastern Asia to the USA still relied on farmers' knowledge and experience, from the 1920s scientists linked to the US Department of Agriculture began to lead the way.[5]

Since then, agricultural science has become a cornerstone of the modern world, with soy playing a pivotal role. Although the transition from a biologically old regime to the high-tech soybean nevertheless endured many incongruencies, trade-offs and alternatives, the pathway for its global modernisation mainly relied on chemical- and gene-based bio-technologies.[6] The dawn of the soybean age since the 1970s – also known as the *Soyacene* – witnessed the scientification of agriculture in several geographical and political contexts.[7] Soil improvement programmes, breeding of staples, geopolitical manoeuvres, increased economic demand and the introduction of new cultivars formed the bedrock of the alliance between agrarian scientists and political elites.[8] At the subtropical and tropical latitudes of the Latin American Southern Cone, soybeans' experiences developed disjointedly and

G. Machado, M. Cunha, A. Walter, A. Faai and J.J.M. Guilhoto 'Biobased economy for Brazil: Impacts and strategies for maximizing socioeconomic benefits', *Renewable and Sustainable Energy Reviews* **139** (2021): 1–14.

4   See J. von Liebig, 'Chemische Briefe', *Landwirtschafltliche Monatsschrift* **6** (1) (1857): 329–43 and J.R. McNeill and V. Winiwarter, 'Breaking the sod: humankind, history, and soil', *Science* **304** (5677) (2004): 1627–29.

5   J.J. Kloppenburg, *First the Seed: The Political Economy of Plant Biotechnology, 1492–2000* (Madison, Wis.: University of Wisconsin Press, 2004), p. 78.

6   About this transition in general, see R.B. Marks, 'The (modern) world since 1500', in J.R. McNeill and E.S. Mauldin (eds), *A Companion to Global Environmental History*, pp. 57–78 (Chichester: Wiley-Blackwell, 2012), p. 58. Some examples of actors involved in this transition are the German *Rationelle Landwirtschaft* and the French Physiocracy from the late 18th and early 19th centuries. About the former, see J. Radkau, *Nature and Power: A Global History of the Environment* (Cambridge: Cambridge University Press, 2008), pp. 206–08; and about the latter, Y. Charbit, and A. Virmani, 'The political failure of an economic theory: Physiocracy', *Population* **57** (6) (2002): 855–83.

7   See C.M. da Silva and C. de Majo, 'Genealogy of the soyacene: The tropical bonanza of soybean farming during the Great Acceleration', *International Review of Environmental History* (forthcoming).

8   See, for the European case, J. Drews, *Die "Nazi-Bohne": Anbau, Verwendung und Auswirkung der Sojabohne im Deutschen Reich und Südosteuropa (1933–1945)* (Münster: Lit Verlag, 2004). About the Global South, see da Silva and de Majo, 'Genealogy of the soyacene'.

7. A Soyacene of Fire

combined with developmental solutions and political arrangements tracing back to the nineteenth century.[9]

Since the onset of the Great Acceleration, political polarisation engendered in the so-called Third World authoritarian regimes eager to implement revenue above all potential environmental and political externalities.[10] In this context, soybean offered one of the most viable options to push the advance of profitable monocultures along the agricultural frontier, deforesting native ecological formations and expanding over indigenous reservations and traditional farming lands.[11] Fire was one of the main instruments of such destructive power.

In this chapter we intend to discuss the historical link between the tropicalisation of soybeans and burning practices. In an increasingly heating world, it is worth remarking on the paradoxical alliance that has propelled the rise of the Soyacene. As game-changing technologies such as genetic manipulation and chemical fertilisers transformed tropical agriculture, they were hinged on primordial fire management technologies. The latter were in turn converted from environmental management practices pioneered by indigenous populations through millennia of trial and error to an agent of massive ecological destruction. The mixed temporalities engendered by this paradoxical encounter reveal the complexity of current environmental scenarios related to intensive soybean farming. As the behemoth of agrarian capitalism assimilated traditional practices, this led to political polarisation between global and national actors and blurred the margins between developmental discourses and environmental activism. Because the tropicalisation of soybean occurred amid rising environmental consciousness and the unprecedented politicisation of global ecological issues, this crop quickly became one of the ultimate villains.[12] As local communities, indigenous peoples, and

9   D. R. Headrick, *Humans Versus Nature: A Global Environmental History* (New York: Oxford University Press, 2020). For a Brazilian perspective, see: A. Acker, *Volkswagen in the Amazon: The Tragedy of Global Development in Modern Brazil* (Cambridge: Cambridge University Press, 2017).

10   See J.R. McNeill and P. Engelke, *The Great Acceleration: An Environmental History of the Anthropocene since 1945* (Cambridge, MA: Belknap Press, 2016), p. 123 and C.M. da Silva and C. de Majo, 'Towards the soyacene: Narratives for an environmental history of soy in Latin America's Southern Cone', *Historia Ambiental Latinoamericana Y Caribeña (HALAC)* **11** (1) (2021): 347.

11   G. de Oliveira and S. Hecht, 'Sacred groves, sacrifice zones and soy production: globalization, intensification and neo-nature in South America', *The Journal of Peasant Studies* **43** (2) (2016): 251–85.

12   For more details, see E.M. Pereira, 'A década da destruição da Amazônia: José Lutzenberger e a contrarreforma agrária em Rondônia (Anos 1980)', *História Unisinos* **21** (1) (2017): 26–37; and K. Niebauer, *Regenwald und ökologische Krise: Die Globalisierung Amazoniens im 20. Jahrhundert* (Berlin: Campus Verlag, 2021).

civil society continue to suffer and denounce the social-ecological hazards related to soybean monocultures, fire continues to constitute the tool par excellence of this demoniac process.[13] Yet, a brief historical outline of the agricultural use of fire over the last three centuries is needed to understand the complex interplay of igneous technologies and soybean monocultures.

Fire and Agriculture: Igneous Geographies in the Modern World and the Soybean

Fire has been part of the agronomic experience of mankind since the very inception of the Neolithic. Without dwelling on the details of ancient agricultural practices, one could maintain that they have marked the omnipresent 'agrarian systems of swidden cultivation, across different temporalities of geographies'.[14]

Between the eighteenth and nineteenth centuries, foresters and reformers from Central Europe acknowledged the importance of fire in making European landscapes. However, they were also prone to consider this igneous inheritance a synonym for ignorance and backwardness.[15] The European Enlightenment despised fire-based agricultural techniques, flooding several fields with biased assumptions. For example, German agronomist Carl Sprengel maintained that only American and Neoamerican peoples could still use fire for farming, given limited resources, irregular demography, and low-levelled technologies.[16] Befriended by Albrecht Thaer and Johann von Schwerz, Sprengel set clearly the igneous borders within 'rational farming' (*Rationelle Landwirtschaft*). In this context, fire was only acceptable either on the outskirts of Europe or outside the continent and should be erased from local landscapes by any available means.

Overall, as agricultural research increasingly galvanised in the nineteenth century, the remnants of fire activities in Central Europe indicated the stub-

13 S. J. Pyne, *World Fire: The Culture of Fire on Earth* (Seattle: University of Washington Press, 1997): 61.

14 M. Mazoyer and L. Roudart, *História das agriculturas no mundo: Do neolítico à crise contemporânea* (São Paulo/Brasilia: UNESP/NEAD, 2009/2010).

15 S. J. Pyne, *Vestal Fire: An Environmental History, Told through Fire, of Europe and Europe's encounter with the World* (Seattle: University of Washington Press, 2012), p. 167.

16 C. Sprengel, *Die Lehre von den Urbarmachungen und Grundverbesserungen: oder Beschreibung und Erklärung aller Urbarmachungen und Grundverbesserungen, welche die Sümpfe, Brüche, Hochmoore, Teiche, Haiden, Wüstungen, Wälder, Sandschollen, Dünen, felsigen Gründe, Aecker, Wiesen und Weiden betreffen* (Leipzig: Baumgartner's Buchhandlung, 1846), pp. 409–10.

bornness of recalcitrant peasants who were incapable of seeing the future. In the Prussian western provinces along the Rhine, conflicts between peasants and public officials over traditional cultivation practices often led to social unrest.[17] Many families from these regions later migrated to subtropical South America, where they reinvented new geographies of fire, juxtaposing European and Amerindian agronomic inheritances.[18]

However, fire persisted not only in the European fields but in the natural sciences themselves. Carl Linnaeus' *Oeconomia Naturae* (1749) described slash-and-burn cultivation techniques as a perfect cycle of the natural economy, with fire providing nutrients for the social reproduction of the peasantry and revenues for the State. Linnaeus also observed that fire accelerated the transfer of nutrients from one species to another and recommended its use.[19] Whilst Scandinavia and Russia constituted the bulwark of European fire ecologies, similar practices also involved parts of France, Austria, Switzerland and the Mediterranean basin.[20]

The pressure exerted on fire ecologies also mirrored an age of frequent warfare and revolution. The French Revolution, the Napoleonic wars and other liberal revolutions across Europe caused unprecedented damage to forests.[21] Just as importantly, the enclosure of the commons all over Europe unleashed novel and old tensions among the peasantry.[22] Because sylvan arsons became a recurrent tool of angry armies, mobs and groups in demand, the enactment of 'forest crimes' equated to political radicalism, arsons and irrationality.[23] Foresters,

17   U.E. Schmidt, *Der Wald in Deutschland im 18. und 19. Jahrhundert: Das Problem der Ressourcenknappheit dargestellt am Beispiel der Waldressourcenknappheit in Deutschland im 18. und 19. Jahrhundert ; eine historisch-politische Analyse* (Conte. Forst, Saarbrücken: Conte-Verlag, 2002), pp. 134–35.

18   E. Relly, 'A agricultura e floresta dos alemães no Brasil: mobilidade, conhecimentos e transfers no Urwald (século XIX)', *Estudos Ibero-Americanos (Online)* **46** (1) (2020): 1–16.

19   M.R. Dove, 'Linnaeus' study of Swedish swidden cultivation: Pioneering ethnographic work on the "economy of nature"', *Ambio* **44** (3) (2015): 239–48.

20   J.G. Goldammer, S. Montag and H. Page, 'Nutzung des Feuers in mittel- und nordeuropäischen Landschaften Geschichte, Methoden, Probleme, Perspektiven', *Alfred Toepfer Akademie für Naturschütz, Schneverdingen* **10** (5) (1997): 18–38.

21   K. Matteson, *Forests in Revolutionary France* (Cambridge: Cambridge University Press, 2015), p. 228.

22   Karl Marx documented conflicts related to the privatisation of common forests. See 'Proceedings of the Sixth Rhine Province Assembly. Third Article. Debates on the Law of the Theft of Wood', in *Marx & Engels: Collected Works, Vol. I. Karl Marx 1835–43*, pp. 224–65 (London: Lawrence & Wishart, 2010).

23   B.S. Grewe, *Der versperrte Wald: Ressourcenmangel in der bayerischen Pfalz (1814–1870)* (Köln, Weimar, Wien: Böhlau, 2004), p. 69.

bureaucrats and landowning elites welcomed this assumption, celebrating the progress of "scientific" knowledge in forestry and agriculture for the national economy, while fire ecologies were negatively stigmatised.[24]

European imperialism added a new layer to the ill-view of global fire and its dynamics. In Europe and the United States, political and scientific discourses on natural resources' rationalisation and conservation condemned non-Western fire-based management practices as synonyms for backwardness and environmental devastation.[25] In addition, the enforcement of anti-burning legislation in colonial lands often turned traditional shifting cultivation practices into crime.[26] For example, in colonial India, The Indian Forest Act (1927) subjected shifting cultivation to control, restriction and abolition by the state government.[27]

Negative views of fire cultures also permeated the political rhetoric of newly-established Southern American nation-states. For example, Brazilian policymakers sparked a national debate on improving fire-free farming methods. In contrast, Argentineans established national parks in peripheral areas (*Tierra del Fuego* and *Patagonia*) to ensure natural conservation.[28] However, in Chile, Brazil, Paraguay and Argentina, European colonists raised their firesticks, shaping fire landscapes along the whole perimeter of the subtropical Atlantic forests, against the expectations of political and scientific elites.[29]

24   J. Radkau, *Wood: A History* (Cambridge: Polity Press, 2012), p. 176.

25   C. Ross, *Ecology and Power in the Age of Empire: Europe and the Transformation of the Tropical World* (Oxford: Oxford University Press, 2017), pp. 302–03. Western imperialism also triggered a rampant ethnographic-style literature about the irrationality of igneous cultures. See F. Sigaut, 'Swidden cultivation in Europe. A question for tropical anthropologists', *Social Science Information* **18** (4–5) (1979): 681.

26   See, for example, for French West Africa, A. Bertrand, J. Ribot and P. Montagne, 'The historical origins of deforestation and forestry policy in French-speaking Africa: from superstition to reality?' in D. Babin (ed.), *Beyond Tropical Deforestation: From Tropical Deforestation to Forest Cover Dynamics and Forest Development*, pp. 451–64 (Paris: UNESCO, 2004); for South Africa see J. Carruthers, *National Park Science: A Century of Research in South Africa. Ecology, Biodiversity and Conservation* (Cambridge: Cambridge University Press, 2019), p. 242; for the East Dutch Indies, C. Geertz, *Agricultural Involution: The Processes of Ecological Change in Indonesia* (Berkeley: University of California Press, 1963), p. 13.

27   Sachchidananda, *Shifting Cultivation in India* (New Delhi: Concept Publishing Company, 1989), p. 76.

28   About the Brazilian case, see J.A. Pádua, *Um sopro de destruição: Pensamento político e crítica ambiental no Brasil escravista, 1786-1888* (Rio de Janeiro: Zahar, 2004), p. 342; on the Argentinean one, see O. Kaltmeier, *National Parks from North to South: An Entangled History of Conservation and Colonization in Argentina* (New Orleans: University of New Orleans Press, 2020), p. 85.

29   See, for example, L. Otero, *La huella del fuego: Historia de los bosques nativos: poblamiento y cambios en el paisaje del sur de Chile* (Santiago del Cile: Pehuén, 2006), p. 87; S.B. de Holanda, *Raízes do*

7. A Soyacene of Fire

While the tropical sciences contributed to the *damnatio memoriae* of tropical burning, contemporary scholars have reassessed slash-and-burn cultivation in cultural, demographic and ecologic terms.[30] Notwithstanding these efforts, research on subtropical/tropical igneous ecologies has remained mostly confined to local and traditional farmers' customary agricultural practices.[31] These smallholders adopt fire-based practices emerging from millennial coevolutionary processes to work their lands, generally located on the fringe of large plantations. Although their environmental knowledge is solid, their land rights are fragile, and their revenues are dependent on seasonal crash crop opportunities. In this context, fire constitutes a residual form of traditional agriculture in the face of an ever-expanding hi-tech farming system.[32] On the other hand, fire has also been adopted as a tool of environmental devastation by agro-business to steal territories occupied by smallholders and indigenous groups – a process known as 'land grabbing' or 'green grabbing'.[33]

This paradox constitutes the underlying scenario of subtropical and tropical soybean production. Whilst soybean is not the only crop to face similar contradictions, its significant biotechnological amelioration and economic demand have turned it into a global mediatic symbol of environmental destruction.[34] Thus, fire and soybean hi-tech agriculture go hand in hand with

Brasil (Šao Paulo: Companhia das Letras, 2007), pp. 66–67.

30   On traditional tropical sciences condemning fire, see P. Gourou, *Les pays tropicaux: principes d'une géographie humaine et économique* (Paris: Presses universitaires de France, 1948). On studies reevaluating fire-based management practices, see H.C. Conklin, 'The study of shifting cultivation', *Current Anthropology* **2** (1) (1961): 27–61; C. Geertz, *Agricultural Involution*; E. Boserup, *Evolução agraria e pressão demográfica* (São Paulo: Hucitec, 1987); M.R. Dove, 'Theories of swidden agriculture, and the political economy of ignorance', *Agroforestry Systems* **1** (1983): 85–99.

31   See, for examples, B Schmook et al., 'Persistence of swidden cultivation in the face of globalization: A case study from communities in Calakmul, Mexico', *Human Ecology* **41** (1) (2013): 93–107, M.C. Silva-Forsberg and P.M. Fearnside, 'Brazilian Amazonian caboclo agriculture: effect of fallow period on maize yield', *Forest Ecology and Management* **97** (3) (1997): 283–91.

32   See, for example, H. França, M.B. Ramos Neto and A. Setzer, *O fogo no Parque Nacional das Emas* (Brasília: Ministério do Meio Ambiente, 2007); and R.C.R. Abreu et al., 'The biodiversity cost of carbon sequestration in tropical savanna', *Science Advances* **3** (8) (2017): e1701284.

33   M. Backhouse, 'Green grabbing', in J. Brunner et al. (eds), *Wörterbuch Land- und Rohstoffkonflikte*, pp. 122–26 (Bielefeld: Transcript Verlag, 2019), p. 125.

34   Palm oil, for instance, is a driving force in (fire-induced) deforestation in Southeast Asia. See M.E. Cattau et al., 'Sources of anthropogenic fire ignitions on the peat-swamp landscape in Klimantan, Indonesia', *Global Environmental Change* **39** (2016): 217. About the public outcry associated with soy, see Friends of the Earth Germany, *Soja Report: Wie kann die Eiweißpflanzenproduktion der EU auf nachhaltige und agrarökologische Weise angekurbelt werden?* (Berlin: BUND, 2019).

present-time deforestation scenarios, expanding along tropical and subtropical agricultural frontiers.[35]

Soybean, Fire and Non-forested Ecozones: Cerrado and the Pampa

After World War II, the relationship between fire and the soybean came along with tropical soy farming. Before *Glycine max* became a viable crop in the tropics, varieties adjusted to the longer photoperiods and higher temperatures of the Brazilian savannah bioregion known as Cerrado needed to be developed.[36] The combination of international partnerships, developmental policies and the foundation of national institutes for agricultural research in the Global South led to the transformation of soybean into a seemingly tropical/subtropical crop in the 1970s.[37]

As soybean cultivation established along new agrarian frontiers, it touched the fringes or remnants of forested subtropical/tropical biomes, especially in the Brazilian states of Mato Grosso, Mato Grosso do Sul and Paraná, and in eastern Paraguay.[38] This process came along with arsons flaring the Atlantic Rainforest and Amazon bioregions. In contrast, the Argentinean case initially showcased a different pathway since the first expansion of the crop occurred in the subtropical Pampas as a complement to winter wheat. As soybean began to demand lands beyond the Pampas' social and ecological carrying capacities, transgenic crops propelled the expansion of the agricultural frontier in the northern forests of the Argentinean Chaco, intersecting arsons, glyphosate fumigations and deforestation.[39]

35   D. Kaimowitz and A. Angelsen, 'Forest cover and agricultural technology', in D. Babin (ed.), *Beyond Tropical Deforestation: From Tropical Deforestation to Forest Cover Dynamics and Forest Development*, pp. 431–38 (Paris: UNESCO, 2004), p. 437.

36   J.R.B. Farias, 'Requisitos climáticos', in FAO (ed.), *El Cultivo de la Soja en Los Trópicos: Mejoramiento Y Producción*, pp. 13–17 (Rome: FAO, 1995).

37   Some examples included the Brazilian Research Agricultural Coroporation (EMBRAPA) and the Argentinean National Agricultura Technology Institute (INTA). See C. M. da Silva, 'Modernizar é preciso. Pensamento social e mudança no Brasil rural (1944–1954)', *Iberoamericana* **17** (64) (2017): 207–09 and J.L. Rodríguez, 'Consecuencias económicas de la difusión de la soja genéticamente modificada en Argentina, 1996–2006', in A.L. Bravo et al. (eds), *Los señores de la soja. La agricultura transgénica en América Latina*, pp. 155–259 (Buenos Aires: Ediciones CICCUS, 2010).

38   E. Langthaler, 'Ausweitung und Vertiefung. Sojaexpansionen als regionale Schauplätze der Globalisierung', *Österreichische Zeitschrift für Geisteswissenschaften* **30** (3) (2019): 119.

39   C. Reboratti, 'Un mar de soja: la nueva agricultura en Argentina y sus consecuencias', *Revista de Geografía. Norte Grande* **45** (2010): 68.

7. A Soyacene of Fire

Cerrado

Genetically modified soybeans were designed to adapt to a wide array of environments.[40] Travelling across different ecozones, they encountered different fire regimes. While fire has maintained its role as a tool to maximise soil nutrition in forested areas, its application has been restrrained in the Brazilian Cerrado, negatively affecting traditional low-scale fire regimes.[41] Conversely, other non-forested bioregions with a massive presence of *Glycine max*, such as the Pampa grasslands, have constructed a more nuanced relationship with fire.

As one of the most extensive soybean landscapes globally, the Cerrado savannah possesses a long history of fire use.[42] In particular, traditional populations used controlled fire to open space for animal farming against constant shrub encroachment and fertilise the soil for small-scale cultivation. In time, local arboreal species have also adapted to fire regimes and base their reproductive cycles on anthropogenic or natural fires.[43] The traditional indigenous slash-and-burn farming practices characterising the moist soils of forested ecozones such as the Amazon or the Atlantic Rainforest were relatively absent in the Cerrado.[44] According to some theories, the whole neotropical savannah bioregion developed its ecological infrastructure between thirty and forty million years ago, creating an endemic ecosystem. Introduced exotic grasses and livestock later modified these native ecologies.[45] Thus, as a bioregion, the Cerrado has formed on an igneous border, separating it from other moister forested bioregions such as the Amazon

---

40 Toledo et al., 'Genética y mejoramiento', 22.

41 L.T. Kelly et al., 'Fire and biodiversity in the Anthropocene', *Science* **370** (6519) (2020): 3. On the positive impact of fire and the negative impact of reduced fire regimes in the Cerrado biome, see G. Durigan, 'Zero-fire: not possible nor desirable in the Cerrado of Brazil', *Flora* **268** (2020): 151612.

42 A.S. Barbosa et al., *O piar da juriti pepena: Narrativa ecológica da ocupação humana do cerrado* (Goiânia: Editora PUC Goiás, 2014), pp. 114–26.

43 See M.F. Simon and R.T. Pennington, 'Evidence for adaptation to fire regimes in the tropical savannas of the Brazilian Cerrado', *International Journal of Plant Sciences* **173** (6) (2012): 711–23 and M.F. Simon et al., 'Recent assembly of the Cerrado, a neotropical plant diversity hotspot, by in situ evolution of adaptations to fire', *Proceedings of the National Academy of Sciences* **106** (48) (2009): 20359–64.

44 A. Prous, *O Brasil antes dos brasileiros: A pré-história de nosso país* (Nova biblioteca de ciências sociais (Rio de Janeiro: Zahar, 2012), p. 34.

45 S. Dutra e Silva and A.S. Barbosa, 'Paisagens e fronteiras do Cerrado: ciência, biodiversidade e expansão agrícola nos chapadões centrais do Brasil', *Estudos Ibero-Americanos* (Online), **46** (1) (2020): 4–5; S. Dutra e Silva, 'Challenging the environmental history of the Cerrado: Science,

Eduardo Relly and Claudio de Majo

and the Atlantic Rainforest. In addition, its mixed composition – open *Cerrado*, *Cerradão* and some deciduous forests of limited extent – has induced multifaceted fire regimes.[46]

The technocratic colonisation of the Cerrado by pasture and soy between the 1950s and 1970s has significantly altered local fire regimes.[47] Since the inception of soybean monocultures, fire regimes have changed considerably, becoming the primary tool to clear the landscape and raising environmental concerns. Specifically, two major fire regimes have characterised this bioregion. The first is linked to the expansion of exotic grasses for animal pasture, while the second is related to the creation of soy monocultures.[48] In both cases, endemic fire regimes were reshaped by the need to clear the land for intensive farming, affecting local biodiversity. Since the 1970s, attempts to recover depleted lands have also played a significant role.[49] In this context, the agribusiness sector has embraced no-till farming as an allegedly sustainable system, allowing cultivators to avoid fire-based land fertilisation. However, no-till agriculture has not managed to curb arson linked to soybean expansion along the Cerrado's farming frontier.[50] Moreover, the accumulation of dry biomass waste linked to the total absence of fires can enhance the risks of catastrophic firestorms.[51]

biodiversity and politics of the Brazilian agricultural frontier', *Historia Ambiental Latinoamericana Y Caribeña (HALAC)* **10** (1) (2020): 82–116.

46 Pyne, *World Fire*, p. 62.

47 On this process, see C.M. da Silva, 'Nelson Rockefeller, a Associação Americana Internacional (AIA) e a ideologia da modernização em busca de novas fronteiras 1946–1961', *Tempos Históricos* **17** (1) (2013): 171–84; R. Nehring, 'Yield of dreams: marching west and the politics of scientific knowledge in the Brazilian Agricultural Research Corporation – Embrapa', *Geoforum* **77** (2016): 206–17; S. Dutra e Silva, *No Oeste, a terra e o céu: a expansão da fronteira agrícola no Brasil Central* (Rio de Janeiro: Mauad X, 2017); C.M. da Silva and C. de Majo, 'The making of a pastureland biome: American scientists, miracle grasses and the transformation of the Brazilian Cerrado', *Environment and History* (online first 2020): 21.

48 See V.R. Pivello, 'The use of fire in the Cerrado and Amazonian rainforests of Brazil: Past and present', *Fire Ecology* **7** (2011): 24–39; E.R. da Silveira et al., 'Controle de gramíneas exóticas em plantio de restauração do Cerrado', in G. Durigan and V. Soares Ramos (eds), *Manejo adaptativo: primeiras experiências na restauração de ecossistemas*, pp. 9–14 (São Paulo: Páginas & Letras Editora, 2013); and G. Durigan et al., 'Pastoreio controlado para a restauração de Cerrado invadido por braquiária', in *Manejo adaptativo*, pp. 47–49.

49 EMBRAPA, 'Integrar para conquistar o Cerrado. A abertura do bioma para a agricultura na década de 1970 motivou a criação dos primeiros sistemas de produção', 15 July 2015: https://www.embrapa.br/busca-de-noticias/-/noticia/3622209/integrar-para-conquistar-o-cerrado (Accessed 6 Aug. 2021).

50 L.L. Rausch et al., 'Soy expansion in Brazil's Cerrado', *Conservation Letters* **12** (6) (2019): 4.

51 See Durigan, 'Zero-fire'.

7. *A Soyacene of Fire*

Pampa

Recently, soy monocultures have significantly increased in the Pampa bioregion, an international ecozone (grasslands) encompassing Uruguay, parts of northern Argentina and southern Brazil. Like the Cerrado, this bioregion has endured a rapid transformation due to the expansion of industrial forestry and soybean monoculture, associated with inadequate environmental protection measures. Between 2000 and 2015, soy in the region has steeply increased in concurrence with enhanced igneous regimes.[52] However, unlike the Cerrado, the Pampa presents a remarkable temperature gradient and a non-distinguishable dry season. Moreover, endemic fires for renewing grass stocks depend on increased temperatures and do not occur with seasonal regularity. In Brazil and Argentina, the wetlands that punctuate the macroregion – the so-called *banhados* and *humedales* – bring additional complexities to the widespread use of fire. While research on the relationship between fire and soybean monocultures is still insubstantial, enhanced fire regimes have been detected in conservation areas over the last decades.[53] The fire conditions in the southern Brazilian Pampa roughly apply also to the expanding Uruguayan soybean frontier, especially in Rio Negro and Soriano. While local researchers have just begun connecting the dots, this process directly relates to the 'foreignisation' of lands occupied by foreign corporations from different parts of the world.[54]

Crossing the border towards Argentina, fire patterns have also recently changed in the vast local grasslands of non-forested Pampa. This territory constitutes the country's central soybean production hub, especially its humid wetlands, whose fertility has sustained the expansion of agribusiness. Soybeans are primarily produced in Santa Fé, Buenos Aires, Córdoba, Entre Rios and La Pampa.[55] Inversely to the processes that involved the Brazilian Cerrado in the 1960s, the inception of soybeans in Argentine happened in already consolidated farming areas, complementing wheat and sunflower cultivation. It emerged from the synergic effort of public and private actors

52   T.M. Kuplich, et al., 'O avanço da soja no bioma Pampa', *Boletim Geográfico do Rio Grande do Sul* **31** (2018): 89.

53   See J. Batista de Jesus et al., 'Análise da incidência temporal, espacial e de tendência de fogo nos biomas e unidades de conservação do Brasil', *Ciencia Florestal* **30** (1): 176–91.

54   D.E. Piñeiro, 'Land grabbing. Concentration and "foreignisation" of land in Uruguay', *Canadian Journal of Development Studies / Revue canadienne d'études du développement*, **33** (4) (2012): 483–84.

55   P.M. Bender, 'O complexo de soja argentino, análise de sua configuração espacial e rendas diferenciais', *Caminhos de Geografia* **18** (62) (2017): 220.

who cooperated in research, financial support and technological transfer.[56] Without the need to clear wildlands for cultivation, initially fire did not play the same role as in Brazil. In addition, the absence of a different dry season and precarious drainage in the Pampa hindered fire regimes and created a lack of interest compared to forested areas.[57]

Finally, just as in the case of Brazil, fire regimes have been openly opposed, at least rhetorically, by local agribusiness' carbon-free and sustainability rhetoric.[58] In this context, advanced soybean farming supposedly harmonises production and climate goals, combining no-till techniques and technological appliances that increase productivity while sparing land clearing and fire-based fertilisation.[59]

However, the bioregion has been lately ravaged by enormous and uncontrolled fires. Within a decade (2005 to 2015), 1,509,997 hectares were burned, especially in the ecozones between Pampa and Monte.[60] Severe fire episodes also hit the Pampa in 2017 and 2020, turning it into one of the most iconic fire landscapes.[61] The net increase in arson incidents is directly linked to the rise of soybean monocultures as the main symptom of an illegally expanding agricultural frontier. Moreover, evictions of smallholders have accelerated

56   D.L.M. Alvarez, 'Historia de la soja en la Argentina. Introducción y adopción del cultivo', in H.J. Baigorri and L.R.S. Navarro (eds), *El cultivo de soja en Argentina*, pp. 11–31 (Buenos Aires: Agroeditorial, 2012), p. 29.

57   For more information on the debate grasslands *versus* forests and the respective roles of fire in the Brazilian, Argentinean and Paraguayan cases see H. Wilhelmy, 'Probleme der Urwaldkolonisation in Südamerika', in H. Wilhelmy and G. Kohlhepp (eds), *Geographische Forschungen in Südamerika. Ausgew. Beitr*, pp. 36–47 (Berlin: Reimer, 1980). For the Chilean case, see Otero, *La huella del fuego*, pp. 77–79.

58   A.L. Cerdeira et al., 'Agricultural impacts of glyphosate-resistant soybean cultivation in South America', *Journal of agricultural and food chemistry*, **59** (11) (2011): 5799.

59   A. Titor, 'Towards an extractivist bioeconomy? The risk of deepening agrarian extractivism when promoting bioeconomy in Argentina', in M. Backhouse et al. (eds), *Bioeconomy and Global Inequalities. Socio-Ecological Perspectives on Biomass Sourcing and Production*, pp. 309–30 (Berlin: Springer, 2021), p. 317.

60   Ministerio del Medio Ambiente y Desarollo Sustentable, *Áreas afectadas por incendios forestales y rurales en la región pampeana y noreste de la región patagónica durante la temporada 2016– 2017*, Informe Técnico 13 (Esquel, Chubut, 2018): https://www.argentina.gob.ar/sites/default/files/ambiente-it13_incendios_2016-2017.pdf,

61   See M. Centenera, 'Los incendios arrasan más de 600.000 hectáreas en la pampa argentina', *El País*, 1 Feb. 2018: https://elpais.com/internacional/2018/02/01/argentina/1517505093_051169.html (Accessed 7 July 2021); and A. Klipphan, 'En los últimos 14 meses los incendios forestales carbonizaron una superficie que equivale 59 veces a la Ciudad de Buenos Aires', *Infobae*, 15 Feb. 2021: https://www.infobae.com/politica/2021/02/15/en-los-ultimos-14-meses-los-incendios--forestales-carbonizaron-una-superficie-que-equivale-59-veces-a-la-ciudad-de-buenos-aires/ (Accessed 1 Aug. 2021).

this trend, as their land management practices usually prevented episodes of uncontrolled fire combustion.[62]

Forested Ecozones, Soybean and Fire

There are three major forested ecozones where intensive soybean farming has consolidated, augmented by fire regimes: the Gran Chaco's dry forests, the Atlantic Rainforest and the Amazon Forest.

Gran Chaco

Soybean farming in the Gran Chaco roughly comprises northern Argentina, western Paraguay and south-eastern Bolivia. In the Argentinean portion, arson and deforestation phenomena directly related to soybean expansion in the Pampa have pushed the agricultural frontier westward, reaching the northern Chaco's tropical dry forests and savannahs.[63] This late advancement has burned down parts of the Argentinian Mesopotamia humid forest, reaching the protected *Misiones* jungle bordering southern Brazil. From the 1990s onwards, Argentinean environmentalists have considered increasing deforestation and arson in these bioregions as direct soybean externalities.[64]

Similarly, the Bolivian and Paraguayan Chaco regions have registered more frequent fires directly associated with deforestation since the introduction of genetically modified soy crops during the 1990s.[65] In Bolivia, Mennonites, Japanese and Brazilian farmers have been enriched through the soybean business, as early signs of land exhaustion and erosion in the Bolivian Chaco pushed soybean monoculture northwards, encroaching on the fringes of the Amazon Forest.[66] Inversely to the Brazilian case, soybean expansion has enjoyed less financial and governmental support, leaning on

---

62   About the expanding agrarian frontier, see Fundación Ambiente y Recursos Naturales (FARN), *Argentina incendiada: Lo que el fuego nos dejó* (Buenos Aires: FARN, 2020), pp. 8–9. Available at https://farn.org.ar/wp-content/uploads/2020/12/DOC_ARGENTINA-INCENDIADA_links. pdf (Accessed 30 Jul 2021). On the eviction of smallholding farmers, see H.A. Urcola et al., 'Land tenancy, soybean, actors and transformations in the pampas: A district balance', *Journal of Rural Studies* 39 (2015): 5.

63   V. Fehlenberg et al. 'The role of soybean production as an underlying driver of deforestation in the South American Chaco', *Global Environmental Change*, 45 (2017): 24–25.

64   P. Lapegna, *Soybeans and Power. Genetically Modified Crops, Environmental Politics, and Social Movements in Argentina* (Oxford: Oxford University Press, 2016), p. 32.

65   Fehlenberg et al., 'The role of soybean production', 32.

66   B. McKay, 'Agrarian extractivism in Bolivia', *World Development* 97 (2017): 207.

coca industry infrastructures.[67] Notably, Mennonites who migrated from the Russian steppes to the Paraguayan and Bolivian Chaco between the late nineteenth and early twentieth centuries could have refrained from using fire for land fertilisation, unlike German, Italian and Polish *colonos*.[68] However, the specific circumstances of soybean farming persuaded them to use fire.

Overall, the interaction between soybeans and fire presents dissimilarities in Bolivia and Paraguay, regardless of localised ecological circumstances. This is mainly due to the technological gap resulting from the lack of national incentives compared to neighbouring countries like Brazil and Argentina. As the lack of agricultural credit hampered the development of no-tillage systems, burning fodder remnants became an option for soil nutrition, even in small land tenures. In addition, the lack of conservation strategies has paved the way for widespread arson, in a context of little public outcry.[69]

Overall, the landscape homogenisation brought by arson combined with soybean and cattle farming has led to the transformative process known as *pampeanización*.[70] According to estimates, in the local wetlands (Chaco Húmedo) alone, around four million hectares have been burnt every year during the 2000s, mostly in densely forested areas.[71] In Bolivia, the driving forces of the burning follow unstable market trends, favouring soybean farming during the 1990s and cattle ranching since the mid-2000s.[72]

Atlantic Rainforest

The Atlantic Rainforest ecozone in which soybean production occurs includes southern Brazil, the western part of the Brazilian State of São Paulo, part of Mato Grosso do Sul and eastern Paraguay. The Atlantic Rainforest is Brazil's most populous bioregion and was the first soybean farming hub of Latin America in the 1930s.[73] However, soybean expansion in southern Brazil had

---

67   S. Hecht, 'Soybeans, development and conservation on the Amazon frontier', *Development and Change* **36** (2) (2005): 378–81.

68   Wilhelmy, 'Probleme', 40–41.

69   Reboratti, 'Un mar de soja', 67.

70   W.A. Pengue, *El vaciamiento de las Pampas La exportación de nutrientes y el final del granero del mundo* (Buenos Aires, Santiago de Chile: Heinrich Böll Foundation, 2017), p. 38.

71   N.J. Carnevale, C. Alzugaray, N. Di Leo 'Evolución de la deforestación en la cuña boscosa santafesina', in J.H. Morello and A.F. Rodriguez (eds), *El chaco sin bosques. la pampa o el desierto del futuro*, pp. 203–28 (Buenos Aires: Orientacion Grafica Editora, 2009), p. 205.

72   McKay, 'Agrarian extractivism', 206.

73   Although several soyfarming experiences were carried out in Brazil from the late nineteenth century, Polish immigrants were the first group to stably adopt the crop in the *colonias* of Rio

to compete with other industries and traditional smallholding tenures, which impaired the development of large-scale monocultures.[74] From the mid-1960s, the national military regime fostered mechanisation and agricultural research, creating the ideal conditions for expanding soybean monoculture in the Atlantic Rainforest. This process was especially evident in western Paraná, where producers substituted coffee for soybean plantations. By the mid-1970s, this process also involved the State of Mato Grosso do Sul.[75] Beyond the Brazilian borders, Paraguay promoted intense modernisation of its agriculture under the military regime of Alfredo Stroessner (1954–1989), attracting farmers from southern Brazil.[76] Soon, the so-called *brasiguaios* created a porous frontier along the Paraná River, fuelled by arson on both sides, as fire became a tool of social mobility, allowing deforestation to clear the land and implement soybean monocultures.[77]

Agriculture in the Atlantic Rainforest has been traditionally fire-based, turning local igneous cycles into part of the region's socio-environmental heritage.[78] Given the lack of an ideal dry season, burnings usually happen between early October and late November. However, moving southward, fire regimes increase in abundance and frequency as low temperatures demand deciduous species that spread their dry leaves on the ground. In this context, soybean farming enjoyed optimal conditions, especially in the north-western Rio Grande do Sul, the first area of permanent cultivation through mechanised agriculture, and western Paraná.[79]

Grande do Sul since the 1930s. See R.T. Zaleski Trindade, 'A soja e os colonos poloneses no sul do Brasil: o caso de Ceslau Biezanko e outros personagens (1930–1934)', *História Unisinos* **22** (2) (2018): 254–63.

74   P.A. Zarth, 'Terras de uso comum nos ervais do Rio Grande do Sul', in M.A.B. da Silva and P.J. Koling (eds), *Terra e poder. abordagens em história agrária*, pp. 57–72 (Porto Alegre: FCM Editora, 2015), pp. 57–59.

75   G. Kohlhepp et al., *Colonização agrária no Norte do Paraná: Processos geoeconômicos e sociogeográficos de desenvolvimento de uma zona subtropical do Brasil sob a influência da plantação de café* (Maringá: Editora da Universidade Estadual de Maringá, 2014), p. 157.

76   L.A. Galeano, 'Paraguay and the expansion of Brazilian and Argentinian agribusiness frontiers', *Canadian Journal of Development Studies* **33** (4) (2012): 458–59.

77   J. Blanc, 'Enclaves of inequality. Brasiguaios and the transformation of the Brazil-Paraguay borderlands', *The Journal of Peasant Studies* **42** (1) (2015): 147.

78   D.d.C. Cabral, *Na presença da floresta: Mata Atlântica e história colonial* (Rio de Janeiro: Garamond, 2014), p. 112. Also see W. Dean, *With Broadax and Firebrand: The Destruction of the Brazilian Atlantic Forest* (Berkeley: University of California Press, 1995).

79   In contrast, in the southern Paraná plateau, dominated by the dominant pine species *Araucaria angustifolia*, low-lying fires are more difficult, as endemic gymnosperm arboreal species retain their

158

Eduardo Relly and Claudio de Majo

In western Paraná, eastern Paraguay and Mato Grosso do Sul, flatlands and warmer climates set the agends for soybean monoculture, especially in territories claimed from indigenous reserves for national development.[80] Furthermore, mega-infrastructures like the Itaipu Dam, a Brazilian-Paraguayan combined effort, accelerated modernisation and contributed to the monocultural conversion of local lands.[81] In this context, intensive farming ecologies requiring fire for clearance replaced traditional activities such as fire-free harvesting of yerba mate. On the other hand, the recent diffusion of no-tillage farming, green manure and genetically modified seeds paradoxically prevented physiological fire cycles.[82] Likewise, in the Argentinean case, fire-free soybean production has been claimed as a technological conquest by the Brazilian agribusiness sector to promote sustainable agriculture with zero carbon emissions.[83]

Amazon

Further north lies the most extensive soybean frontier: the Amazon Forest. Fire has also been an enduring feature of local ecologies for millennia. Soybean production and cattle ranching have notably altered natural succession processes and traditional swidden cultivation systems.[84] In pre-Columbian times, fire return intervals ranged from 400 to 1,000 years. In contrast, burning has become a common feature of the Amazon basin over the last fifty years, as climatic factors such as El Nino southern oscillation and the Atlantic Multi-decadal Oscillation have been coupled with forest

leaves. See M.A. Conterato, S. Schneider, P.D. Waquil, 'Estilos de agricultura: uma perspectiva para a análise da diversidade da agricultura familiar', *Ensaios FEE* **31** (1) (2010): 180.

80   P. Antunha Barbosa and F. Mura, 'Construindo e reconstruindo territórios Guarani: dinâmica territorial na fronteira entre Brasil e Paraguai (séc. xix–xx)', *Journal de la société des américanistes* **97** (2) (2011): 302–03.

81   J. Blanc, 'A turbulent border: geopolitics and the hydreletric development of the Paraná river', in J. Blanc and F. Freitas (eds), *Big Water: The Making of the Borderlands Between Brazil, Argentina, and Paraguay*, pp. 211–41 (Tucson: University of Arizona Press, 2018), p. 227

82   J. Wilkinson and P. Perreira, 'Sojaanbau in Brasilien. Neue Formen der Finanzierung und Regulierung', in M. Ramírez and S. Schmalz (eds), *Extraktivismus. Lateinamerika nach dem Ende des Rohstoffbooms*, pp. 119–38 (München: Oekom Verlag, 2019), pp. 121–22.

83   K. Lorenzen, 'Sugarcane industry expansion and changing rural labour regimes in Mato Grosso do Sul (2000–2016)', in M. Backhouse et al. (eds), *Bioeconomy and Global Inequalities. Socio-Ecological Perspectives on Biomass Sourcing and Production*, pp. 217–38 (Cham: Palgrave Macmillan, 2021), p. 222.

84   See R. Carmenta et al., 'Shifting cultivation and fire policy: Insights from the Brazilian Amazon', *Human Ecology* **41** (2013): 603–14.

degradation.[85] Soybean production and arson are especially pronounced in the Brazilian and Bolivian Amazon.

Forest fires linked to deforestation and agricultural expansion are pivotal elements of global environmentalism, epitomising the main concerns of our times. On the other hand, Amazonian natural wealth has placed the bioregion at the core of developmental projects.[86] In this context, fire has played the symbolic role of purging transition from an 'extreme world periphery' towards an integrated capitalistic world region.[87]

Setting the Amazon ablaze is no easy task, as the region remains humid throughout the year, its soils are moist and few deciduous species inhabit it. Unlike the Atlantic Rainforest, the dry season occurs earlier (from June to November) during the so-called 'Amazonian summer' (*verão amazonico*).[88] Although intensive agro-farming practices are bound to alter agricultural settings everywhere, the Amazon's fire calendar occurs during traditional seasons with increased intensity because of land clearing. The new agents of Amazonian fire are descendants of European immigrants who expanded the agrarian frontier from the Atlantic Rainforest, moving upstream along the Paraná River and finally reaching the Amazon. This phenomenon is ongoing as agribusinesses continue to fuel the frontier myth, claiming a pioneer *ethos*.[89] Just as European immigrants established new fire regimes in the Atlantic Rainforest, overshadowing millennial indigenous practices, present-day agribusinesses are now shifting towards the Amazon.[90]

As of the 1950s, colonisation schemes targeting southern Brazilian farmers were implemented along with roads, facilities and other infrastructure.[91] However, this process gained pace in the mid-1970s as the military regime attempted to curb agrarian tensions in the south of Brazil. More specifically, the construction of an inland motorway system better connecting the Amazon to the rest of the country has led to the establishment of an 'arc of

---

85   A.A. Alencar et al., 'Landscape fragmentation, severe drought, and the new Amazon forest fire regime', *Ecological Applications: A Publication of the Ecological Society of America* **25** (6) (2015): 1493.

86   P.I. Vieira, *States of Grace: Utopia in Brazilian Culture* (Albany NY: SUNY Press, 2018), p. 33.

87   Acker, *Volkswagen in the Amazon*, p. 122.

88   R.B. Lima e Silva, J.E. de Souza Vilhena and J. da Luz Freitas, *Climatologia do Amapá: Quase um século de história* (Rio de Janeiro: Gramma, 2018), p. 53.

89   Oliveira and Hecht, 'Sacred Groves', 268.

90   Relly, 'A agricultura', 8.

91   V. Dubreuil et al., 'Evolução da fronteira agrícola no Centro-Oeste de Mato Grosso: municípios de Tangará da Serra, Campo Novo do Parecis e Diamantino', *Cadernos de Ciência & Tecnologia* **22** (2) (2005): 465.

fire' or 'arc of deforestation' in the southern and eastern fringes of the Amazon Forest.[92] As a result, aside from agricultural expansion, noble tropical timbers have been added to other coveted raw materials from the region, supplying both domestic and international markets. Furthermore, thanks to an improved road network, tractors could access more remote territories and cut down heavy trunks, augmenting the forest's flammability.[93] Since the 1980s, environmentalist José Lutzenberger has warned about shifting fire patterns between the Atlantic Rainforest and the Amazon. Lutzenberger attributed the transfer of southern Brazilian igneous agricultural practices to the Amazon to the POLONOROESTE road project. His environmentalist effort even inspired the documentary *A Década da Destruição* (1980–2009), which portrayed fire as a powerful mediatic broker epitomising the brutal developmental effort carried out by national authorities in the Amazon.[94]

Although fire gained momentum from the 1980s onwards, local deforestation phenomena between 1985 and 2018 were primarily related to logging instead of fire. Burning has engulfed the dry deciduous forest fragments where the bioregion intersects with the Gran Chaco and the Cerrado.[95] Nevertheless, soybean monoculture has steadily expanded in the Amazon, increasing more than tenfold between 2000 and 2019, as part of a deforestation process starting with fire and followed first by cattle farming and then by highly capitalised soybean plantations.[96]

Despite such trends, two recent developments that occurred during the 2000s have brought fire to new patterns. The first is the so-called 'soy moratorium', a joint initiative of the Brazilian Association of Vegetable Oil Industries and the National Association of Cereal Exporters, forbidding the expansion of soybean farming as a vehicle of deforestation. The moratorium was established in 2006 and was extended indefinitely in 2016.[97] The second initiative is the Action Plan for the Prevention and Control of Legal

---

92   P.M. Fearnside, 'Brazil's Cuiabá - Santarém (BR-163) Highway. The environmental cost of paving a soybean corridor through the Amazon', *Environmental Management* **39** (2007): 604.

93   Oliveira and Hecht, 'Sacred groves', 252.

94   E.M. Pereira, '"A década da destruição" da Amazônia: José Lutzenberger e a contrarreforma agrária em Rondônia (Anos 1980)', *História Unisinos* **21** (1) (2017): 30.

95   V. Zalles et al., 'Rapid expansion of human impact on natural land in South America since 1985', *Science Advances* **7** (14) (2021): 2.

96   X.-P. Song et al., 'Massive soybean expansion in South America since 2000 and implications for conservation', *Nature Sustainability* **4** (2021): 784–92.

97   B.F.T. Rudroff et al., 'The soy moratorium in the Amazon biome monitored by remote sensing images', *Remote Sensing* **3** (1) (2011): 185–202.

7. A Soyacene of Fire

Amazon Deforestation (PPCDAm), launched in 2004 to slow deforestation in the region. Although these policies managed to curb deforestation rates in the Amazon, partially breaking the local fire cycle, this downward trend has perilously reversed in recent years as the cattle frontier advances, fuelled by the expanding soybean frontier in the Cerrado (the so-called 'spillover effect').[98] Just as significantly, the election of right-wing president Jair Bolsonaro has dramatically increased deforestation trends, generating public outcry. In this context, fire has once again constituted the tool and symbol of environmental devastation. On 10 August 2019, the northern Brazilian newspaper *Novo Progresso* announced the 'day of fire' (*dia do fogo*).[99] This unashamed statement officialised the return of the uncontrolled arson that had swept the Amazon during the military regime, bringing land grabbing, environmental plunder and human rights violations. This igneous controversy generated public outcry and reached the upper spheres of world politics, leading French President Emmanuel Macron to take a clear stance on the geopolitical weight of France in the Amazonian question – namely, to pledge rational management of environmental policies and resources in the region via French Guyana – and to denounce Brazilian environmental-climate politics publicly.[100] In the political background, Bolsonaro's supporters and allied politicians downplayed Macron's harsh words on Amazonian arson as part of his vested interests in safeguarding French farmers after the penalising trade agreements between the Mercosur and the European Union. While this discussion could potentially evolve in the future, raising the issue of igneous environmental depletion and preaching in favour of the observation of minimal environmental standards responded to a clear political agenda. The French statements implicitly intended to acquit the European

---

98   About the soy moratorium, see N.G.R. de Mello and P. Artaxo 'Evolução do Plano de Ação para Prevenção e Controle do Desmatamento na Amazônia Legal', *Revista do Instituto de Estudos Brasileiros* **66** (2017): 127. On the spillover effect, see N. Kuschnig et al. 'Unveiling drivers of deforestation: evidence from the Brazilian Amazon', *Ecological Economic Papers 32 WU Vienna University of Economics and Business* (2019); and R. Rajão et al., 'The rotten apples of Brazil's agribusiness. Brazil's inability to tackle illegal deforestation puts the future of its agribusiness at risk', *Science* **369** (6501) (2020): 246–48.

99   L. Machado, 'O que se sabe sobre o "dia do fogo", momento-chave das queimadas na Amazônia', *BBC News Brasil*, 27 Aug. 2019: https://www.bbc.com/portuguese/brasil-49453037 (Accessed 8 Aug. 2021).

100   'Incendies en Amazonie: Bolsonaro exige que Macron "retire sus insultes" avant de discuter de l'aide du G7'. *Le Monde*, 27 Aug. 2019: https://www.lemonde.fr/international/article/2019/08/27/le-bresil-rejette-l-aide-du-g7-pour-combattre-les-incendies-en-amazonie_5503166_3210.html (Accessed 8 Aug. 2021).

Union from any responsibility in expanding the agribusiness sector in South America, selling an image of the old continent as a fire-free continent with a deep environmental sensibility. With Amazon arson conveying the imagery of an ill-conducted agribusiness, environmentalist discourses aimed to accomplish the double intent of generating political gains for European farmers and environmentalists alike.

A Soyacene of Fire

Over recent decades, the combined expansion of commodities such as soybean monoculture and cattle ranching along the agricultural frontier has framed patterns of fire regimes in the whole Southern American Cone. This process came alongside the destruction of millennial coevolutionary practices between indigenous human groups and other endemic natural species, who acquired resistance to controlled fire regimes and even based their reproduction cycles on the latter. The expansion of the commodity frontier has paradoxically turned fire from a valuable ally into the epitome of global environmental destruction, climate change and mass extinction. This is problematic in relation to the negative ecological impact of altered fire regimes and as far as fire-based knowledge practices are concerned. Whilst nineteenth-century European colonisers condemned essential fire-based practices as backward and harmful, they were proven wrong by their fellow countrymen, who both at home and abroad integrated these fires into their agricultural practices. However, because of the relentless misappropriation of fire regimes by technocrats and agribusinesses as part of the expanding agricultural frontier, present-day public debates continue to see fire almost solely as an actor of ecological devastation. The projected expansion of soybean and cattle farming in the coming years is likely to further exacerbate these issues, despite the proliferation of scientific research demonstrating the importance of controlled fire regimes in endangered bioregions such as the Cerrado and the Pampa. The contemporary fire conundrum faced by policymakers, scientists and environmental activists certainly adds further nuance to an already complex picture of ecological damage and tentative amelioration policies. As the Soyacene relentlessly advances, assimilating practices and temporalities at different geographical scales, the diverging impacts of controlled versus uncontrolled fire regimes pose the arduous challenge of careful understanding changing socioeconomic circumstances and their localised ecological implications. In a scenario of such complexity, fire regimes are here to stay and will continue to uniquely influence the region's ecological balance, for better and for worse, long after the dawn of intensive anthropogenic activities.

CHAPTER 8.

SOYBEAN CULTIVATION AND ITS ENVIRONMENTAL HEALTH AND WATER IMPACTS IN BRAZIL AND THE UNITED STATES SINCE THE 1950S: AN ANALYSIS OF RECENT LITERATURE

Larissa de Lima Trindade, John Hoornbeek, Mutlaq Albugmi, Joshua Filla, Rodrigo Fortunato de Oliveira

Introduction

There are more than 7.5 billion people on earth, and agriculture plays a major role in providing nutrition. More than 570 million farms generate the food these people consume, producing a wide range of crops. However, four crops are responsible for providing half of the world's food calories: corn, wheat, rice and soybeans. Despite the importance of these crops to human sustenance, environmental impacts are created by both their production and the food supply chain needed to produce them. For example, Poore and Nemecek note that 13.7 billion tons of carbon dioxide, or 26 per cent of anthropogenic emissions from greenhouse gases, are traceable to agriculture, as is 32 per cent of global terrestrial acidification. They also note that 67 per cent of the world's available freshwater is used for agricultural irrigation and that 78 per cent of water eutrophication results from agricultural production.[1]

These environmental impacts of agricultural production are significant and suggest anthropogenic influences consistent with McNeill and Engelke's book, *The Great Acceleration*.[2] These authors discuss the onset and development of the 'Anthropocene period', a geologic period of global history initially

---

[1]   J. Poore and T. Nemeck, 'Reducing food's environmental impacts through producers and con-sumers', *Science* **360** (6392) (2018): 987–992.

[2]   J.R. McNeill and P. Engelke, *The Great Acceleration: an Environmental History of the Anthropocene since 1945* (Cambridge, MA: Belknap Press of Harvard University Press, 2016).

THE AGE OF THE SOYBEAN: 164–184 **doi:** 10.3197/63800040695086.ch08

8. Soybean Cultivation and its Water-related Impacts

conceived by Crutzen and Stoermer,[3] which began in the late eighteenth century with the growing use of fossil fuel energy sources. They argue that the impacts of the Anthropocene period – or the 'Human Age' – began accelerating mainly from the 1950s, as economic growth expanded at the expense of social, environmental and economic impacts on the planet. For McNeill and Engelke, therefore, the Great Acceleration is defined as a period in which there was an exponential increase in energy use and population growth, resulting in technological and cultural impacts on both society and the planet. However, as it is highly dependent on natural resources, McNeill and Engelke suggest that this time interval cannot last very long. They recognise humankind as a geological agent interfering with the biosphere, yielding harmful effects – including emission of greenhouse gases, reduction of the ozone layer, acidification of oceans and loss of biodiversity. Perhaps not coincidentally, we have seen substantial growth in soybean cultivation since the 1950s to support increasingly global commerce systems in order to meet the sustenance needs of a growing world population.[4]

While growth in soybean production in the US and Brazil since the middle of the twentieth century is in part attributable to expansion in farmland devoted to it, increasing soy yields clarify that increased efficiency in production has also contributed to its expansion. Some of this change is traceable to scientific and technological advances. In the 1990s, advances in genetic engineering revolutionised soybean production, as Monsanto patented genetically modified (GM) soybeans and began providing them for widespread use.[5] Genetically Modified Organisms (GMOs), or transgenics, change DNA, causing characteristics to be modified by transferring genes from one organism to another.[6] What is sought with these GMOs is mass production through the developed resistance of GM seeds to herbicides. What is sought more broadly is food with more nutritional value than otherwise

3   P. Crutzen and E. Stoermer, 'The Anthropocene', *IGBP Global Change Newsletter* **41** (2000): 17–18.

4   C.M. da Silva and C. de Majo, 'Towards the Soyacene: Narratives for an environmental history of soy in Latin America's Southern Cone', *Historia Ambiental Latinoamericana Y Caribeña (HALAC) Revista De La Solcha* **11** (1) (2021): 329–56. https://doi.org/10.32991/2237-2717.2021v11i1. p329-356.

5   Antonio Regalado, 'As patents expire, farmers plant generic GMOs', *MIT Technology Review*, 30 July 2015: https://www.technologyreview.com/2015/07/30/166919/as-patents-expire-farmers-plant-generic-gmos/.

6   A.P. Gravioli and J. da S. Nunes, 'A soja transgênica no Brasil e suas influências à saúde e ao meio ambiente', *Revista Científica da Faculdade de Educação e Meio Ambiente* **6** (2) (2015): 1–16.

Larissa Trindade et al.

would have been possible, as GM processes are thought to produce more soybean plants of better quality and nutritional potential.[7]

The transgenic market in soybean agriculture has been expanding rapidly due to the influences mentioned above. For every 100 hectares planted with soybeans on the planet, eighty come from seeds with altered genes. In the past two decades, transgenic crops have risen 100-fold, from 1.7 million hectares to 175.2 million.[8] In the US, more than ninety per cent of planted acres of soybeans are of the GM variety;[9] in Brazil, almost 100 per cent of the soy used is genetically modified.[10]

While influences of the great acceleration can be seen in the evolution of agricultural practices generally and across the globe, this chapter focuses on soybean production and its water and environmental health impacts in Brazil and the United States (US). The text begins by outlining the research approaches we use to understand soybean cultivation and its water and environmental health impacts in Brazil and the US. It then summarises recent trends in soybean production in these two countries and offers examples of the impacts soy agriculture appears to be having on water and environmental health. The US and Brazil are the two largest producers of soybeans in the world, so the insights presented here should provide a valuable picture of the production of this crop and how its cultivation may generate environmental health and water impacts.[11] We hope that this picture will prove helpful to researchers, policymakers and practitioners with work and responsibilities relating to agriculture, health, and the environment.

Research Approach

To collect information for the chapter, we reviewed existing peer-reviewed and professional literature on soybean production and ways in which soy

---

7   Ibid.

8   Empresa Brasileira de Pesquisa Agropecuária (EMBRAPA), *Transgênico* [data from Embrapa's website]: https://www.embrapa.br/tema-transgenicos/perguntas-e-respostas.

9   United States Department of Agriculture (USDA), 'Recent Trends in GE Adoption', *Economic Research Service*: https://www.ers.usda.gov/data-products/adoption-of-genetically-engineered-crops-in-the-us/recent-trends-in-ge-adoption.aspx (Accessed 17 April 2020).

10   A. Dall'Agnol, *A Embrapa Soja no contexto do desenvolvimento da soja no Brasil: Histórico e contribuições*, (Brasilia: Embrapa Soja, 2016).

11   Empresa Brasileira de Pesquisa Agropecuária (EMBRAPA), *Soja em números* (2018/19) [data from Embrapa's website] 7 June 2019: https://www.embrapa.br/soja/cultivos/soja1/dados-economicos (Accessed 8 May 2020).

cultivation may lead to environmental health and water impacts in Brazil and the US. Our review included independent literature searches in databases available through university libraries in Brazil and the US, web-based searches and targeted reviews of government agency websites. Our initial baseline review efforts involved independent searches of databases of literature in Brazil and the US, using similar keywords. The keywords used included (but were not limited to) combinations of these terms: soybeans, water, water use, water supply, water quality regulation, financial incentives and public policy. These searches began in late 2019 and continued until Spring 2020 and focused on literature published between 2000 and 2019. In Brazil, we searched Capes Journals Reports and other available databases. In the US, we explored various databases, including PubMed, the Public Affairs Index, Biological and Agricultural Index, Agricola, Greenfile and Environment Index. In reviewing the articles uncovered through these searches, we focused on pieces that described or analysed the impacts of soybean cultivation or agriculture on water, the environment or health. We used the products of these searches to identify other helpful information, including information on soybean cultivation and environmental health and water impacts of soybean agriculture. To obtain this information, we conducted follow up web searches and consulted websites of government and non-profit organisations. While we found relevant material through these searches, we found no systematic analysis of soybean cultivation and environmental, health and water impacts in Brazil and the US.

Soybean Cultivation and Use in Brazil and the US

Soybean production has been increasing rapidly across the globe since the middle of the twentieth century.[12] In the decades since then, soy production has grown particularly fast in Brazil and to a significant degree in the US as well.[13] In the 1950s, the decade marked by some as the beginning of 'Great Acceleration', the US was the world's leading soybean exporter.[14] By 1967/68, global soy production totalled 36.454 million metric tons (MMT), and about

---

12   A. Barrett, *Long-Term World Soybean Outlook* (article from US Soy's website), 21 Nov. 2019: https://ussoy.org/long-term-world-soybean-outlook/.

13   Barrett, *Long-Term World Soybean Outlook*.

14   J. Seven, 'How a Chinese crop became an American winner': https://www.history.com/news/soybean-china-american-crop-tariffs

168

Larissa Trindade et al.

73 per cent of this production occurred in the US.[15] At that time, China was the second leading soybean producer, having cultivated 6.95 MMTs, or about 19 per cent of the global total.[16] Brazil produced just 0.654 MMT in that year, less than 2 per cent of global soybean production output. By 1985/86, global production had increased to 96.88 MMT, with the US (57.113 MMT, or 59 per cent), Brazil (14.1 MMT, or 15 per cent) and China (10.5 MMT, 11 per cent) producing the vast majority of the global total[17].

Over the decades since, global soybean production growth has continued at expeditious rates, with an increasing proportion of global production occurring in Brazil. Across the globe, soy production has increased from 104 MMT in 1990 to more than 300 MMT in recent years.[18] This increasing soybean production has roots in growing global demand for soy, and demand is traceable to increasing world population, increasing consumption of meat and accelerating demand for soy in Asia and China.[19] Brazil and the US have both increased soybean production to meet this growing global demand, although soy production growth rates in Brazil (7.9 per cent) have outstripped growth rates in the US (3.1 per cent) between 1990 and 2016.[20]

Since 1960, US farmers have more than tripled both the acreage devoted to soy and their harvests of soybean crops. In 1960, US farmers planted 24,440,000 acres of soy and harvested 23,655,000 of those acres; by 2017, they planted 90,162,000 acres and harvested 89,542,000 acres.[21] During this same period, yields on the soybean acres planted also increased. In 1960, US soybean acreage yielded 23.5 bushels per acre, producing a total of 555,085,000 bushels across the country. By 2017, these figures had increased to 49.3 bushels per acre and 4,411,633,000 bushels, respectively.

The growth in soy production in Brazil has been even more spectacular. From the 1960s onwards, soybean production expanded in Brazil, with

---

15   J. Schaub, W.C. McArthur, D. Hacklander, J. Glauber, M. Leath and H. Doty, 'The U.S. soybean industry', *Economic Research Service, U.S. Department of Agriculture, Agricultural Economic Report Number 588* (May 1988).

16   Ibid.

17   Ibid.

18   D. Widmar, 'Trends in global soybean production', *Agricultural Economic Insights*, 24 July 2017: https://aei.ag/2017/07/24/global-soybean-production/.

19   Barrett, *Long-Term World Soybean Outlook.*

20   Widmar, 'Trends in global soybean production'.

21   USDA, 'Table 2—Soybeans: Acreage planted, harvested, yield, production, value, and loan rate, U.S., 1960-2019, Economic Research Service': https://www.ers.usda.gov/webdocs/DataFiles/52218/Soy.xlsx?v=6235.8 (Accessed 25 July 2021).

growing demand in the oil industry and the international market.[22] According to Agrolink's 2017 data, soybean production proliferated from 1960 to 2017, globally and particularly in Brazil. Global soy production grew 7.98 times between 1960 and 2017 (44 Mt to 351 Mt).[23] In Brazil, production growth was even more substantial, increasing by a factor of 76 (1.5 Mt in 1970 to 114 Mt in 2017). And in 2020, Brazil 'overtook the United States as the leading soybean producing country with a production volume of some 126 million metric tons in 2020/21', in part due to increased conversion of Amazon rainforest lands to agricultural production of soybeans and other crops.[24]

Consistent with McNeill and Engelke's analysis, Brazil and the US export as well as consume substantial proportions of the soybeans they produce. In 2018, Brazil exported 83.6 MMT of grain (U$ 33.2 billion), 16.9 MMT of bran (U$ 6.7 billion) and 1.4 MMT of oil (U$ 1.0 billion) while consuming 44 MMT of grain internally.[25] By contrast, in the 2018/19 year, the USA exported about 1,748 million bushels of soybean (47.6 MMT), equal to roughly 40 per cent of soybeans it produced.[26] It used 2,092 million bushels of soybeans (56.9 MMT) to produce 'crush' for soybean oil, feedstock for animals and bio-diesel fuel, and 132 million bushels for seed (3.59 MMT), feed, and residual.[27] Overall, over 70 per cent of US soybean production is used for animal feed, 15 per cent is used for human consumption (mainly in cooking oils), 5 per cent for biodiesel fuel, and the remainder is used for a variety of purposes, including various industrial applications and organic foods.[28] Table 1 compares the values of the 2018/2019 soybean harvest in Brazil and the US.

22   Dall'Agnol, *A Embrapa Soja*.

23   Agrolink, 'A saga da soja no Brasil e no Mundo', 2017: https://www.agrolink.com.br/colunistas/coluna/a-saga-da-soja-no-brasil-e-no-mundo_400724.html. (Accessed 8 Sept. 2020).

24   M. Shahbandeh, 'Soybean production worldwide 2012/13-2020/21, by country', *Statista.com*: https://www.statista.com/statistics/263926/soybean-production-in-selected-countries-since-1980/#:~:text=From%202015%2F16%20to%202018,metric%20tons%20in%202020%2F21 (Acessed 21 Mar. 2021).

25   EMBRAPA, *Soja em números*.

26   USDA, 'Table 3—Soybeans: Supply, disappearance, and price, U.S., 1980/81–2020/21', Economic Research Service: https://www.ers.usda.gov/webdocs/DataFiles/52218/Soy.xlsx?v=6235.8 (Accessed 25 July 2021).

27   Ibid.

28   C. Ingraham, 'Soybeans, explained for the agriculturally impaired', *The Washington Post*, July 2018: https://www.washingtonpost.com/business/2018/07/26/soybeans-explained-agriculturally-impaired/.

170

Larissa Trindade et al.

Variable (MMT)	Brazil	USA
Production	117.0	120.5
Consumption	44.8	60.5
Export	74.6	47.6
Stocks	30.5	24.7

Table 1.

Soybean Harvest 2018/2019 Brazil and USA in MMT. Data source: Federação das Indústrias de São Paulo – (FIESP, 2020).[29]

In terms of geography, both Brazil and the US produce soybeans across large portions of their territories, with cultivation concentrated in specific states and regions. In Brazil, soy production occurs throughout the country. However, much production is concentrated in certain key Midwest and Southern Brazilian states: Mato Grasso (32,445 million tons), Rio Grande do Sul (19,187 million tons) and Paraná (16,253 million tons).[30] In recent years, soy production has expanded in Northern states proximate to the Amazon Forest, primarily due to production infrastructure investment (in seeds, pesticides, and ports) made by major multi-national corporations such as Archer Daniels Midland (ADM) and Bunge, Dreyfus and Cargill. Together, these corporations are reported to finance approximately sixty per cent of Brazilian soy production.[31] Figure 1 shows a map of the geographic distribution of soy producing establishments, by state, in Brazil, according to data from the last agricultural census.[32]

In the US, soybeans are produced in at least 31 states, with production concentrated across the central US, particularly in the states of Illinois (10,600 thousand acres), Iowa (10,000 thousand acres), Minnesota (8,150 thousand acres) and North Dakota (7,100).[33] However, there is also significant

29   Federação das Indústrias do Estado de São Paulo (FIESP), 'Informativo Março 2020: Safra Mundial de Soja 2019/2020', March 2020: https://www.fiesp.com.br/indices-pesquisas-e-publicacoes/safra-mundial-de-soja/.

30   Instituto Brasileiro de Geografia e Estatística (IBGE), 'Censo Agropecuário 2017'. Rio de Janeiro, 2017: https://censoagro2017.ibge.gov.br/templates/censo_agro/resultadosagro/index.html.

31   M.S. Domingues and C. Bermann, 'O arco de desflorestamento na Amazônia: Da pecuária à soja', *Ambiente & Sociedade* **15** (2) (2012): 1–22 doi: 10.1590/S1414-753X2012000200002.

32   IBGE, 'Censo Agropecuário 2017'.

33   United States Department of Agriculture (USDA), 'Crop Production 2019 Summary', Washington DC, 2020: https://www.nass.usda.gov/Publications/Todays_Reports/reports/cropan20.pdf.

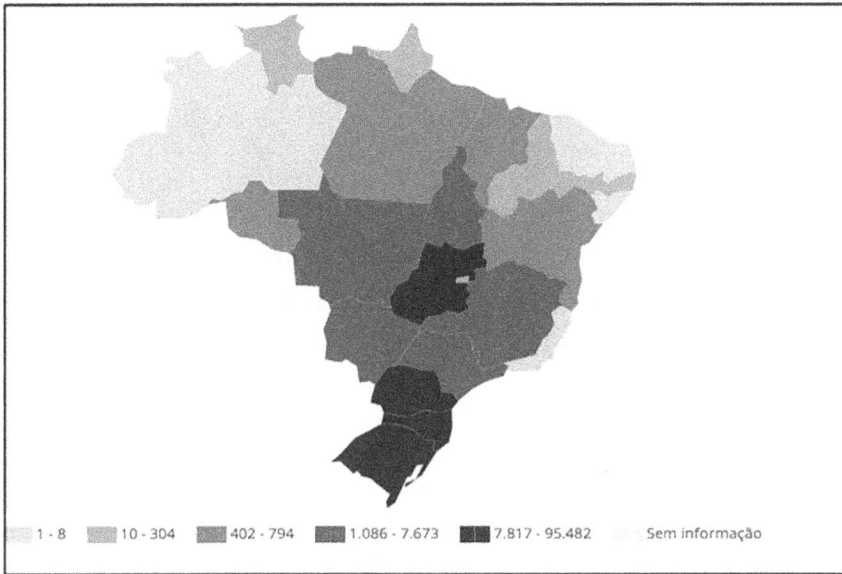

Figure 1.

Soy cartography in Brazil by number of establishments (2017). Source: Instituto Brasileiro de Geografia e Estatística – IBGE, 2017.

soybean production in the southern Mississippi River valley and states on the eastern coast of the US and in the Great Plains region generally. Based on data from the USDA, Figure 2 provides a map of US soybean production by acreage for 2017.[34]

Water and Environmental Health Impacts of Soybean Production in Brazil and the US

Soy cultivation requires water, and it also has potential environmental health and water quality impacts. This section discusses these impacts in Brazil and the US, based on our literature review and searches for information. More specifically, we discuss water use relating to agriculture and soybean production, the use and impacts of pesticides in agriculture and soybean production, and water quality impacts.

34   Ibid.

Larissa Trindade et al.

Area planted United States in 2017 in 1000 acres

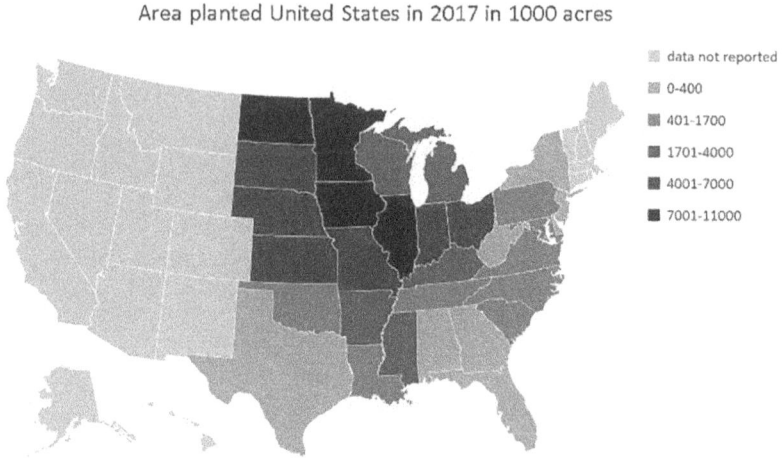

- data not reported
- 0-400
- 401-1700
- 1701-4000
- 4001-7000
- 7001-11000

Figure 2.

Soybean cultivation in the United States, by state and 1,000 acres. Data source: USDA, Crop Production Summary, January 2020.[35]

However, before we focus on these areas, it is worth recognising that an essential concern with soybean agriculture relates to climate change and the impacts of clearing land for soy production. Because natural ecosystems absorb carbon, they play a central role in combating anthropogenic releases of carbon that contribute to global climate change. In the US, the forests of the upper Midwest and Great Plains states were largely cleared decades ago and agricultural land for soybean cultivation is now plentiful in those states. In Brazil, however, efforts to expand soybean agriculture are pushing northward from the central and southern states that have long provided the foundation for Brazilian agriculture. Increasingly soybean agriculture is expanding to states and areas abutting the Amazon Forest in the north. And efforts to build ports and roads, aided by major multi-national corporations, such as ADM and Cargill, Dreyfus and Bunge, support the transportation needed to expand agricultural production in northern Brazil.[36]

The use of Amazon Forest areas for soybean agriculture met obstacles during the first decade and a half of the twenty-first century, as the Brazilian government limited conversion of forest lands to agriculture and sought to enforce those limits with some aggression. These policies resulted in decli-

35   Ibid.
36   Domingues and Bermann, 'O arco de desflorestamento na Amazônia'.

ning deforestation rates in the Amazon region during this period. However, it should be noted that despite government measures, forest deforestation has increased, albeit at a slower pace.[37] As a result, some of the most critical environmental impacts of the global expansion of soybean agriculture may now be occurring through deforestation in the Amazon region of Brazil. However, this chapter focuses on other potential environmental health and water impacts, so we return to these areas of focus in the following paragraphs.

Growing soy requires water, and both Brazil and the US have abundant water supplies compared to many other nations. Brazil accounts for twelve per cent of all freshwater available for consumption globally, and its uses are varied.[38] The US is also relatively water-rich, except for some of the arid regions in the western part of the country. In its latest report on the state of Brazilian water resources, the Brazilian National Water Agency documented significant sources of water consumption and quantified them. These estimates are shown in Table 2, along with water use estimates from the US Geological Survey (USGS) for the USA.

Water Consumption Type	Brazil Percentages*	US Percentages**
Irrigation	209.2 (68.4%)	118 (36.7%)
Mining sector	2.4 (0.8%)	4 (1.2%)
Urban supply/Public supply	26.3 (8.6%)	39 (12.1%)
Animal supply/Livestock	33 (10.8%)	2 (.6%)
Thermoelectric	0.6 (0.2%)	133 (41.4%)
Industry sector	26.9 (8.8%)	14.8 (4.6%)
Rural supply/Self supplied	7.34 (2.4%)	3.26 (1%)
Aquaculture	-	7.55 (2.3%)
TOTAL	**305.91 BGPD (%)**	**326.61 BGPD (%)**

Table 2.

Main sources of water consumption. Data sources: ANA, 2018 and USGS,[39] 2018 (2015 figures), in billions of gallons per day (BGPD).

37   D. Boucher, 'How Brazil has dramatically reduced tropical deforestation', *Solutions* 5 (2) (2014): 66–75.

38   Agência Nacional de Águas (ANA), 'Conjuntura dos Recursos Hídricos no Brasil 2018: Informe anual', Brasília 2018: http://arquivos.ana.gov.br/portal/publicacao/Conjuntura2018.pdf.

39   ANA, 'Conjuntura dos Recursos Hídricos no Brasil 2018: Informe anual'; United States Geological Survey (USGA), 'Summary of Estimated Water Use in the United States 2015' (United States Geological Survey, 2018): https://pubs.usgs.gov/fs/2018/3035/fs20183035.pdf.

Larissa Trindade et al.

A couple of observations on the data in this table are warranted. First, both Brazil and the US use large amounts of water for various purposes, as both countries use more than 300 billion gallons per day. And second, large proportions of water used in Brazil and the US are for irrigation of farmlands – more than two thirds of the water used in Brazil and more than a third of water used in the US are for this purpose. Because soybeans are among the most prevalent crops grown in both countries, significant amounts of water are used to grow soybeans.[40]

Pesticide use

Both Brazil and the US use pesticides in agriculture, including soybean agriculture. The literature reveals that pesticide use has increased significantly in Brazil during the twenty-first century, while reported trends in Brazil and the US suggest growing attention to the impacts of the increasing use of GM soybean crops on pesticide use.

In Brazil, Almeida et al. reported pesticide use based on analyses using data from the Brazilian Institute of Geography and Statistics (IBGE).[41] They suggest that pesticide use has increased significantly in Brazil since the legal introduction of GM crops in the early 2000s and report that the use of pesticides in Brazil increased 1.6 times between the years 2000 and 2012. During this same period, Almeida and colleagues also noted that pesticides to grow soybeans increased more than three times.[42] Pignati et al. reinforced this claim, adding that 63 per cent of pesticides used in Brazil in 2016 were to grow soybeans.[43] They also reported that some states make greater use of pesticides than others. For 2016, they reported that the State of Mato Grosso planted 13.9 million hectares and consumed 207 million litres of pesticides, followed by Paraná with 10.2 million hectares, consuming 135 million litres of pesticides and Rio Grande do Sul with 8.5 million hectares planted, using 134 million litres of pesticides. Almeida et al. (2017) also point

---

40   The soybean plant needs 450 to 800 mm of water throughout its cycle – an average of 620 mm. The volume of water required by the plant depends a lot on the cultivar chosen for planting and, consequently, on the length of its cycle.

41   V.E.S. de Almeida, K. Friedrich, A.F. Tygel, L. Melgarejo and F.F. Carneiro 'Uso de sementes geneticamente modificadas e agrotóxicos no Brasil: Cultivando perigos', *Ciência & Saúde Coletiva* **22** (10), (2017): 3333–39.

42   Almeida et al., 'Uso de sementes'.

43   W.A. Pignati, F.A.N. de S Lima, S.S. de Lara, M.L.M. Correa, J.R. Barbosa, L.H. da C. Leão and M.G. Pignati, 'Distribuição espacial do uso de agrotóxicos no Brasil: uma ferramenta para a Vigilância em Saúde', *Ciência & Saúde Coletiva* **22** (10) (2017): 3281–93.

out that, in 2014, the year in which pesticide sales were highest, the area of Brazil planted with GM crops reached 42.2 million hectares, an increase of 1306.67 per cent compared to the 3 million hectares registered in 2003. Almeida and colleagues (2017) also suggest that multiple studies report direct associations between the global consumption of pesticides and the use of genetically modified (GM) crops.[44] These findings of positive relationships between GM crop production and pesticide use run counter to early claims that GM crops – which are modified genetically to resist glyphosate pesticide impacts – would reduce the need for pesticide use.[45]

In the US, pesticide use has grown significantly between the mid-1960s and 2010,[46] although recent pesticide uses estimates have focused on differing aspects of broader trends. Drawing on USDA data, Osteen and Cornejo report that overall pesticide use in the US grew substantially between the 1960s and early 1980s and stabilised some through 2010. They also report that the trends revealed by their USDA data align well with USEPA estimates showing total pesticide use of 366 million pounds (of active ingredient) in 1964, growing to a high of 843 million pounds in 1979, and then levelling somewhat to 666 million pounds in 1987, 767 million pounds in 1997 and 684 million pounds in 2007.[47] They also present data showing that soybeans account for a significant portion of pesticide use in the US – just over 20 per cent (about 113 million pounds of the 543 million pounds they estimate were used in 2010 using USDA, not USEPA, data). They also report that more than 95 per cent of soybean acreage planted was treated with herbicides during the first half-decade of the twenty-first century.[48]

Benbrook (2012) also focuses on pesticide use in the US in his analysis of

44   See P.J. Landrigan and C. Benbrook, 'GMOs, herbicides, and public health', *New England Journal of Medicine* **373** (8) (2015): 693–95. doi: 10.1056/NEJMp1505660; R. Peshin and W. Zhang, 'Integrated pest management and pesticide use', in D. Pimentel and R. Peshin (eds), *Integrated Pest Management* (Dordrecht: Springer, 2014); C.M. Benbrook, 'Impacts of genetically engineered crops on pesticide use in the U.S.: The first sixteen years', *Environmental Sciences Europe* **24** (2012), doi: 10.1186/2190-4715-24-24; B.G. Young, 'Changes in herbicide use patterns and production practices resulting from glyphosate-resistant crops', *Weed Technology* **20** (2) (2006): 301–07 doi: 10.1614/WT-04-189.110; W.A. Pengue, 'Transgenic crops in Argentina: The ecological and social debt', *Bulletin of Science Technology & Society* **25** (4) (2005): 314–22.

45   Benbrook, 'Impacts of genetically engineered crops'.

46   C.D. Osteen and J. Fernandez-Cornejo, 'Economic and policy issues of U.S. agricultural pesticide use trends', *Pest Management Science* **69** (9) (2013): 1001–25. https://pubag.nal.usda.gov/download/57149/PDF

47   Ibid.

48   Ibid., 1006.

USDA data between 1996 and 2011, but his research focuses primarily on pesticides for GM crops. These crops are herbicide-resistant (HR) and – as noted above – were thought likely to reduce overall aggregate pesticide risks by replacing what has been considered to be a relatively benign herbicide, glyphosate, for a range of potentially more toxic herbicides that were used more frequently before the introduction of GM crops in the mid-1990s.[49] However, Benbrook suggests that the use of GM crops led to an increase of about 404 million pounds in pesticide used between 1996 and 2011, an increase of about seven per cent. He further suggests that this increase is primarily driven by the development of glyphosate-resistant weeds, which have led farmers to supplement their increasing use of glyphosate (from 2,500 to 30,000 tons per year between 1995 and 2002 alone) with the use of other herbicides to kill the weeds that have become resistant to glyphosate over time. In this regard, Benbrook's study echos findings offered by Altien (2012) for Brazil, which suggest that GM crops there have necessitated greater use of the pesticide Emamectin Benzoate to kill weeds that have become resistant to glyphosate.

Based on the analyses summarised above, it appears likely that the growing use of GM formulated soybeans is influencing pesticide application practices over time. At least some of these changes in practice seem to have implications for aggregate pesticide use in Brazil and the US. With continuing changes in GM seeding practices over time, likely future changes in pesticide use trends and patterns are not as clear as they might be.

Environmental health impacts of pesticides

While herbicide-resistant forms of soybeans and other crops raise questions about their effects on pesticide use, there are also questions and concerns about the impacts of pesticides on human health and the environment. This subsection addresses some of these questions, including those relating to pesticides on foods, potential carcinogenic effects and potential impacts on pregnant women and their children.

In Brazil, studies have documented pesticide residues on food, adverse health impacts from pesticide exposure and correlations between geographic locations of high agricultural use and negative effects of pesticide exposure. In 2015, the Brazilian Association of Public Health reported that about a third of all the foods consumed by Brazilians daily are contaminated in some fashion

49   Benbrook, 'Impacts of genetically engineered crops'.

by pesticides.[50] Several years later (2019), Canineu reported that a similar study by the Brazilian National Health Surveillance Agency (ANVISA) conducted between August 2017 and June 2018 found that nearly a quarter of 4,600 food samples in Brazilian grocery stores were contaminated with 'dangerous traces of pesticides'.[51] Moreover, in 2018, Gross reported that data from the Brazilian Notification of Injury Information System (*Sistema de Informação de Agravos de Notificação*, SINAN) documented more than 4,000 cases of pesticide poisoning in Brazil in 2016, 355 of which resulted in fatalities.[52]

Pignati et al. suggest that these adverse health impacts are associated with the widespread use of pesticides in agriculture, including soybean cultivation.[53] They present spatial analyses of areas planted with crops, pesticide consumption and related health problems. Soybean, corn and sugarcane crops, which accounted for 76 per cent of the planted area in Brazil in 2015, were predominant, and nearly 900 million litres of pesticides were reported to have been sprayed on these crops. The states of Mato Grosso, Paraná and Rio Grande Sul were reported as having used the most significant quantities of these pesticides. In Mato Grosso, the authors identified the concentration of pesticide consumption in municipalities (148 municipalities of Mato Grosso State) located mainly in the centre (3.3 to 14.6 million litres) and the south of Brazil (744 million to 3.3 million litres), where agricultural production is more extensive. Overall, results from Pignati et al. show a positive correlation between pesticide usage and acute poisoning and incidents of fetal malformation and mortality from childhood cancer.[54]

There has also been attention paid to pesticides used to grow soybeans and other crops in the US, including their fate and impacts. Indeed, recognised concerns about pesticides and their potentially toxic effects have contributed to a general receptiveness to the use of glyphosate, as it has been thought to have fewer negative health and environmental impacts than many other

50   F.F. Carneiro, L.G.o da Silva Augusto, R.M. Rigotto, K. Friedrich and A.C. Búrgio (eds), *Dossiê ABRASCO: Um alerta sobre os impactos dos agrotóxicos na saúde* (Rio de Janeiro: Expressão Popular, 2015): https://abrasco.org.br/dossieagrotoxicos/.

51   M.L. Canineu, *Brazil Needs More Pesticide Regulation, Not Less*, Human Rights Watch website, 23 Dec. 2019. https://www.hrw.org/news/2019/12/23/brazil-needs-more-pesticide-regula-tion-not-less.

52   A.M. Gross, 'Brazil's pesticide problem poses global dilemma, critics say', Mongabay, 27 Aug. 2018: https://news.mongabay.com/2018/08/brazils-pesticide-poisoning-problem-poses-glob-al-dilemma-say-critics/ (Accessed 1 May 2020).

53   Pignati et al., 'Distribuição espacial'.

54   Ibid.

pesticides in the US market. However, more recent studies and public debates about pesticides and their impacts give rise to further questions and concerns, particularly as they relate to cancer and pregnancy-related health impacts.

While Hongbing points out that multiple studies have linked pesticides to cancer and diabetes in farmworkers, he offers data and analyses suggesting a connection between pesticides in the natural environment and elevated colorectal cancer rates in counties in the Mississippi River Basin floodplain.[55] He points out that eighty per cent of US agricultural production comes from the Mississippi River basin in the US and roughly two-thirds of pesticide application occurs in that region as well, resulting in potentially significant human exposures to pesticides in the floodplain areas of that basin. His analysis suggests that the incidence of colorectal cancer in 86 counties in the Mississippi River embayment is 29 per cent higher than that of counties in the 48 contiguous states of the US and that colorectal fatalities were also higher in 63 of those counties.

Recent studies also raise questions about the widely used pesticide glyphosate and its carcinogenicity. Based on an assessment from the World Health Organization's (WHO) International Agency for Research on Cancer (IARC), the State of California determined that glyphosate is a probable human carcinogen and – in 2017 – required labelling of glyphosate-based pesticides to include information to this effect.[56] At least one subsequent study has corroborated the WHO-IARC finding.[57] Even so, the USEPA has expressed disagreement with the WHO-IARC finding and issued a decision in 2019 prohibiting those registered to sell glyphosate-based pesticides from labelling their products as instructed by the State of California. While federal preemption under US laws means the USEPA decision takes precedence, tort liability decisions made by juries may ultimately decide the fate of these pesticides in the US.[58] Moreover, in June 2020, Bayer Corporation, which recently bought Monsanto, the manufacturer of glyphosate, agreed to pay out more than $10 billion in payments to settle tens of thousands of claims

55   S. Hongbing, 'Pesticide in the Mississippi River floodplain and its possible linkage to colon cancer risk in the US', *Toxicological & Environmental Chemistry* **100** (8–10) (2018): 794–814.

56   E. Schkloven, *EPA, State of California Debate Labeling Glyphosate as Carcinogen*, Medtruth, 5 Sept. 2019. https://medtruth.com/articles/news/epa-california-glyphosate-carcinogenic-prop-65/

57   L. Zhang, I. Rana, R.M. Shaffer, E. Taioli and L. Sheppard, 'Exposure to glyphosate-based herbicides and risk for non-Hodgkin lymphoma: A meta-analysis and supporting evidence', *Mutation Research/Reviews in Mutation Research* 781 (2018): 186–206.

58   C.H. Brown, 'EPA moves to block California's Roundup cancer warning', The Counter, 12 Aug. 2019: https://thecounter.org/epa-california-cancer-glyphosate-monsanto-prop-65/

while selling the pesticide without safety warnings.[59] While tort settlements are being made, scientific uncertainties and associated debates regarding the carcinogenicity of glyphosate-based pesticides remain.[60]

There are also questions about the impacts of glyphosate-based pesticides on pregnant women and the children to whom they give birth.[61] Parvz et al. analysed urine samples and drinking water for 71 pregnant women in central Indiana to assess the potential impact of glyphosate exposures on pregnant women and their offspring. They found evidence of glyphosate exposure in the urine of more than ninety per cent of the women, but no evidence of exposure through drinking water. Overall, while Parvez and colleagues found no evidence of glyphosate impact on infant growth variables (Low Birthweight rates, for example), they did find evidence of shortened gestational periods.[62] While this study should probably be viewed as preliminary and in need of follow up research, it nevertheless adds to questions about the impacts of exposure to glyphosate-based pesticides.

Soybean agriculture and water quality

Soybean agriculture can also affect water quality in rivers, streams, lakes and other water bodies, as rainwater runoff and evaporation may disseminate pollutants through water. Our research revealed some concerns about pesticide contamination and excess nutrient enrichment in Brazilian and US waters.

Oliveira et al. draw attention to Brazilian water quality control system discrepancies. Unlike other countries, including the USA, Brazil does not have a public programme to report on the control and quality of Brazilian waters.[63] Despite stipulations in legislation, such information is not yet made available by the water agency. What exists are only isolated and occasional investigations. This fact represents one of the significant weaknesses of the Brazilian water management system.

59   P. Cohen, 'Roundup maker to pay $10 billion to settle cancer suits', *New York Times*, 24 June 2020: https://www.nytimes.com/2020/06/24/business/roundup-settlement-lawsuits.html, (Accessed 20 Mar. 2021).

60   Benbrook, 'Impacts of genetically engineered crops'.

61   S. Parvez, R.R. Gerona, C. Proctor et al., 'Glyphosate exposure in pregnancy and shortened gestational length: a prospective Indiana birth cohort study', *Environmental Health* **17** (23) (2018). https://doi.org/10.1186/s12940-018-0367-0.

62   Ibid.

63   J.R.A. de Oliveira, L. Vilela and M.A. Ayarza, 'Adsorção de nitrato em solos de cerrado do Distrito Federal', *Pesquisa Agropecuária Brasileira* **35** (6) (2000): 1199–205. https://doi.org/10.1590/S0100-204X2000000600017.

Larissa Trindade et al.

However, studies have documented pesticides in surface water, rainwater and artesian wells serving urbanised areas in Brazil. Moreira et al. analysed water quality in rivers and streams in a city in the Brazilian Midwest and found pesticide residues from Endosulfan, Futriafol and Metolachlor.[64] More than one of these pesticides were found in combination in several samples. They also found pesticide residues of Endosulfan alpha and beta, flutriafol, and metolachlor in 83 per cent of the sampled artesian wells serving urbanised areas of the city and pesticide residues in more than half of their rainwater samples. In another study, Belo et al. provided further corroboration of the potential for pesticides in rainwater and the possibility of exposure through air deposition.[65] They conducted a preliminary descriptive exploratory study in 2008 and 2009 in Mato Grosso, one of Brazil's most intensive soy-producing regions. Based on triangulated methods using analyses of an agricultural database, biological indicators of exposure to pesticides and analyses of rainwaters, they found evidence of pesticide accumulation in rainwater and exposure to pesticides among workers and residents in proximity to planting areas where pesticides were used.

Excessive nutrification is also a concern with agricultural and soybean production. Human activities, including large-scale agriculture, affect the cycling of nutrients in the Earth System, as highlighted by Gaffney and Steffen.[66] There are estimates that 72 per cent of the nitrogen that reaches bodies of water comes from agricultural activity, including synthetic fertilisers.[67] The impacts of excess nutrient loadings (involving phosphorus and/ or Nnitrogen) include eutrophication, pH changes and dissolved oxygen and toxicity, producing imbalances in ecosystems and potentially making productive and recreational uses of affected water bodies more difficult.

However, while crops' agricultural production – including that of soybeans – holds the potential to contaminate surrounding water bodies when nitrogen

---

64   J.C. Moreira, F. Peres, A.C. Simões, W.A. Pignati, E. de C. Dores, S.N. Vieira, C. Strusssmann and T. Mott, 'Contaminação de águas superficiais e de chuva por agrotóxicos em uma região do estado do Mato Grosso', *Ciência & Saúde Coletiva* 17 (6) (2012): 1557–68. https://www.scielo. br/scielo.php?script=sci_arttext&pid=S1413-81232012000600019&lng=pt&nrm=iso&tlng=pt. doi: 10.1590/S1413-81232012000600019

65   M.S.P. Belo, W.A. Pignati, E.F.G. De C. Dores, J.C. Moreira and F. Peres, 'Uso de agrotóxicos na produção de soja do Estado do Mato Grosso: Um estudo preliminar de riscos ocupacionais e ambientais', *Revista Brasileira de Saúde Ocupacional* **37** (125) (2012): 78–88.

66   O. Gaffney and W. Steffen, 'The Anthropocene equation', *The Anthropocene Review* 4 (1) (2017): 53–61.

67   Ibid.

8. Soybean Cultivation and its Water-related Impacts

and/or phosphorus from fertilisers run off croplands to nearby surface waters during rain events, we identified only one study documenting these kinds of effects in Brazil.[68] Bertol et al. studied a no-tillage agricultural operation in Parana State in Brazil and found that liquid swine manure used for fertiliser increased total phosphorus concentrations, particulate phosphorus and reactive dissolved phosphorus by substantial amounts in average rains. They also reported increases in phosphorus loads that grew with the intensity of the rainfall.[69] These findings suggest some concern regarding excess nutrification associated with Brazilian agriculture, including soybean agriculture.

Several reports and studies have documented the potential impact of agriculture on surface water and rainwater quality in the US, including impacts from nutrients through both over-land flows and air deposition. The USEPA, for example, reports that agriculture is the number 1 'known' source of impairment in rivers and streams in the US ('unknown' ranks slightly above agriculture in the figures reported), based on information submitted by states in their biennial section 303/305 combined reports on the quality of water within their borders.[70] Based on these reports, the USEPA indicates that nutrients are the number 3 cause of water quality impairment in the US behind pathogens and sediment, as state 303/305 combined reports document more than 118,000 miles of rivers and streams that are threatened or impaired in the US due to excess nutrients. In some cases, these nutrients can lead to harmful algal blooms and 'dead zones' in receiving water bodies, as have occurred in Lake Erie, the Long Island Sound and Tampa Bay areas of the US.[71] While nutrients come from various sources, including point source wastewater discharges, agricultural operations are reported to be a common culprit. And, given that soy is a commonly grown crop in the US, there seems little doubt that fertilisers and runoff used in soy cultivation contribute

---

68   J. Mateo-Sagasta, S.M. Zadeh, H. Turral and J. Burke, *Water Pollution from Agriculture: A Global Review* (Rome: Food and Agriculture Organization of the United Nations, 2017) http://www.fao.org/3/a-i7754e.pdf

69   O.J. Bertol, N.E. Rizzi, N. Favaretto and M. do C. Lana. 'Phosphorus loss by surface runoff in no-till system under mineral and organic fertilization', *Scientia Agricola* **67** (1) (2010): 71–77.

70   United States Environmental Protection Agency –USEPA, 'National Summary Causes of Impairment in Assessed Rivers and Streams Office of Water', Washington DC, 2020: https://ofmpub.epa.gov/waters10/attains_nation_cy.control#total_assessed_waters.

71   J. Hoornbeek, F. Joshua and Y. Soumya, 'Watershed based policy tools for reducing nutrient flows to surface waters: Addressing nutrient enrichment and harmful algal blooms in the United States', invited article for Symposium Issue on America's Water Crisis: An Issue of Environmental Justice', *Fordham University Environmental Law Review* XXIX (1) (2017).

to at least some of the reported nutrient-based water quality impairments.

Our literature review also identified other evidence of the impacts of agriculture and/or soybean cultivation on water quality in the USA. In 2014, Stone and colleagues released findings from water sampling and pesticide analyses done at more than 100 river and stream sites across the US by the USGS between 1992 and 2011.[72] They excluded glyphosate from their sampling and analyses but found pesticide concentrations exceeding aquatic life standards for more than half of the sites in both agricultural and urban areas and almost half the sites in mixed (urban-rural) areas. They also found that human health criteria for pesticides analysed in their study were rarely exceeded.

In 2017, Nowell et al. published results of their more comprehensive study of more than 100 midwestern streams in May–August 2013 and found large numbers of pesticides interacting to produce potentially harmful conditions in the streams sampled.[73] More specifically, Nowell et al. found a median of 25 different pesticides per sample and 54 pesticides per site, and they predicted – based on their results – that the combinations of pesticides encountered were likely toxic to invertebrate populations in the environments in which they were found.[74]

In addition, in 2000, Majewski et al. documented the accumulation of pesticides (other than glyphosate) in rainwater samples in three states in the highly agricultural Mississippi River Basin (Minnesota, Iowa, and Mississippi) and one background site (near Lake Superior in Michigan).[75] Their study demonstrated the tendency of pesticides to accumulate in rainwater and deposit on land and water where they can lead to human health and ecological exposure.

These studies have since been supplemented by another study conducted using USGS data which documented glyphosate accumulation in surface waters in the US. In early 2020, the journal *Science of the Total Environment* published a study by Medalie et al. which focused on the presence of glypho-

72   W.W. Stone, R.J. Gilliom and K.R. Ryberg, 'Pesticides in U.S. streams and rivers: Occurrence and trends during 1992–2011', *Environmental Science and Technology* **48** (19) (2014): 11025–30.

73   L.H. Nowell, P.W. Moran, T.S. Schmidt, J.E. Norman, N. Nakagaki, M.E. Shoda and M.L. Hladik, 'Complex mixtures of dissolved pesticides show potential aquatic toxicity in a synoptic study of Midwestern U.S. streams', *Science of the Total Environment* **613** (614) (2017): 1469–88.

74   Ibid.

75   M.S. Majewski, W.T. Foreman, D.A. Goolsby, 'Pesticides in the atmosphere of the Mississippi River Valley, Part 1 – rain', *Science of the Total Environment* **248** (2–3) (2000): 201–12.

sate and its degradate, Aminomethylphosphonic acid (AMPA), in seventy streams across the US.[76] The study suggested that glyphosate and AMPA were present in most streams in their sample. However, their reported concentrations were 'far below' human health and ecological benchmarks. They also found that proximate land uses were a better predictor of glyphosate and AMPA concentrations than larger basin characteristics and that glyphosate was more often present in smaller streams and rivers with AMPA relatively more prevalent in larger water basins. They concluded that their study results provide a foundation for further research to illuminate our understanding of glyphosate and AMPA transport and ecological fate.

Conclusions

This section presents some final thoughts on soybean cultivation in Brazil and the US and its potential environmental health and water impacts. First, soybean cultivation is essential for the economies of both countries. Brazil and the US are the world's two largest producers of soy, and they also both export and consume soybeans in multiple forms. Second, while existing studies do not yield conclusive results on the overall impacts of soybean cultivation on water and pesticide use or its effects on human health, the environment and water quality, they do raise questions and suggest reasons for concern.

While we found relatively few recent studies on soybean production and its impacts on health and the environment, we did find evidence suggesting soybean cultivation impacts to water and environmental health in both Brazil and the US:

i) Both countries display intensive water use for irrigation in agriculture, and – because soybean cultivation accounts for substantial portions of agricultural land use in Brazil and the US – growing soybeans probably accounts for a notable amount of this water use.

ii) Pesticide use is growing in Brazil and remains substantial in the US. While GM crops in general – and GM soybeans in particular – have been thought to hold the potential to diminish pesticide use overall, there is growing concern that the development of resistance to glyphosate in surrounding crops may be leading to the use of other pesticides in greater quantity.

iii) There is growing – although still not complete – evidence that soybean

---

76   L. Medalie, N.T. Baker, M.E. Shoda, W.W. Stone, M.T. Meyer, E.G. Stests and M. Wilson, 'Influence of land use and region on glyposate and aminomethylphosponic acid in streams in the USA', *Science of the Total Environment* **707** (2020): 1360008.

cultivation negatively impacts human and environmental health in Brazil and the US. In regions of Brazil where there is intensive agriculture, especially soybean, corn and wheat – as in the State of Mato Grosso – there is assertion and evidence that pesticide use has negative human health impacts on rural and urban workers and potentially others as well. There is concern about the release of pesticides tino US surface waters and the potential effects of pesticide use on human health in the US, as evidenced by the 2020 litigation settlement by Bayer Corporation (which owns Monsanto), with thousands claiming that glyphosate has impacted their health. There is also evidence of concern in the US regarding the impacts of excess nutrient flows from generally agricultural operations, including soybean cultivation.

Glyphosate and its impacts also appear to warrant attention, as it is widely used in Brazil and the US. As noted above, the WHO's Cancer Research Agency declared in 2015 that glyphosate is a probable cause of human cancer. This finding has led several European countries to ban its use. However, Brazil and the US have not recognised this conclusion as valid and continue to enable widespread use.

More broadly, it seems likely that soy agriculture has become a mechanism through which the 'Great Acceleration' of environmental stresses proliferating in the Anthropocene era is influencing human health, the environment and long-term sustainability – particularly in Brazil and the US, where its production has been substantial and accelerating. In these countries, as in others, there is a need to better understand the use and impacts of large-scale soy production, so soy cultivation and its potential side-effects can be managed in ways that protect human beings and the environment. In the meantime, as national and subnational efforts to address soybean cultivation and its impacts are undertaken, prudent steps to guide soy cultivation and minimise its potentially harmful environmental health and water-related effects would seem to be appropriate.

CHAPTER 9.

FIVE DECADES OF SOYBEAN AGRICULTURE: SOIL NITROGEN EXPORTS AND SOCIAL COSTS IN THE ARGENTINE PAMPAS, 1970–2021

Enrique Antonio Mejia

Despite unprecedented yields, and food output that has lifted some of the most vulnerable communities of the world out of food poverty and malnourishment, contemporary industrial agriculture has also contributed to a wide variety of cross-scale social-ecological imbalances; evidently playing a major role in the transgression of several planetary boundaries.[1] The alteration of the global nitrogen cycle is one such transgression that comes as a result of massive anthropogenic nitrogen loading from the acceleration in agricultural output.[2] This is due to the fundamental role of nitrogen in the synthesis of proteins vital for biological life.[3] Anthropogenic nitrogen loading from agricultural activities takes place via the fixation of nitrogen from the atmosphere during the production of synthetic fertilisers, cultivation of legumes and fossil fuel combustion as well as the mobilisation of nitrogen from long-term ecological storage pools through biomass burning, clear cutting and deforestation and non-regenerative agriculture.[4]

However, any analysis of the nitrogen cascade – the process by which

---

1   W. Steffen et al. 'Planetary boundaries: Guiding human development on a changing planet', *Science* **347** (6223) (2015).

2   Both the acceleration in agricultural output and the massive levels of anthropogenic nitrogen loading are fixtures of a broad range of rapidly transforming social-ecological indicators referred to as The Great Acceleration which is argued to be both historical and idiosyncratic: J.R. McNeill and P. Engelke, *The Great Acceleration* (Cambridge: Harvard University Press, 2016).

3   Biochemically, not all nitrogen is the same. Inert atmospheric nitrogen is inaccessible to most life on Earth, although it makes up 78% of our atmosphere's composition; while reactive nitrogen is the most biologically useful. Reactive nitrogen will be simply referred to as nitrogen in this chapter, as inert nitrogen is not central to the analysis.

4   P.M. Vitousek et al. 'Technical report: Human alteration of the global nitrogen cycle: sources and consequences', *Ecological Applications* **7** (3) (1997): 737–50, at 738–39.

THE AGE OF THE SOYBEAN: 185–203 **doi:** 10.3197/63800040695086.ch09

nitrogen atoms flow and transform through social-ecological systems – at the 'planetary' level masks local and historical distributions of causes and consequences, power relations and place-specific contexts.[5] For example, it is now empirically evident that the structure of global agrofood trade is rooted in a political economy of asymmetric biophysical resource flows characterised by ecologically unequal exchanges, not least concerning embodied nitrogen.[6] This is because the pecuniary valuation of a commodity does not, and cannot, accurately reflect the unaccounted-for costs borne by local actors, communities and ecosystems within places of production. Furthermore, these unaccounted-for costs are articulated in different ways depending on institutional and ecological contexts.

In order to demonstrate these critical aspects, I focus this chapter on the case of export-oriented soybean monoculture in the Argentine Pampas where the mobilisation of nitrogen has degraded soil quality and bears significant unaccounted-for costs for local actors, communities and ecosystems. First, I situate the Pampean case within the greater historical political economy of asymmetric embodied nitrogen flows and use a simple mass balance of nitrogen as well as annual production data between the years 1970–2021 to assess to what degree soil nitrogen is extracted and exported as a result of local soybean monoculture. Second, influenced by K. William's Kapp's theory of social costs, I use an institutional economic lens to differentiate two orders of unaccounted-for costs of soil nitrogen depletion in the Argentine Pampas: those directly resulting from soil nitrogen depletion and those resulting from so-far-unsuccessful attempts at overcoming the ecological limitations posed by diminished soil nitrogen.[7] Therefore, this chapter represents an original attempt at merging nitrogen related research with social science by linking exogenous pressures to the place-based social-ecological disruptions occurring in the Pampas.

5   J.N. Galloway et al. 'The nitrogen cascade', *BioScience* **53** (4) (2003): 341–56; F. Biermann and R.E. Kim. 'The boundaries of the planetary boundary framework: A critical appraisal of approaches to define a "safe operating space" for humanity', *Annual Review of Environment and Resources* **45** (2020): 497–521, at 502.

6   See D.A. Díaz de Astarloa and W.A. Pengue, 'Nutrients metabolism of agricultural production in Argentina: NPK input and output flows from 1961 to 2015', *Ecological Economics* **147** (2018): 74–83; A. Oita et al. 'Substantial nitrogen pollution embedded in international trade', *Nature Geoscience* **9** (2) (2016): 111–15; L. Lassaletta et al. 'Nitrogen embedded in global food trade', in P. Ferranti, E.M. Berry and J.R. Anderson (eds), *Encyclopedia of Food Security and Sustainability* (Oxford: Elsevier, 2019), pp. 105–09.

7   K.W. Kapp, *The Social Costs of Business Enterprise* (Nottingham: Russell Press, 1978).

Soil Nitrogen Depletion in the Argentine Pampas

Over the last century, soybeans have become a significant input within the global agrofood system, showing marked growth in production volumes and geographical expansion. While the United States has held the position as a global producer for the greater part of this history, South American countries have become some of the world's largest producers and traders of soybeans since the 1970s.[8] For example, in 2019, the export baskets of South American countries demonstrated stark specialisation in the export of whole soybean grain, soybean meal and soybean oil as well as all processed derivatives – i.e. the soybean complex – compared to those of the United States.[9] This is despite the fact that the crop had little significance or presence on the continent before the middle of the twentieth century.

While originally driven by a global demand for oil in the early and mid-twentieth century, soybean cultivation is now being driven by an increasing demand for animal-based protein which has markedly accelerated since the 1960s. This is because soybean meal is prized on the global market for its high protein concentration – forty per cent of the calories in soybeans are derived from protein, as opposed to 25 per cent for most other feed crops – and has therefore become intricately tied to livestock production chains; mainly poultry and pork.[10] Though European countries were the leading net-importers of soybeans throughout the twentieth century, the bulk of global soybean exports today land in China and India, thus fuelling a rapid transformation in national diets.[11] Additionally, use in biofuel production has begun to rapidly increase and may become a contributing driver for soybean expansion in the near future.[12]

Both production weight and area harvested of soybeans have drastically grown in South America, rising from just over 300,000 metric tons harve-

8   FAO, FAOSTAT Online Database (Rome: Food and Agriculture Organization of the United Nations, 2021): http://www.fao.org/faostat/en/#data (Accessed 8 Aug. 2021).

9   In 2019, the soybean complex accounted for 25% of Argentina's export basket, 15% of Brazil's, 33% of Paraguay's and 8% of Uruguay's versus 2% of the United States': OEC, The Observatory for Economic Complexity (2021): https://oec.world/en (Accessed 20 Jul. 2021).

10   M. Turzi, *The Political Economy of Agricultural Booms: Managing Soybean Production in Argentina, Brazil and Paraguay* (Cham: Springer, 2016), p. 3; M. Baraibar Norberg, *The Political Economy of Agrarian Change in Latin America* (Cham: Springer, 2020), pp. 117–63.

11   M.A. Sutton et al. *Our Nutrient World. The Challenge to Produce More Food and Energy with Less Pollution* (Edinburgh: Centre for Ecology and Hydrology, 2013), p. 12.

12   W.A. Pengue, 'Agrofuels and agrifoods: Counting the externalities at the major crossroads of the 21st century', *Bulletin of Science, Technology & Society* **29** (3) (2009): 167–79.

sted over roughly 280,000 hectares in 1961 to 184 million metric tons over about 60 million hectares in 2019.[13] Argentina is the third largest soybean producer in the world, second to Brazil in South America. However, the soybean complex accounted for 25 per cent of Argentina's total export basket in 2019, as opposed to roughly fifteen per cent of Brazil's.[14] Argentina has also maintained a categorically large physical trade balance deficit since the early twentieth century, accelerating in the postwar period.[15] Further, the country has become a net-exporter of protein, indicative of South American countries specialising in the production of soybeans.[16]

The Argentine Pampas play a significant role in the country's soybean narrative. The area under soybean monoculture has increased roughly 1,500 times over between 1970 and 2021. Soybean cultivation has become a common feature of the region as not only has a process of expansion occurred, but the yield of soybeans per hectare has seen significant intensification, evidenced by the drastic growth in yields between 1970 and 2021 – at 1.2 metric tons per hectare and 2.5 metric tons per hectare respectively.[17] As result of this expansion and intensification, the Pampas have experienced a process of agrarian conversion in which other historical forms of land-use and native ecosystems have been subsumed under the treadmill logic of *sojización*. This agrarian conversion is facilitated by the advent of the soybean technological package, a high-input, capital-intensive form of production centred on Monsanto's[18] patented genetically modified Roundup Ready soybean that has become hegemonic in Argentina: since its introduction, there has been practically a 100 per cent adoption rate.[19] The technological

---

13   FAOSTAT Online Database 2021.

14   OEC 2021.

15   See Díaz de Astarloa and Pengue, 'Nutrients metabolism of agricultural production'; P.L.P. Manrique et al. 'The biophysical performance of Argentina (1970–2009)', *Journal of Industrial Ecology* **17** (4) (2013): 590–604; J. Infante-Amate et al. 'Las venas abiertas de América Latina en la era del Antropoceno: Un estudio biofísico del comercio exterior (1900–2016) [The open veins of Latin American in the era of the Anthropocene: A biophysical study of foreign trade]', *Diálogos Revista Electrónica de Historia* **21** (2) (2020): 177–214.

16   Lassaletta et al., 'Nitrogen embedded in global food trade', 105.

17   MINAGRI, Estimaciones agrícolas [Agricultural statistics]. (Ministerio Agricultura, Ganadería y Pesca Argentina, 2021): https://www.magyp.gob.ar/datosagroindustriales (Accessed 21 Oct. 2021).

18   Bayer, a German multinational pharmaceutical and life sciences company, purchased Monsanto and the rights to their products in 2018.

19   ISAAA, 'Global status of commercialized biotech/GM crops in 2018' (Ithaca, NY: International Service for the Acquisition of Agri-biotech Applications, 2018) Brief No. 54, p. 18.

package is designed to manage weed populations through the use of glyphosate (Roundup) and no-tillage farming – also argued to improve soil health and increase yields as it does not overturn the topsoil.[20]

While the specialisation in soybean production and export has been considered a success story in facilitating economic growth in some South American countries, the accelerated growth in the Argentine Pampas has not come without severe social-ecological consequences. Degradation of renewable and non-renewable resources, cultural and biodiversity loss, disruption to livelihoods and diminishing development alternatives are but some of the significant costs unaccounted for in the market price reflected in international trade and in some cases have no pecuniary value. Because these costs go unaccounted for, structural patterns emerge in the globalised trade systems that tend to cumulatively benefit one actor over another. These unequal exchanges between actors result in a variety of costs that tend to change in circular and cumulative ways, making them difficult to grasp. Furthermore, some costs may not manifest immediately, instead rearing their heads in the future, becoming more consequential.

The idea that unequal exchanges are inherent within globalised trade is not new and these inequalities are typically argued to be represented by the asymmetric trade of embodied labour, land or money; i.e. Polanyian fictitious commodities.[21] The underlying assumptions of these arguments is that the pecuniary valuation of a commodity does not, and cannot, accurately reflect the costs resulting from productive processes and that the asymmetric trade of either of these fictitious commodities benefits more economically and politically powerful actors within the production chain. In this way, value as expressed in monetary terms is not only subjected to the preferences, valuations and speculations of powerful vested interests but also results from the axiological fallacy that the value of anything can be converted into an exchangeable value towards everything. Therefore, money 'conditions us to abstraction, interchangeability and disembeddedness, which tends to alienate us not only from fellow humans but also from our natural environment'.[22]

Cross-national trade comparisons have demonstrated that, in the wake of the Great Acceleration, South American countries have become net-exporters

---

20   Bayer, 'Sustainability report' (2020): https://www.bayer.com/en/media/sustainability-reports (Accessed 31 Oct. 2021), 25–35.

21   K. Polanyi, *The Great Transformation* (New York: Farrar and Rinehart, 1944).

22   A. Hornborg, *Nature, Society, and Justice in the Anthropocene: Unraveling the Money-Energy-Technology Complex* (Cambridge: Cambridge University Press, 2019), p. 6.

Enrique Antonio Mejia

of biophysical resources.[23] Additionally, studies have found that South American countries are net-exporters of nitrogen embodied in the protein content of food and feed.[24] In fact, eight of the top ten countries exporting protein are also among the top ten producing soybean, half of which are in South America.[25] This is because export-oriented soybean monoculture is one of the principle pathways for nitrogen to enter agricultural production chains on the continent.[26] This is evidenced by soybean trade accounting for 56 per cent of the continent's total nitrogen exports in 2016.[27] However, while the coupled effects of synthetic fertiliser use and biological fixation from soybean production account for the majority of nitrogen entering agricultural systems for the continent as whole, the different institutional and ecological context of the Argentine Pampas has led to a situation where soil nitrogen is being depleted due to historically-low fertiliser use and reduced biological nitrogen fixation.

The nitrogen needed to create proteins is made available to soybean agriculture from a limited number of sources. It is either already present in the soil through the mineralisation of organic matter, deposited by weather events or animal droppings or fixed from the atmosphere biologically via a symbiotic relationship with rhizobium bacteria or through the addition of synthetic fertiliser – fixed industrially through various energy and material-intensive processes. It is typically believed that soybeans biologically fix their required nitrogen supply, and this has been demonstrated in mixed-crop rotations; however, as a monoculture with the explicit intent of increasing yields with as little input as possible, soybeans fail to meet their nitrogen requirements through biological fixation.[28] For example, in Argentine soybean agriculture, nitrogen uptake from biological fixation can range from 12–90

---

23   See Infante-Amate et al., 'Las venas abiertas'; L. Rivera-Basques et al., 'Unequal ecological exchange in the era of global value chains: The case of Latin America', *Ecological Economics* **180** (106881) (2021).

24   Lassaletta et al., 'Nitrogen embedded in global food trade'.

25   FAOSTAT Online Database 2021.

26   A. T. Austin et al. 'More is less: Agricultural impacts on the N cycle in Argentina', *Biogeochemistry* **79** (1) (2006): 45–60, at 48.

27   R. Guareschi et al. 'Balanço de nitrogênio, fósforo e potássio na agricultura da América Latina e o Caribe [Nitrogen, phosphorus and potassium balance in Latin American and Caribbean agriculture]' *Revista Terra Latinoamericana* **37** (2) (2019): 105–19, at 111.

28   For mixed-crop rotation data, see F.J. Bergersen et al. 'Natural abundance of 15N in an irrigated soybean crop and its use for the calculation of nitrogen fixation', *Australian Journal of Agricultural Research* **36**(3) (1985): 411–23.

per cent, averaging 60 per cent, while in Brazil the average is 81 per cent between a range of 69–94 per cent.[29]

In the Argentine Pampas, this figure ranges between 20–55 per cent, with a mean value of 40 per cent. Thus, Pampean soybeans uptake less nitrogen from biological fixation than other continental averages. This happens because soybeans tend to reduce biological fixation when planted in nitrogen-rich soils, since nodulation and biological fixation are expensive in terms of other macronutrients and energy.[30] While the soils of the Pampas are heterogenous, the majority of soybean production has taken place on the fine soils of the central and eastern Pampas. The soil quality of these areas has made them world-renowned and they are typically credited for how Argentina became a breadbasket of the world in the early twentieth century and maintains this position today. Because of this, the application of synthetic fertilisers on Pampean soil has been historically low, especially in the case of nitrogen fertilisers on soybean crops. The remainder of the plant's required nitrogen must then come from the soil. Therefore, unless managed by economically-disincentivised inputs – such as fertilisers, cover crops or permaculture, export-oriented soybean monoculture tends to mine soils of nitrogen over time.

Based on localised models, results show that of the sixty kilograms of nitrogen contained per metric ton of harvested grain dry matter, 52 kilograms of nitrogen are biologically fixed.[31] Assuming that nitrogen fertiliser application is negligible, a simple mass balance of nitrogen finds that on average in Pampean soybean agriculture a total of eight kilograms of nitrogen per hectare is extracted from the soil per metric ton of harvested grain dry matter.[32] This is based on nitrogen concentration in above-ground biomass, i.e. grain and shoots, litter and roots, as well as the percentage of biological nitrogen fixation for the soybean crop as a whole and for its different compartments. Since increasing crop yield is tied to an increasing amount of nitrogen yields derived from biological fixation and increased return of nitrogen to the soil, average production data per year of Pampean soybean

29 S. Tamagno et al. 'Interplay between nitrogen fertilizer and biological nitrogen fixation in soybean: Implications on seed yield and biomass allocation', *Scientific Reports* **8** (17502) (2018): 1–11, at 3.

30 F. Salvagiotti et al. 'Growth and nitrogen fixation in high-yielding soybean: Impact of nitrogen fertilization', *Agronomy Journal* **101** (4) (2009): 958–70.

31 C. di Ciocco et al. 'Nitrogen fixation by soybean in the Pampas: Relationship between yield and soil nitrogen balance', *Agrochimica* **55** (6) (2011): 305–13.

32 Ibid.

192

Enrique Antonio Mejia

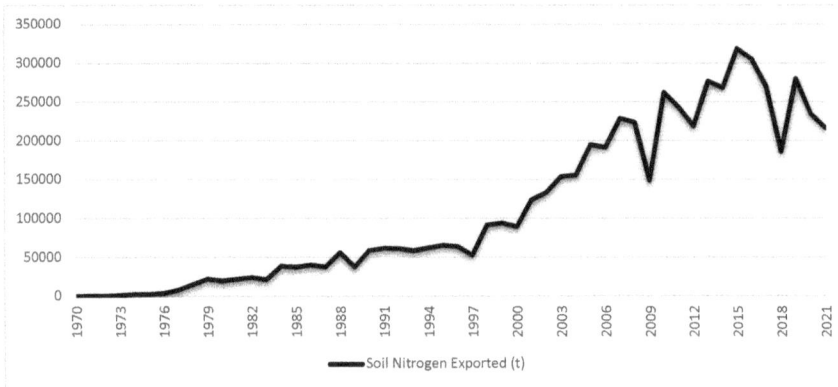

Figure 1.

Extracted soil nitrogen exported from Argentine Pampean soil per year between 1970–2021 in metric tons. Graph produced by author with data compiled from di Ciocco et al., 'Nitrogen fixation by soybean in the Pampas'; and MINAGRI, 'Estimaciones agrícolas'.

agriculture between 1970–2021 is assessed. The year 1970 is chosen as a starting point, both as a practicality due to the availability of data at the departmental level and because, in the 1970s, Argentina began exporting soybeans as a global commodity. While these calculations show the average amount of nitrogen extracted from Pampean soils per year, they do not represent how much of this extracted soil nitrogen is exported to the world market. As Argentina has had little domestic use for soybeans and soybean meals in the commodity's short history, the majority of what is produced is exported. To calculate the total extracted nitrogen destined for export, a national average of 6.2 per cent is subtracted and represents the proportion of soy production used domestically.[33] The results demonstrate the total amount of nitrogen in metric tons per year exported from Pampean soils has significantly risen over time (Figure 1).

While other mass balance calculations for nitrogen have been conducted on Pampean soybean monoculture, they have mainly focused on annual averages for the early 2000s and have not calculated the total soil nitrogen

33   This average is based on annual national figures proposed by Trase. While the limitation of using this average is that it cannot reflect fluctuations per year, domestic use has remained low and does not significantly affect the results: Trase, 'SEI-PCS Argentina soy v1.0.1 supply chain map: Data sources and methods' (2020): www.trase.earth (Accessed 15 Sept. 2021).

exported to the global agrofood system.[34] Furthermore, research has taken place that has been found to overestimate soil nitrogen extraction, suggesting that upwards of 126 kilograms of nitrogen per hectare is extracted from the soil per metric ton of harvested grain dry matter.[35] This is perhaps due to a lack of access to departmental-level production data or the limitation of analysing the results of only two local experiments. Despite this, there is consensus that soybean monoculture is associated with negative nitrogen balances where soils are mined.[36]

The results presented in this chapter show that an acceleration in the amount of soil nitrogen exported per hectare begins in 1996, a significant year as it marks the introduction of the soybean technological package. Therefore, there is a clear relation between sojización and soil nitrogen mining. At the national-level, export-oriented agrofood production in general is responsible for an annual average of 26 kilograms of nitrogen per hectare extracted from soils. As Pampean soybean monoculture has an annual average loss of 17 kilograms of nitrogen per hectare, it is a major contributor to soil mining nationally.

In addition, the average annual extraction of soil nitrogen per year since 1996 has been roughly 195,000 kilograms per year, which seems marginal compared to the 1.5 million kilograms that are biologically fixed and even more so considering biological fixation from Pampean soybean agriculture is responsible for roughly 0.002 per cent of the planetary threshold for anthropogenic nitrogen loading: averaged at 80 teragrams per year.[37] However, while soil nitrogen extraction constitutes a small contribution to overall annual anthropogenic nitrogen loading, it is important to highlight that this is the amount lost from but one local social-ecological system. Moreover, poor nitrogen use efficiencies in commodity chains and the nitrogen cascade mean that roughly 85 per cent of this nitrogen will be lost to the environment before it reaches consumers.[38] Thus, although the scale of global anthropogenic nitrogen

34   di Ciocco et al., 'Nitrogen fixation by soybean in the Pampas'; R. Alvarez et al. 'A regional audit of nitrogen fluxes in Pampean agroecosystems' *Agriculture, Ecosystems and Environment* **184** (2014): 1–8.

35   Austin et al., 'More is less', 50.

36   Alvarez et al., 'A regional audit', 7, Austin et al., 'More is less', 52; Díaz de Astarloa and Pengue, 'Nutrients metabolism of agricultural production', 81.

37   W. de Vries et al. 'Assessing planetary and regional nitrogen boundaries related to food security and adverse environmental impacts', *Current Opinion in Environmental Sustainability* **5** (3) (2013): 392–402, at 399.

38   J.N. Galloway and E.B. Cowling, 'Reflections on 200 years of nitrogen, 20 years later', *Ambio* **50** (4) (2021): 745–49, at 746.

loading is colossal, the planetary-level analysis of this social-ecological crisis masks the significant unaccounted-for costs for local actors, communities and ecosystems not only in the present, but also in the future.

This represents an interesting aspect of ecologically-unequal exchange, in that the pecuniary valuations of the soybean complex sold on the market neither reflect the present unaccounted-for costs associated with soil nitrogen depletion nor the costs that are likely to occur. Attempts to arrest this depletion incur further costs, creating inertia whereby soybean agriculture becomes cumulatively specialised, which is characteristic of the types of productive processes that lead to ecologically-unequal exchanges as they tend to limit the development potential of extractive economies.[39] Thus, while the productivity rates of soybean agriculture are significantly rising over a relatively short period of time, the social-ecological disruptions that take place remain absent from mainstream accounting. To capture the unaccounted-for costs associated with soil nitrogen depletion, I propose an innovative institutional framework inspired by Kapp's theory of social costs that differentiate two orders in which these costs occur.[40]

Social Costs of Soil Nitrogen Depletion and Institutional Roots

The *social* in social costs does not limit the damages to society as such, but is rather to demonstrate that the costs are borne by those who do not receive a profit. Social costs are then defined as harmful consequences and damages which various stakeholders or ecosystems sustain as a result of productive processes, and for which business enterprises are not held accountable. Furthermore, social costs must have two qualities, it must be possible to avoid them and they must be part of productive processes that can be shifted to third persons, communities or ecosystems. This does not however imply that the business enterprise is strictly private. In fact, Kapp revised his earlier writings on social costs to account for state enterprises in the Soviet Union and elsewhere that operated under similar economic calculations as private enterprises.[41] The theory of social costs is especially

---

39   S.G. Bunker, 'Modes of extraction, unequal exchange, and the progressive underdevelopment of an extreme periphery: The Brazilian Amazon, 1600–1980', *American Journal of Sociology* **89** (5) (1984): 1017–64.

40   Kapp, *The Social Costs of Business Enterprise*.

41   Ibid., p. 14.

relevant as productive processes associated with sojización have become increasingly dominated by vastly integrated transnational corporations that both right- and left-wing governments have allowed to flourish; albeit for disparate purposes. For example, Peronists in Argentina have maintained export taxes on the soy complex.[42] The taxes are seen as a proverbial 'cash cow' and are in practice intended to develop social welfare projects. While this approach taken by left-wing governments has been argued as demonstrating a Polanyian institutional recalibration, the contradictions that are embedded in this institutionalisation of neo-extractivism, not in the least the resulting environmental degradation, are too stark to ignore.[43]

Social costs are left as a broad category to not only account for those costs with a market value, but also those that are difficult to articulate and may have no satisfactory pecuniary valuation. This means that social costs can derive from diverse productive processes such as deforestation, automation, the privatisation of health care, land-use change or export-oriented mono-culture. In neoclassical economics, social costs are considered externalities; as in external to economic activity. However, social costs as not only internal to economic activity, but are necessary for business enterprises to maximise profit via cost-shifting.

The negative effects of social costs resulting from productive processes are sometimes felt immediately, but often remain hidden for considerable periods of time. These effects can be circular and cumulative; thus, changes that result from primary changes tend to progress in the same direction, i.e. better gets better and worse gets worse.[44] Further, mismatches between social and ecological scales compound this; not in the lease concerning spatial, temporal and institutional dimensions. Therefore, analysing the social costs of soil nitrogen depletion requires a fundamental understanding of the biophysical context in which it occurs. The complex interdependencies involved with agriculture mean that historically-formed institutions and ecological context matter and

---

42   For example, Peronist president Albert Fernández of Argentina during his tenure expressed interest in acquiring the failing soymeal giant, Vicentin. This was later rescinded after opposition from the private sector and conservative politicians: H. Bronstein and M. Heath, 'Argentina ditches takeover plan for soymeal giant Vicentin', *Reuters* 2020: https://www.reuters.com/article/us-argentina-grains-vicentin-idUSKCN24W39C (Accessed 22 Oct. 2021).

43   Cf. C. Berndt et al. 'Postneoliberalism as institutional recalibration: Reading Polanyi through Argentina's soy boom', *Environment and Planning A: Economy and Space* **52** (1) (2020): 216–36.

44   As Myrdal has described in his theory of circular and cumulative causation: G. Myrdal, 'The meaning and validity of institutional economics', in K. Dopfer (ed.), *Economics in the Future* (London: Macmillan Education, 1976), pp. 82–89.

changes create further changes: evolutions and adaptations arise. For example, it has been repeated *ad nauseum* that a main social cost of using chemical pesticides is the genesis and explosive growth of superweeds.[45]

Still, concerning soil nitrogen depletion, the practice of 'soybean-on-soybean' monocropping has become a recurring pattern in the Argentine Pampas, coming to a height in the 2015/16 growing season. The institutional structure of land tenure in Argentina has compounded the effects of this depletion by incentivising sojización, and limiting regenerative agriculture practices. This is explained by high land values and land lease agreements lasting rarely over one year, facilitating a behaviour of profit-maximisation for both land-owners and tenants. Because leases are not guaranteed to be renewed for the following season – though it is common that they are, the extra costs associated with planting cover crops or using fertilisers are di-sincentivised. Further, the market price of single soybeans, which are grown earlier in the growing season is higher than second soybeans – planted after rotation of a winter crop, such as wheat, and planted later in the season. As leases are only a year in length, the profit motive leads soybean producers to cultivate single soybeans and leave the soil bare over the winter; thus, leaving it susceptible to elemental erosion – a threat that is not without historical precedent and especially affects the western ends of the Pampas. In doing so, the costs of maintaining and remedying the soil are shifted to future lease holders, who under this pattern continue to shift the costs.[46] The lack of crop rotation also makes agriculture in the Pampas vulnerable to cyclical droughts. This is evidenced by the 2009 and 2018 harvest years where yields were 39 per cent and 29 per cent less than the previous year, respectively.[47]

The result of this is that the soil is continuously mined of nitrogen and never replenished because the nitrogen is converted into protein contained within the soybean grain that is either sent to processing plants or exported whole. Higher land values have also pushed livestock producers and other small farmers out into the margins, and are a leading driver for a number

45   See F. Zorzoli, '¿Límites ecológicos y fronteras tecnológicas en el negocio agrícola? Agricultura y ambiente en los sectores agrarios medios del noroeste argentino [Ecological limits and technological frontiers in the agricultural business? Agriculture and environment in the medium agricultural sectors of north-western Argentina]', *Población y Sociedad* **25** (1) (2018): 163–95.

46   P. Arora et al. 'Ownership effect in the wild: Influence of land ownership on agribusiness goals and decisions in the Argentine Pampas', *Journal of Behavioral and Experimental Economics* **58** (2015): 162–70.

47   The droughts of 2009 and 2018 are also reflected in Fig. 1: since average yield declined, so did total nitrogen exported.

of social-ecological impacts and related social costs such as deforestation, biodiversity loss and the alteration of soils in disparate ecosystems. This leads to a differentiation of social costs: those directly resulting from soil nitrogen depletion and those resulting from the so-far-unsuccessful attempts at overcoming soil nitrogen depletion.

Social costs of the first order

Soil is not merely dirt. It is a complex ecosystem of microbes, nutrients, fungi, plant roots and decomposing organic matter. When soil is healthy, it is a renewable resource; however, when a critical threshold has passed, and the soil has become severely degraded, reviving the soil may be outside any reasonable human timescale – if it ever returns. Therefore, returning to the *status quo ante* within a reasonable human timescale requires technological innovation and cultural practices to change, as well as significant capital investment and time, both of which counter the profit-maximising *modus operandi* of the business enterprise. Due to the cumulative and circular changes associated with environmental disruption, a multitude of slow and delayed feedbacks can manifest later in time, compounding already existing unaccounted-for costs or incurring new ones.

Soil nutrient depletion, and its ultimate form – soil erosion – represents one of the strongest cases for the theory of social costs. Under the logic of private property rights and with the spectre of the tragedy of the commons looming, the neoclassical assumption holds that productive land entered into the market system should guarantee the preservation of soil health. However, despite strong private property rights and a well-established foothold within the global capitalist agrofood economy, 36 per cent of Argentine territory – roughly 100 million hectares – has become eroded over the past 25 years.[48]

The combined effects of soil nitrogen extraction and the simplification of the agrarian landscape into a continuous cropping system have had marked effects on the biodiversity of the region. Native species have experienced a reduction in geographical range and abundance with regional extinctions taking place, while non-native species and some pollinators have thrived.[49]

---

48   INTA, 'El 36% del suelo argentino sufre procesos de erosion [36% of Argentine soils suffer erosion processes]', (Argentina: Instituto Nacional de Tecnología Agropecuaria 2019): https://www.argentina. gob.ar/noticias/el-36-del-suelo-argentino-sufre-procesos-de-erosion (Accessed 27 Aug. 2021).

49   D. Medan et al. 'Effects of agriculture expansion and intensification on the vertebrate and invertebrate diversity in the Pampas of Argentina', *Biodiversity and Conservation* **20** (13) (2011): 3077–100.

This is not only the case with larger fauna and flora of the region, but also the macrofauna found within soils.[50] The loss of these genetic reservoirs, the erosion of the soil and the dialectic between productivity and extraction are driving the Pampean agroecosystem towards an irreversible regime shift. This process draws significant parallels with an earlier ecological crisis in the 1940s that witnessed extensive soil erosion, crop failure and diminished productivity rates, famine and death of livestock and a mass rural exodus.[51] This was spurred by agricultural expansion into fragile soils, cyclical droughts and the mismanagement of soil.

Despite the lack of institutional incentives and historically low levels of synthetic nitrogen fertiliser use on soybeans, the practice has steadily grown in Argentina. However, studies have shown that current fertilising practices have led to a reduction of biological nitrogen fixation in soybeans and have no impact on increasing protein concentration in seeds.[52] In order to apply fertilisers and not limit biological fixation, producers must place controlled-release nitrogen fertilisers below the soil surface at the roots of the soybean, which involves additional costs and practice changes.[53] As soil nitrogen levels decline, yields increase and land becomes scarcer, it is likely that synthetic fertiliser use will continue to rise.

In order to circumvent the ecological limitations of depleting soil nitrogen and meet the increasing demand for protein, incredible amounts of capital, labour and energy have been put into researching and creating genetically modified varieties suitable for production in otherwise marginal lands and harsh climatic conditions, improving transportation networks and constructing the necessary infrastructure to store and process soybeans. The implementation of high-yielding varieties has been a clear strategy towards overcoming declining protein concentrations, increasing land values and finite productive land.

---

50   A. Domínguez et al. 'Soil macrofauna diversity as a key element for building sustainable agriculture in Argentine Pampas', *Acta Oecologica* **92** (2018): 102–16.

51   G. Covas, 'Evolución del manejo de suelos en la región pampeana semiárida [Evolution of soil management in the semi-arid Pampas region]', *Actas de las Primeras Jornadas de Suelos en Zonas Áridas y Semiáridas* (Buenos Aires: Instituto Nacional de Technología Agropecuaria, 1989), pp. 1–11.

52   N. Cafaro La Menza et al. 'Is soybean yield limited by nitrogen supply?' *Field Crops Research* **213** (2017): 204–12; Salvagiotti et al., 'Growth and nitrogen fixation'.

53   Ibid.

Social costs of the second order

Because of soil nitrogen depletion, soybean agriculture has struggled to achieve the demanded protein output necessary to satisfy foreign demand. Further, it has been shown that the social-ecological context matters in that protein concentration varies considerably across Argentina.[54] Over time, Argentine soybeans have continuously declined in their protein concentration; therefore, producers have adopted the strategies of cultivating higher-yielding varieties and increased processing to meet requirements set by traders.[55] In cases where protein concentrations were not up to standard – 47.5–49 per cent after being processed into soymeal – prices fetched by processors were substantially reduced. To compensate for missed revenue, processors have reduced wages, resulting in strikes by labourers.[56]

While there can be a multitude of social-ecological causes for this decline in protein concentration, a fundamental reason is described as the inverse yield nitrogen law, that increased yields will be followed by a decrease in protein concentration.[57] The pursuit of higher yields to meet protein demand is paradoxically driving the decline in protein concentration in soybeans. This is compounded by the fact that producers are paid by yield and weight rather than protein concentration, a disconnect from the valuation of soybean meal placed on processors. Therefore, a positive feedback is instilled where higher yields will over time beget the need for higher yields.

Additionally, high-yielding varieties often require a number of complementary inputs that are scarce and costly when first implemented. The cultivation methods of such varieties are based upon large quantities of water,

---

54   L.B. Bosaz et al. 'Management and environmental factors explaining soybean seed protein variability in Central Argentina', *Field Crops Research* **240** (2019): 34–43; D.M. Maestri et al. 'Seed composition of soybean cultivars evaluated in different environmental regions', *Journal of the Science of Food and Agriculture* **77** (4) (1998): 494–98.

55   CAC, 'Campaña soja 2018 relevamiento calidad industrial – Junio', (Cámara Arbital de Cereales de Rosario, 2018): http://www.cac.bcr.com.ar/es/arbitraje-y-calidad/informes-de-calidad/informe-calidad-de-cosecha-soja/campana-soja-2018 (Accessed 13 July 2021).

56   For example, at the end of 2020, A 20-day strike over wages paralysed exports from international agro-giants such as Cargill Inc, Bunge Ltd. and Louis Dreyfus Co. As a result of the strike, the loading of 162 ships was delayed, valued at the market price of US$1.5 billion in exports (of course not taking into consideration all other social costs). Eventually, an increase in salaries was achieved: H. Bronstein and M. Heath, 'UPDATE 3- Argentine soy crushers sign deal with oilseed workers, ending strike', *Reuters* 2020: https://www.reuters.com/article/argentina-grains-strike-idUSL1N-2J90YS (Accessed 23 Aug. 2021).

57   O.W. Willcox, 'Factual base of the inverse yield-nitrogen law', *Agronomy Journal* **41** (11) (1949): 527–30.

200

energy, nutrients, pesticides and context-specific machinery. For example, it is common practice to inoculate the soil with cultures of symbiotic rhizobium bacteria to promote nodulation within the roots of soybean plants to increase biological nitrogen fixation. However, the use of herbicides such as Roundup has been shown to affect the rhizobium and lessen the efficiency of biological fixation. In this way, the inputs become contradictory and over time costlier to maintain.[58]

The transaction costs of switching to high-input, mechanised agriculture and increasing land values and rents are exclusionary to those without significant capital. Thus, this specific form of agricultural production leads to a concentration of ownership in land and the means of production by increasingly demanding higher economies of scale to operate. In this climate, sowing pools – network firms that pool resources from foreign finance and trust funds, amongst other sources – have obtained a competitive advantage in their purchasing capacity and are able to return high profits to investors.[59] For example, it is estimated that about a quarter of small farms in Argentina were acquired by large-scale farmers between 1998–2002 alone.[60] The resulting dispossession of tenants and small farmers represents a substantial social cost by contributing to the exodus of landless workers to urban centres of Argentina. As history has shown, not just in Argentina, but in South America in general, this type of development has the circular and cumulative effect of contributing to greater inequalities, poverty and social instability, in that dispossession often represents a drastic and sometimes violent rift in social relations, livelihoods and people's relation to land. For example, this process has polarised the social structure of agriculture in the region with increasing monopolistic, capitalist enterprises on one end and a decreasing plurality of small, medium and peasant producers on the other. The tensions are becoming more evident as vandalism to silo bags – which can be 2,000 metres long and contain 1,000 tons of harvested soybeans – have become a form of resistance towards large producers.[61]

58   J.B. dos Santos et al. 'Tolerance of Bradyrhizobium strains to glyphosate formulations', *Crop Protection* **24** (6) (2005): 543–47.

59   Baraibar Norberg, *The Political Economy of Agrarian Change*, p. 8.

60   M. Altieri and W.A. Pengue, 'GM soybean: Latin America's new colonizer', *Seedling* (2006): http://www.grain.org/es/article/entries/588-gm-soybean-latin-america-s-newcolonizer (Accessed 16 Oct. 2021).

61   M. Reinke, '"Ultrajados": Rompieron cuatro silobolsas y el delito no para de crecer ["Outraged": They broke four silo bags and crime does not stop growing]', *La Nación* (2021): https://www.

9. Five Decades of Soybean Agriculture

While sojización appears to be all-consuming, recent changes in export tax regulations by former President Macri, such as the elimination of tax on wheat and maize yet maintaining the thirty per cent export tax on soybeans, have affected agricultural production in some locales. For example, though the initial costs are higher and the requirement for synthetic fertilisers adds to this, maize has maintained higher profit margins than soybean production, thus incentivising producers to rotate crops by planting second soybeans followed by wheat and bringing maize back into the shuffle. This shift has made some allude to the potential of maize to become Argentina's new star crop.[62] The role of institutional change in shaping Argentina's agrarian landscape is then hard to ignore. And although maize is beginning to receive special attention, recent negotiations around the long-anticipated Mercosur-EU trade agreement, may see the abolition of the soybean export tax, thus lowering the costs placed on producers and re-establishing soybean as the leading cash crop. Whether maize becomes the new star crop or sojización continues to subsume the Pampean agrarian landscape, the productive process of capital intensive, high-input, export-oriented agrofood production in Argentina is a fixture of not only continuous soil mining, but also an escalating global nitrogen problem. In this way, the soybean itself is not the singular issue, but rather the structural and institutional arrangements that surround it leading to sojización.

Conclusion

This chapter has focused on the case of export-oriented soybean agriculture in the Argentine Pampas where the mobilisation of nitrogen has degraded soil quality, bearing significant unaccounted-for costs for local actors, communities and ecosystems. By situating the Pampean case within the greater historical political economy of asymmetric embodied nitrogen flows, the soybean narrative of this region is seen to be one of neo-extractivism, myopia and specialisation.

The assessment of the amount of soil nitrogen per year resulting from local soybean agriculture over a historical period presented not only empirical evidence of the net-export of soil nitrogen from the Argentine Pampas, but

lanacion.com.ar/economia/campo/ultrajados-rompieron-cuatro-silobolsas-y-el-delito-no-para-de-crecer-nid25082021/ (Accessed 26 Aug. 2021).

62   J. Hiba, 'Is Argentina's soy boom over?', *Dialogo Chino* (2021): https://dialogochino.net/en/agriculture/44411-is-argentinas-soy-boom-over/ (Accessed 23 Aug. 2021).

Enrique Antonio Mejia

also that this has accelerated since 1996 onward. This is linked to sojización, where the adoption of the soybean technological package has facilitated both agricultural expansion and intensification not only in the Pampas but across Argentina in general. The social costs of soil nitrogen depletion are manifold and multi-scalar. Therefore, influenced by K. William's Kapp's theory of social costs, I have used an institutional economic lens to differentiate two orders of unaccounted-for costs in the Argentine Pampas: those directly resulting from soil nitrogen depletion and those resulting from so-far-unsuccessful attempts at overcoming soil nitrogen depletion.

Biodiversity loss, the simplification of the agrarian landscape and the threat of soil erosion are substantial social costs that are all too real. The patterns that contributed to the ecological crisis of the 1940s are re-emerging, and the threat of an irreversible regime shift in the Argentine Pampas should be taken seriously. These issues are not only impacting society and ecosystems in the present, but will have cumulatively worse effects in the future.

Furthermore, cultivation of higher-yielding varieties, inoculating soils with Rhizobium, processing soybean into meal cakes with globally competitive protein concentrations, operating at greater economies of scale and expanding into marginal lands all represent attempts at overcoming the ecological constraints of limited soil nitrogen and decreasing protein concentration in soybean grain. However, these are not without unaccounted for costs. For example, the cultivation of high-yielding varieties paradoxically drives the need to cultivate high-yielding varieties because increased yields lead to decreased protein concentrations. Also, the accumulation of land under sowing pools and large-scale farming operations has dispossessed small and medium producers spurring rural migration and expanding the rentier economy. Strong private property rights and land leases lasting rarely over a year are institutional constraints on regenerative agriculture, since the profit motive leads to a behaviour of cost-shifting where future leasers bear the costs of depleted soil.

Despite the clear social costs to society and ecosystems, conditions suggest that export-oriented soybean monoculture in the Argentine Pampas will continue. For one, the versatility of the soybean as an input for various industries and the aberrant quality of protein contained within soybeans have interlinked the production into vast, cross-scalar commodity chains. Secondly, the incredible investments into research programmes, breeding, infrastructure, intellectual copyright patents, technology and land represent forms of inertia that contribute towards a path dependency of specialisation and reprimarisation of the Argentine economy. Lastly, the structure of land tenure in Argentina and strong attitudes towards private property disincentivise the practice of regenerative agriculture. Therefore, the declining soil nitrogen quality of the Pampas and the cultural and biodiversity loss of the region have simplified the agrarian landscape, narrowing windows of opportunity for alternative development and economic pathways. Unless dramatic shifts occur in crop rotation, land tenure reform and the ways in which export commodities are valued, social costs will continue to accumulate well into the future and be borne by third parties and ecosystems that have no agency in the decisions taken today. All in the name of producing just a little bean.

Section IV
Soybean Nutrition and Technology

CHAPTER 10.

SOY AND MEAT: A POST-WAR NORTH AMERICAN HISTORY

Anna Zeide

In many parts of North America, if consumers are aware of having eaten soybeans, they may think of the ubiquitous soy sauce on Chinese restaurant tables. Or they order a soy latte at their neighbourhood coffee shop, welcoming the base of beige soymilk as a way to get extra heart-healthy protein. Some consumers might enjoy an occasional tofu dish at a Thai restaurant, remarking, 'You can barely even taste the tofu! It just soaks up the flavors of what's around it.' But most people would rarely think of soy as a mainstay of their diets, or as a central American crop.

In fact, however, soybeans are nearly tied with corn as the leading crop planted in the United States, with wheat and cotton trailing far behind. The US exported $18.7 billion worth of soybeans in 2019, the largest portions going to China and Mexico. Soy is the world's number one source of livestock feed and the number two source of vegetable oil. When consumers eat meat, they are typically eating flesh of an animal fed on soybeans. When they consume anything with vegetable oil or margarine, soy is there.[1]

This disconnect between many consumers' perceptions of soy's importance to American agriculture and the reality lies in the fact that the soybean has become a quintessentially industrial food. Or, rather, industrial *ingredient* might better capture the remarkable position of this bean. During the twentieth century, and especially in the years after World War Two, the soybean experienced a meteoric rise, as industrial producers of all kinds adopted it as a core part of their production processes. These industries broke the small

---

1   USDA National Agricultural Statistics Service, Agricultural Statistics Board, 'Acreage Report', 30 June 2020: https://www.nass.usda.gov/Publications/Todays_Reports/reports/acrg0620.pdf; USDA Foreign Agricultural Service, 'Soybeans' (Accessed 22 Oct. 2020): https://www.fas.usda.gov/commodities/soybeans; USDA Economic Research Service, 'Soybeans & Oil Crops' (Accessed 22 Oct. 2020): https://www.ers.usda.gov/topics/crops/soybeans-oil-crops/; D. Nosowitz, 'Soy is set to become our biggest crop by acreage. But what are we doing with this soy?', *Modern Farmer*, 4 Dec. 2017: https://modernfarmer.com/2017/12/soy-set-become-biggest-crop-acreage-soy/.

THE AGE OF THE SOYBEAN: 207–226 **doi:** 10.3197/63800040695086.ch10

soybean into its constituent parts and reorganised it into a dazzling array of foods and industrial products.[2]

In the second half of the twentieth century, the soybean was adopted by counterculturists, drawing on the groundwork laid by Asian immigrants and the vegetarian Seventh-day Adventists, as a meat and dairy substitute – which itself generated an industry. The soybean was transmogrified into spun proteins and extruded soy bits to serve as a meat extender, especially during the lean years of the 1970s. And at the same time the soybean became a critical animal feed, fuelling the growth of the meat industry and meat consumption in North America and beyond. In all these ways and more, soy has become a ballast of American industry, especially defining the post-World War Two period, characterised by high rates of meat and processed food consumption, with all the attendant environmental and health effects.[3]

In its many variations, soy's most enduring and devastating legacy is in the way it has contributed to the dramatic rise of meat consumption in North America. This is especially true in the United States, but also in Canada and Mexico (and indeed all around the globe, as you can read in other chapters of this volume). In all these places, the rise of soy production enabled the rise in meat consumption. The two are inextricable. But soy's role as a building block of meat is less visible in the public eye. Instead, consumers tend to view soy as a meat *alternative*, something that moves us away from such high rates of meat consumption. These competing identities embody the versatility and wide reach of the humble soybean. This chapter will thus focus on these meat-adjacent uses of soy in post-war America: as animal feed (meat itself), as meat substitute and as meat extender.

Soy Turned into Meat Itself

In the second half of the twentieth century, soy grew to become the chief animal feed in a feedlot system that dominated American – and increasingly global – agriculture. Cheap soy yielded cheap meat, with devastating environmental and health consequences. Between 1950 and 1990, world agriculture produced 300 per cent more soy per capita. And because more than ninety per cent of all soy protein goes to feed livestock, this increase yielded cheaper and more plentiful meat. According to soy scholar Christine

---

2   For an excellent overview of the history of soy in the United States, see M. Roth, *Magic Bean: The Rise of Soy in America* (Lawrence, KS: University Press of Kansas, 2018).

3   Ibid..

10. Soy and Meat: A Postwar North American History

DuBois, 'Defatted soybeans have become so dominant in animal feeds that other protein sources are measured in a unit called "SME," for "Soybean Meal Equivalent."' As meat prices have fallen, meat consumption has risen dramatically, with the world consuming twice as much pork and four times as much chicken in 2018 compared with 1960.[4] Combined with animal breeding efforts and antibiotic research that enabled the confinement and concentration of enormous numbers of animals, cheap soy and corn created the modern factory farm. This American model of the Concentrated Animal Feeding Operation (CAFO) began with chickens by the 1950s, then spread to hog and cattle operations, creating myriad problems: water and land over-use, antibiotic resistance, manure lagoons that contaminate nearby waterways, animal mistreatment, demand for petroleum-based fertilisers and pesticides, and greenhouse gas emissions, among others.[5]

While this story of the transformation of soy into meat began in earnest during the Second World War, some foundations were laid in the years prior. Before the war, some firms in England, Denmark and Germany had begun to crush the soybean, to separate it onto oil and meal, creating what soy chronicler William Shurtleff calls 'the modern Western paradigm' around the use of soy. Shurtleff also reminds us that this early-twentieth century method took off '96% of the way through [the soybean's] history'.[6] That is, although this method soon became the dominant way that soybeans were processed, it broke with the much longer past in which people, especially in China, had eaten the soybean and its products as human food. Depression-era agricultural policies promoted soy production, as farmers were paid to grow less of other crops, like corn, wheat and cotton, while soy was unrestricted. Concerns about the boll weevil pest in Southern cotton fields also led growers to look beyond cotton for oilseed crops.[7] Scientific feeding tests, as at the agricultural experiment stations of Illinois, Ohio and Wisconsin, confirmed that soybean meal, when separated from the oil, provided high-quality feed

4   C.M. Du Bois, *The Story of Soy* (University of Chicago Press, 2018), pp. 10–11.
5   J. Specht, *Red Meat Republic* (Princeton, NJ: Princeton University Press, 2019); Alex Blanchette, *Porkopolis: American Animality, Standardized Life, and the Factory Farm* (Durham, NC: Duke University Press, 2020); P.R. Josephson, *Chicken: A History from Farmyard to Factory* (Cambridge [England]: Polity Press, 2020).
6   W. Shurtleff and A. Aoyagi, *History of Soybean Crushing – Soy Oil and Soybean Meal (980–2016): Extensively Annotated Bibliography and Sourcebook* (Soyinfo Center, 2016), p. 5.
7   Roth, *Magic Bean.*

for cattle, hogs and poultry.[8] Soy industry boosters like D.W. McMillen of Central Soya then promoted soybean meal to veterinarians and livestock feeders to make it 'respectable'.[9]

All of this laid the foundation for the surge in soybean meal as livestock feed during and after World War Two. With the war limiting oil imports from around the world – such as coconut oil from the East Indies and palm oil from the Netherlands Indies – the US government set high production goals for soybeans and other domestic oilseeds.[10] Cargill, which would become an agricultural industry giant, opened the first US soybean crushing plant in 1943, to extract the soybean oil, leaving behind the meal.[11] This availability of soybean meal, along with the demand for more meat, led to a huge increase in meat, dairy and egg production by 1945.[12] A 1942 advertisement in the *Soybean Digest* trade journal, sponsored by eight different companies, read 'Our Workers, Our Armies and Our Allies Depend on the Efficiency of Our Livestock Production'.[13] And efficient livestock production came to depend on soybean meal. By 1942, most of the world's soybeans were sent to crushing mills, with the oil going largely to human food purposes (rather than the earlier industrial uses) and the meal going to feed livestock.[14] In fact, a number of meatpacking companies got into the soybean meal business, in order to link the two even more tightly. A 1947 Swift & Co. advert proclaimed, 'Nothing builds good livestock like good feed … So, it's only natural that Swift has become a leading buyer and processor of soybeans and distributor of fine soybean oil meal.'[15] And in Canada, the country's largest meat packer, Canada Packers Ltd., purchased an Ontario soybean crushing plant, Canadian Vegetable Oil Processing Ltd., in 1950.[16]

8   H. File, 'We can make almost anything from soybeans', *Staley Journal*, Aug. 1936. As cited in Shurtleff and Aoyagi, *Soybean Crushing*, pp. 1188–89.

9   'Honorary life members: D.W. McMillen and Dwayne O. Andreas', *Soybean Digest*, Sept. 1966: 6.

10   R.M. Walsh, 'Soybean production – here and abroad: And possible competition from other oilseeds', *Soybean Digest*, May 1947:18–21.

11   Shurtleff and Aoyagi, *Soybean Crushing*, p. 13.

12   P.R. Record, 'The soybean situation', *Chemurgic Digest* 4 (31 March 1945): 114–15. As cited in Shurtleff and Aoyagi, *Soybean Crushing*, p. 1592.

13   'Each pig costs $5.40 less: "Watch em grow into dough" (Ad.)', *Soybean Digest*, Sept. 1942: 27.

14   Shurtleff and Aoyagi, *Soybean Crushing*, p. 5.

15   Swift & Company, 'Across the soybean belt Swift leads the way! (Ad.)', *Soybean Digest*, Sept. 1947: 83.

16   W. Shurtleff and A. Aoyagi, *History of Soybeans and Soyfoods in Canada (1831–2019): Extensively Annotated Bibliography and Sourcebook* (Soyinfo Center, 2019), p. 7.

10. Soy and Meat: A Postwar North American History

The process of soybean crushing and separation was a complicated multi-step procedure using highly reactive chemicals. Around 1937, the industry shifted from the expeller method to solvent extraction.[17] The primary solvent used was the very flammable hexane, a synthetic, petroleum-based solvent, which is a by-product of gasoline processing. A 1946 description of A.E. Staley Manufacturing Company's new $2 million soy extracting plant in Decatur, IL gives a sense of the industrial process. A power shovel was used to unload soybeans from box cars onto an overhead conveyer belt, which carried the beans to be cleaned, then cracked in vertical three-roll mills, and then passed through a ten-mesh screen, producing bean flakes resembling breakfast cereals. The conveyers then carried the flaked beans through a high enclosed bridge to the extraction building, featuring vapour-tight towers. The flakes were soaked in a hexane bath to separate the oil from the meal, then both parts were evaporated to remove residual hexane. To improve the flavour of the meal, it was then heat-treated, before being cooled and bagged. The oil, meanwhile, was piped to the refinery, for further processing.[18] Despite efforts at control, this process sometimes led to disaster. Hexane fumes leaked out of the extraction buildings and caused explosions.[19] Solvent residues remained in soybean meal, causing haemorrhage and 'extensive death losses' in cattle.[20] These incidents, though decreasing in frequency as the trade evolved, highlighted the industrial nature of soybean processing.

Scientific research and government policies further helped the intertwined soy-livestock industry rise to prominence. The 1948 isolation of vitamin B-12 led to fortification of soy meal to provide this vitamin previously only present in animal proteins. Other studies led to fortification with calcium, phosphorus and other minerals, and later antibiotics and other feed additives.[21] The development of the 'desolventizer-toaster' in 1949–51 helped improve the taste of meal and thus its palatability to livestock.[22] Other research efforts focused on breeding high-oil varieties or varieties adapted to different areas

---

17   'Honorary life members'.

18   'Soybean processing', *Food Industries* **18** (3) (March 1946): 337–41. As cited in Shurtleff and Aoyagi, *Soybean Crushing*, p. 1636.

19   Roth, *Magic Bean*, pp. 87–96.

20   M.J. Twiehaus and E.E. Leasure, 'The presence of a hemorrhagenic factor in soybean pellets extracted with trichloroethylene as a solvent when fed to cattle', *Veterinary Medicine* **46** (11) (Nov. 1951): 428–31.

21   L.E. Hanson, 'Soybean oil meal in livestock and poultry feeds', *Soybean Digest*, Jan. 1958: 17–19.

22   Shurtleff and Aoyagi, *Soybean Crushing*, p. 14.

of the country, mechanising growing and harvesting practices, formulating animal feed rations and more. By 1965, US Secretary of Agriculture Orville L. Freeman was able to describe the 'dramatic soybean success story' that, in the span of 25 years, thanks to research efforts, took soybeans from a 'minor hay crop' to 'rank[in]g first among U.S. oilseed crops and earn[ing] more dollars in farm exports … than any other crop'.[23]

In 1962, no less than President John F. Kennedy himself lauded the soybean industry, calling the soybean a 'miracle crop,' and 'basic to our livestock feeding industry'. He also, notably, referred to his Administration's encouragement of soybean production and exports to create a 'better balance in American agricultural production and to better meet world protein requirements in our Food for Peace program'.[24] Indeed, by the early 1960s, the US soy industry, primarily through the American Soybean Association (ASA), was exporting soybeans abroad to promote meat consumption around the world. By 1973, European consumption of chicken increased nearly 400 per cent, with 23 times more soybean meal being exported to Eastern Europe in 1970 than in 1961 (from 22,947 to 532,772 metric tons).[25] In Japan, chicken consumption skyrocketed 22 times higher between 1960 and 1984, with an attendant 600 per cent increase in imports of US soybeans.[26] Mexico went from not importing any US soybeans at all in 1970 to importing 18.5 million bushels in 1973, supported by duty-free status for imported soybeans granted by Mexico's CONASUPO government agency. An explicit goal of this status was 'to increase the consumption of broiler meat and eggs, and possibly hogs'.[27]

All this international expansion was supported not only by the industry's active promotion, but also, as President Kennedy said, by US federal support. The Food for Peace programme, created as Public Law 480 in 1954, was a way to donate surplus US commodities to balance prices at a home and to develop markets for US grain exports. It was also perceived as a humanitarian food aid project. But American grain companies were the largest beneficiaries, working with the US Department of Agriculture to build demand for US products. One critic in 1980 referred to the program as 'an institutionalized

23   'Freeman credits success of soybeans to research', *Soybean Digest*, Aug. 1965: 30.

24   J.F. Kennedy, 'Greetings from the President of the United States', *Soybean Digest*, Sept. 1962: Front cover.

25   Du Bois, *Story of Soy*, p. 99; J.R. Jones and W.R. Morrison, 'Import demand for soybeans and soybean products in Eastern Europe' (Arkansas Agricultural Experiment Station, March 1976).

26   Du Bois, *Story of Soy*, p. 100.

27   'Limited ASA-ASI Program in Mexico approved', *Soybean Digest*, Oct. 1970: 29.

arm of U.S. imperialism'. The US Feed Grains Council used Food for Peace monies to 'promote the development of local livestock and poultry industries which rely on imported feed grains'. And, as described above, in the years after this programme was launched, livestock industries abroad did indeed grow mightily.[28]

With increasing meat consumption also came an increasing awareness of its environmental impacts. Frances Moore Lappé's *Diet for a Small Planet* raised the red flag about animal agriculture in 1971. She described vividly how animal wastes from factory farms pollute waterways, how growing soy and other livestock feed causes soil erosion and heavy pesticide loads, and how inefficient livestock raising is, in terms of water, land and feed. Because one pound of beef protein requires the input of fourteen pounds of vegetable protein to feed the hungry feedlot cow, any increased demand for meat leads to an outsized demand for grain and soy.[29] In 1981, Lester Brown further decried the world's destruction of our environment for short-term gain, arguing that the three main threats to civilisation – soil erosion; deterioration of forests, grasslands and fisheries; and depletion of oil reserves – all were exacerbated by meat-heavy food systems.[30] In 1987, John Robbins's *Diet for a New America* argued that the meat industry undermined the health of the planet and that reducing meat consumption was the most important thing humans could do to halt the earth's destruction.[31]

The early 1990s brought attention to the way that meat production and consumption create greenhouse gases, contributing to the existential threat of global warming.[32] By the twenty-first century, the environmental impacts of soybean farming itself, along with its effect on meat production, were becoming harder to ignore. The vast majority of soybeans in the US today are genetically modified to be 'Roundup Ready', or resistant to Monsanto's leading pesticide, Roundup. Concerns about genetic engineering – fear of 'super weeds' or 'super pests', health concerns, corporate control – continue

28   R. Burbach and P. Flynn, *Agribusiness in the Americas* (New York: Monthly Review Press, 1980). As cited in Shurtleff and Aoyagi, *Soybean Crushing*, p. 2690.

29   F.M. Lappé, *Diet for a Small Planet* (New York: Ballantine Books, 1971); L.R. Brown, *The Twenty-Ninth Day: Accommodating Human Needs and Numbers to the Earth's Resources* (New York: W.W. Norton & Co., 1978).

30   L.R. Brown, *Building a Sustainable Society* (New York: W.W. Norton & Co., 1981).

31   J. Robbins, *Diet for a New America* (Stillpoint Press, 1987).

32   D. Pimentel, 'Global warming, population growth, and natural resources for food production', *Society & Natural Resources* 4 (4) (1 Oct. 1991): 347–63.

to circulate around soybeans. Finally, soybeans are also implicated in dramatic rates of deforestation, especially in the Amazon, and water usage, as the world's aquifers get funnelled into soybean production.[33] Through all of this, there have been many writers, activists and politicians who have brought attention to the overwhelming environmental problems of soy, both through meat production and on its own.[34] The problems, however, continue, without sufficient political will to implement the solutions.

Soy as Meat Substitute

Even as soy took such a central role in the dramatic expansion of meat production after 1950, its more visible manifestation has been as a meat substitute. Some individuals have even seen soy-based meat substitutes as one solution to the environmental problems of meat production described above. This substitution has deep roots. Soy first arrived in the United States in the eighteenth century, from China. Before the Second World War, it was a marginal crop, largely used for industrial products. But when it was used as a source of human food, following on Chinese practices, it was considered a protein source that could substitute for meat, especially in times of scarcity.[35] Tofu, the cheese-like cake produced through the processes of soaking soybeans, curdling the resulting soymilk and pressing the curds, was known in China as 'meat without bones'.[36] Recognising early the efficiency of getting protein directly from vegetable sources, Yamei Kin, the first Chinese woman to earn a medical degree in the United States, wrote in 1917, 'instead of taking the long and expensive method of feeding grain to an animal until the animal is ready to be killed and eaten, in China we take a short cut by eating the soy bean, which is protein, meat, and milk in itself'.[37] Kin was ahead of her time with these views, presaging arguments like those that Frances Moore Lappé would later make.

---

33   Du Bois, *Story of Soy*.

34   Among many others, see Frances Moore Lappé's daughter following in her footsteps: A. Lappé, *Diet for a Hot Planet: The Climate Crisis at the End of Your Fork and What You Can Do about It* (New York: Bloomsbury USA, 2010).

35   I. Prodöhl, 'Versatile and cheap: A global history of soy in the first half of the twentieth century', *Journal of Global History* **8** (3) (Nov. 2013): 461–82.

36   H.L. Wang et al., 'Soybeans as human food, unprocessed and simply processed', Utilization Research Report No. 5 (USDA Science and Education Administration, January 1979); W. Shurtleff and A. Aoyagi, *The Book of Tofu: Food for Mankind* (Autumn Press, 1975).

37   Roth, *Magic Bean*, p. 56.

10. Soy and Meat: A Postwar North American History

One group in the US that developed soy as a meat substitute, due to religious devotion, was the Seventh-day Adventists. In the 1860s, the church's
founder, Ellen G. White, believed she had a vision from God that turned
her toward vegetarianism and other health beliefs.[38] In order to support their
vegetarianism, Seventh-day Adventists created a wide variety of meat-like
substances. Early products were often made with other vegetable protein
sources, like the peanuts in Protose, created by Ella Kellogg, wife of famed
Adventist John Harvey Kellogg.[39] But by 1920s, soy came to the fore. A
Seventh-day Adventist company in Tennessee, Madison Foods, produced
Soy Bean Meat, the first soy-based meat substitute, in March 1922.[40] The
Adventists founded many of leading companies that made industrial soy
products throughout the twentieth century. John Harvey Kellogg's brother
W.K., at a 1927 meeting of the American Soybean Association, explained
one motivation for Adventist work in this area, echoing the earlier ideas of
Yamei Kin. He said, 'Some day people in the U.S. will realize how foolish
it is to feed one hundred pounds of soybeans to livestock and get back a
very small poundage of meat products which have a protein inferior to the
protein fed to the livestock.'[41] This recurring idea underlay the early soy
proponents' belief in the fundamental logic of eating soy instead of meat,
and in the inevitability of its future dominance.[42]

Throughout the early twentieth century, especially during both world
wars and the Great Depression, there were efforts to make use of soy as
a less expensive protein source, to substitute for meat or milk. One of the
United States' leading industrialists, Henry Ford, was a soy evangelist,
complaining that the cow was 'the crudest machine in the world', and that
'milk' and protein could be produced much more efficiently, as he attempted

38   R.L. Numbers, *Prophetess of Health: Ellen G. White and the Origins of Seventh-Day Adventist Health Reform* (Knoxville, TN: University of Tennessee Press, 1992).

39   A. Shprintzen, 'Modern food as substitute food – Ella Eaton Kellogg's Protose: Fake meat and the gender politics that made American vegetarianism modern', in B.R. Cohen, M.S. Kideckel and A. Zeide (eds), *Acquired Tastes* (Cambridge, MA: MIT Press, 2021).

40   W. Shurtleff and A. Aoyagi, *History of Soybeans and Soyfoods in Mexico and Central America (1877–2009): Extensively Annotated Bibliography and Sourcebook* (Soyinfo Center, 2009), p. 7.

41   G.M. Strayer, 'The battle of the coconut cow. The early days of the American soybean industry and its surprising similarity with today's soyfoods movement', *Soycraft (Colrain, Massachusetts)* 1 (2) (Winter 1980): 50–53. As cited in Shurtleff and Aoyagi, *Soybean Crushing*, p. 2665.

42   For a history of early American vegetarian movements, see A.D. Shprintzen, *The Vegetarian Crusade: The Rise of an American Reform Movement, 1817–1921* (Chapel Hill, NC: University of North Carolina Press, 2015).

to show with his demonstration soy-milk plant, opened in 1934.[43] During World War Two, with rations that extended to meat and dairy, soy foods were promoted as a stand-in.[44]

Beginning in the 1960s and especially in the 1970s, a new group turned their attention to the use of soy as a meat substitute. Influenced by Eastern traditions and by rising concerns about world hunger, counterculturists considered soy a practical and spiritual solution. Learning from Asian immigrants and their descendants, who had continued to produce soy foods in the United States over several generations, the new counterculturists adopted soy products as part of their diets. Natural health food stores began to carry tofu, soymilk and new products like soyburgers.[45] One group of hippies, eager to embody their views in practice, decamped for rural Tennessee, forming a commune called 'The Farm' in the early 1970s. There, the commune members planted and grew their own soybeans – harvesting 162 tons in 1975 – and used them as the foundation for soy sausage, soyloaf, soy yogurt and soy mayonnaise, among many other food inventions.[46] The Farm's founder, Stephen Gaskin, had been inspired by the Zen Buddhism and psychedelic drug use of the Haight-Ashbury neighbourhood of San Francisco in the 1960s. By 1964, he had had a 'psychedelic vision of the soybean, in which he saw it as a great provider for all humankind'.[47]

Two other soy pioneers, William Shurtleff and Akiko Aoyagi, whose staggering Soyinfo Center has today compiled nearly every imaginable source on soy into 65 massive free online bibliographies (which greatly informed this chapter), got their start in this same Japanese-influenced era of the 1960s.[48] After an introduction to Buddhism and macrobiotic diets, the American Shurtleff travelled to Tokyo to study Japanese, where he met his partner Aoyagi. Together, they researched, wrote and published *The Book of Tofu* in 1974, and thereafter became evangelists for soy foods in America, teaching

43   N. Berenstein, 'A brief history of soy milk, the future food of yesterday', *Serious Eats*, 21 June 2018: https://www.seriouseats.com/2018/06/a-brief-history-of-soy-milk-the-future-food-of-yesterday.html.

44   M.M. Lager, *The Useful Soybean: A Plus Factor in Modern Living* (McGraw-Hill Book Company, 1945); Roth, *Magic Bean*.

45   W.J. Belasco, *Appetite for Change: How the Counterculture Took on the Food Industry*, 2nd updated ed. (Ithaca: Cornell University Press, 2007).

46   Wang et al., 'Soybeans as human food', 40; The Farm, *The Farm Vegetarian Cookbook* (Summertown, TN: The Book Publishing Co., 1975).

47   Roth, *Magic Bean*, p. 204.

48   W. Shurtleff and A. Aoyagi, 'SoyInfo Center: Soy from A historical perspective': https://www.soyinfocenter.com/ (accessed 22 October 2020).

classes, giving lectures, writing many books and inspiring legions of creators and entrepreneurs.[49] One of Shurtleff's acolytes in Mexico, Blanca Dominguez, captured the missionary zeal of the soy apostles in those heady days: 'Soya is almost like a conversion. When people adopt it, they turn inside out.'[50] Others who 'converted' opened soyfoods businesses, like Steve Demos's White Wave Foods in 1977 in Boulder, Colorado; launched 1970s soyfoods restaurants or delis, like The Soy Plant in Ann Arbor, MI; and wrote soy cookbooks like Paul and Patricia Bragg's 1975 *Hi-Protein Meatless Health Recipes*.[51]

These acolytes were driven by high ideals: soy wasn't just a food to them – it was a way of life. They were driven by the concerns about environmental degradation, global hunger and animal rights that flowered during the 1960s. The context of proliferating post-war technologies, a rapidly growing population and the rising demand for food around the world created fertile ground for these ideas.[52] These high ideals also led the soy innovators to think beyond their own consumption practices to consider how they could harness the power of soy to alleviate hunger and suffering in other parts of the world. In 1976, a huge 7.5 magnitude earthquake ravaged Guatemala – leaving 22,700 dead, another 76,000 people injured and over $1 billion worth of damage. The Farm commune in Tennessee, through their recently-established non-profit Plenty, sent volunteers to deliver medical supplies and help build homes, schools, clinics, water systems and, notably, to launch the Integrated Soy Project. In partnership with Plenty Canada and the Canadian International Development Agency (CIDA), the volunteers and local Guatemalans identified and planted soybean varieties better adapted to the tropical climate, did extension work to teach Mayan women techniques for making soymilk and tofu and built a soy dairy in 1979. Plenty were in Guatemala for five years before political danger forced them out.[53]

49   Roth, *Magic Bean*, pp. 213–22.

50   B. Dominguez de Diez Gutiérrez, 'Current work with soyfoods in Mexico. Letter to William Shurtleff at Soyfoods Center', 16 Feb. 1982. As cited in Shurtleff and Aoyagi, *Soy in Mexico*, p. 165.

51   Demos Steve, History of White Wave, Inc. Lafayette, California: Soyfoods Center. Unpublished manuscript., interview by William Shurtleff, 1987, as cited in Shurtleff and Aoyagi, *Soy in Mexico*, p. 222. Richard Leviton, 'The soy delicatessen', *Soycraft (Greenfield, Massachusetts)* 1 (1) (Summer 1979): 12–18, as cited in Shurtleff and Aoyagi, *Soy in Mexico*, p. 131; P. Bragg and P. Bragg, *Hi-Protein Meatless Health Recipes: With History and Reasons* (Desert Hot Springs, CA: Health Science, 1975).

52   R. Carson, *Silent Spring* (Houghton Mifflin, 1962); P.R. Ehrlich, *The Population Bomb* (Sierra Club/Ballantine, 1968).

53   R. Fike (ed.), *Voices from the Farm: Adventures in Community Living* (Summertown, TN: Book Publishing Co., 1998); 'Soybean project in Guatemala highlands', *Plenty News* 1 (2) (March 1979):

Anna Zeide

While the countercultural principles of The Farm drove much of this use of soy as meat substitute, their practices were in fact dependent on previous mainstream work. As historian Matthew Roth argues in *The Magic Bean*, much of the alternative soy work depended in large part on preceding projects of the very people who the counterculturists turned against: the establishment work of breeders, farmers, grain companies and other industrial leaders. State and federal agricultural experiment stations had sourced and bred the many varieties of soy that could grow in Tennessee, or in Guatemala. Major companies like Ralston Purina or ADM invested heavily in soy research and technologies. And all of these also built on the work of Asian and Asian-American soy crafters and Seventh-day Adventists.[54]

The more mainstream entities in the soy business used industrial practices to produce high-tech meat substitutes, alongside the hippies' tofu. In 1954, Robert Boyer, who had worked with Henry Ford decades earlier, patented a process of isolating and texturising soy protein, to give it a meat-like consistency. This technology was adopted by many companies to create a wide range of meat substitutes.[55] By 1965, Worthington Foods sold White Chik, Beeflike, Prosage and Holiday Roast; General Mills sold Bac*Os, meatless fried bacon bits from spun soy protein fibre. In 1974, after purchasing Worthington Foods, Miles Laboratories released the Morningstar Farms brand of meat analogues, selling ham-adjacent Breakfast Slices and sausage-like Breakfast Links and Patties.[56] The vice president of Purina in 1963 proclaimed these edible soy products to be 'the most exciting and most promising group of new foods of this decade'.[57]

But consumers were wary of these new products, stymying this hopeful prophecy. For example, Morningstar Farms was predicted to have sales of $100 million a year in 1973, but by 1977, they were earning only $15 million

1–3; Plenty International, *Soy Demonstration Program: Introducing Soyfoods to the Third World. A Step by Step Guide for Demonstrating Soymilk and Tofu Preparation* (Summertown, Tennessee, 1982); 'Citing politics, "The Farm" leaves Guatemala', *Vegetarian Times*, May 1981. As cited in Shurtleff and Aoyagi, *Soy in Mexico*.

54   Roth, *Magic Bean*, p. 205.

55   R. Boyer, High protein food product and process for its preparation, U.S. Patent 2,682,466, filed 6 May 1952, and issued 29 June 1954; G.H Kyd, 'Edible soy-protein fibers promise new family of foods', *Food Processing* **24** (5) (May 1963): 122–26.

56   W. Shurtleff and A. Aoyagi, *History of Meat Alternatives (965 CE to 2014): Extensively Annotated Bibliography and Sourcebook* (Soyinfo Center, 2014), p. 8.

57   Kyd, 'Edible soy-protein fibers', 126.

a year. By 1980, they had cut their advertising budget.[58] Other companies selling industrial meat substitutes had similar experiences. One reason for this disappointing reception, besides the increasingly cheap cost of meat due to soy livestock feed, was that some critics were sceptical of the industrial and opaque nature of this quasi-meats. One 1968 article in the *New Republic* magazine titled 'Unfoods', asked in its subtitle, 'Do you know what you're eating?'[59] These imitation meats at first had few labelling requirements, leaving consumers sceptical and wary. The highly-processed nature of these meat substitutes also meant that there were sometimes-worrying components amid the long ingredient lists. Midland Harvest Burgers produced by multinational conglomerate Archer Daniels Midland (ADM), for example, used methylcellulose – 'wood pulp processed with caustic soda and other chemicals' – as a binding ingredient for all the soy protein.[60]

But soy's fate turned once again by the 1990s and early 2000s, when it became known as a healthful food, low in fat and calories, protective against breast cancer and officially 'heart-healthy' after a 1999 US Food and Drug Administration (FDA) pronouncement.[61] New products continued to come on the market, like the popular Tofurky in 1995.[62] Small meat substitute companies were gobbled up by large corporations – Kellogg purchased Worthington Foods in 1999, Kraft purchased Boca Burger in 2001.[63] These acquisitions attested to the continued value of these products, even as mainstream American sources like *US News and World Report* undermined soy's popularity by writing headlines like, 'soy-based foods are disease fighters, but they can taste pretty weird'.[64]

58   W. Shurtleff and A. Aoyagi, *History of Modern Soy Protein Ingredients – Isolates, Concentrates, and Textured Soy Protein Products (1911–2016): Extensively Annotated Bibliography and Sourcebook* (Soyinfo Center, 2016), p. 11.

59   D. Sanford, 'Unfoods: Do you know what you're eating?', *New Republic* **158** (2) (18 May 1968): 13–15.

60   W. Shurtleff, Talk with Richard Gross, owner of Nature's Oven, Florida, concerning ADM's new Harvest Burger, 3 September 1991. As cited in Shurtleff and Aoyagi, *Soy in Mexico*, p. 255.

61   J.W. Anderson, B.M. Johnstone and M.E. Cook-Newell, 'Meta-analysis of the effects of soy protein intake on serum lipids', *The New England Journal of Medicine* **333** (5) (3 August 1995): 276–82; Department of Health and Human Services, FDA, 'Food labeling: Health claims; soy protein and coronary heart disease', Docket No. 98P–0683 § Federal Register Vol. 54, No. 206 (1999), https://www.govinfo.gov/content/pkg/FR-1999-10-26/pdf/99-27693.pdf.

62   T. Huddleston Jr., 'Tofurky's creator was living in a treehouse when he invented the tofu "bird" that's still a thanksgiving staple', *CNBC*, 18 Nov. 2019: https://www.cnbc.com/2019/11/28/tofurky-creator-lived-in-a-treehouse-before-million-dollar-idea.html.

63   Shurtleff and Aoyagi, *Meat Alternatives*, p. 9.

64   S. Schultz, 'Pass the tofu tacos: Soy-based foods are disease fighters, but they can taste pretty weird', *U.S. News and World Report*, 22 Nov. 1999.

By the second decade of the twenty-first century, soy-based meat sub-stitutes met new challenges and new opportunities. Soy came under attack from all sides: those who feared its hormonal effects, who claimed it increased cancer risks, who fought against it as a genetically modified crop, or who saw its global monocultures as the embodiment of detrimental industrial agriculture.[65] But despite these challenges, meat substitutes as a whole have achieved unprecedented investment and mainstream acceptance, as with the soy-based Impossible Burger. The founder of Impossible Foods, Patrick Brown, set out to eliminate factory farming of animals, which he considered the world's worst environmental problem. The company's Impossible Whopper is now available in Burger King restaurants throughout the United States, a level of penetration that would have been unthinkable to the early meat substitute evangelists a century ago.[66]

Soy as Meat Extender

One additional product that brought soy and meat together, if for a brief historical moment, deserves our attention for the way that it highlights the hopes that soy has held, the malleability of the bean for industrial purposes and the eventual dominance of soy as animal feed above all else.

Around the same time that soy became key to the vegetarian diets of the counterculture, in the 1960s and 1970s, Americans more broadly began to reconsider their meat consumption, not because of spiritual, ethical or environmental concerns, but due to economic considerations. This period saw a large increase in meat prices and food prices across the board. The world food crisis of the 1970s was due to a combination of factors: an energy crisis which raised costs of petroleum-based agricultural pesticides, herbicides and fertilisers; capricious weather; and food policies that limited US production and exports. American housewives – who were most often in charge of their families' food purchases – revolted against these high prices, boycotting meat purchases and complaining to officials in Washington D.C.[67]

Into this arena stepped the soy processing industry, holding up its pro-

65   Du Bois, *Story of Soy.*

66   'Impossible Foods closes $200 million in new funding to accelerate growth', *Business Wire*, 13 Aug. 2020: https://www.businesswire.com/news/home/20200813005733/en/Impossible-Foods-Closes-200-Million-in-New-Funding-to-Accelerate-Growth.

67   'Inflation: Changing farm policy to cut food prices', *Time*, 9 April 1973; A. Eckstein and D. Heien, 'The 1973 food price inflation', *American Journal of Agricultural Economics* **60** (2) (1978): 186–96; 'What happened to world food prices and why?', The State of Agricultural Commodity

ducts as the perfect solution to cheaper meat. The same breeding efforts and soy texturising techniques that had made commercial meat substitutes possible led to the creation of meat extenders. These were combined with ground meats to make a cheaper meat-soy product. In a 1979 issue of the *Processed Prepared Food* trade journal, an ADM advertisement showed a graph comparing the price for meat compared to '24% extended product' between 1967 and 1979. The extended product was consistently around forty per cent cheaper. The caption of the graph read, 'Extending meat with TVP made good sense in 1967. It's making more sense all the time.' In 1973, when meat prices were an all-time high, the list of meat-soy blends included 'Plusmeat from Central Soya … Burger Bonus and Tuna bonus from A. E. Staley … Betty Crocker's Burger Builder from General Mills; Armour-Dial's Burger Savor; Ac'cent ground beef extender from Ac'cent International and Progresso Foods' Extend'n Flavor'.[68]

The technological foundation for this flood of new products began with the same method that produced the meat substitutes of this period: Robert Boyer's process of spinning edible soy protein fibres into meat-like filaments, inspired by research he had done years earlier at Ford Motor Company. When Boyer first developed the method in 1950, the major meatpacker Swift & Co. secretly hired him to create new soy protein products. In a time when the scarcity of World War Two was still fresh on everyone's minds, Swift considered synthetic meats a promising area of development, as a hedge against reduced meat supply. But the company also told Boyer that 'if their Livestock Relations Department found out that Swift was doing research on meat analogs, "all hell would break loose"'. After leaving Swift in 1954, Boyer worked with Worthington Foods, General Mills and Ralston Purina, all of whom produced meat extenders and analogues using his method.[69] In 1970, another soy scientist, W.T. Atkinson of ADM, patented a parallel method for creating soy-based meat extenders by texturising soy flour, producing TVP®, or texturised vegetable protein. One nutritional expert said of this method: 'The ability to produce texture out of soy flour will probably rank

Markets (Food and Agriculture Organization of the United Nations, 2009): http://www.fao.org/3/i0854e/i0854e01.pdf.

68   L. Edwards, 'Soy extenders remain "food of the future" as category sales dwindle', *Advertising Age*, 29 July 1975: 3, 46.

69   R.A. Boyer, Development of meatlike products based on spun soy protein fibers. Part II, interview by William Shurtleff, 11 Oct. 1981, SoyaScan Notes. As cited in Shurtleff and Aoyagi, *Meat Alternatives*, pp. 529–30.

with the invention of bread as one of the truly great inventions of food'.[70]

Both Boyer's and Atkinson's techniques, however, were highly industrialised processes, requiring dramatic transformation in the intermediate steps between the harvested soybean in the fields and the resulting meat-like substance. For Boyer's method, the soybean was first separated into oil and meal, then the soy flakes were soaked in an alkalising vat before being pumped through a centrifuge to isolate the protein. Then, the procedure involved

> adjusting the pH to 10-11, aging the slurry at 40-50°C until the slurry becomes spinnable, and then forcing the 'spinning dope' (slurry) through a spinnerette into a coagulating bath containing acid and salts ... The tow (bundle of fibers) is placed in a second heated bath with additional stretching to reduce the diameter from 200-250μ to 75μ. Coloring, flavoring material, and a binder (usually egg albumen) are added to produce a product of the desired texture, color, and flavor.[71]

Atkinson's method, while simpler, involved moistening soybean flour 'into a "plasticized" mass, bringing it to a high temperature and rapidly forcing it through perforated dies into a chamber of lower temperature and pressure'.[72] These were not the sort of operations individuals would perform in their home kitchens.

While these soy-extended meat blends made some headway among general consumers, one of the biggest markets became the US school lunch programme with the 1971 passage of Food and Nutrition Service (FNS) Notice 219, which allowed up to thirty per cent extenders. One commentator referred to this notice as 'the Magna Carta of textured vegetable protein'.[73] This was a big coup for the soy products industry, which had been pushing for this allowance for years. By 1973, with meat prices climbing, US schools were using up to 40 million pounds of TVP, in foods like 'chili mix, meat loaf or meatballs, patty mix, pizza sauce, sloppy joe, spaghetti sauce, or taco filling'.[74] This widespread market exposed the majority of American children to soy-extended meats, whether they knew it or not.

Another set of concerns, beyond high meat prices, also brought global attention to soy proteins in this period. In the mid-1960s, health officials

70   D.S. Greenberg, 'Slaughterhouse zero: How soybean sellers plan to take the animal out of meat', *Harper's*, Nov. 1973: 40.

71   A.M. Pearson, 'Meat extenders and substitutes', *BioScience* **26** (4) (April 1976): 253.

72   Greenberg, 'Slaughterhouse zero', 39.

73   Ibid., p. 42.

74   USDA Farmer Cooperative Service, 'Appendix – Companies Producing and Distributing Soy Products. Edible Soy Protein: Operational Aspects of Producing and Marketing', pp. 53–82, Jan. 1976.

10. Soy and Meat: A Postwar North American History

around the world identified a so-called 'protein crisis', pointing to protein malnutrition as 'the world's most widespread deficiency disease'.[75] In 1968, after smaller-scale attention to protein malnutrition, a United Nations Advisory Committee published a report entitled 'Feeding the Expanding World Population: International Action to Avert the Impending Protein Crisis'. The report offered one especially promising solution to this crisis: the soybean.[76] Numerous World Soy Protein conferences followed, often with more than 1,000 attendees each, all devoted to soy as an answer to the protein crisis: in Germany in 1973, Singapore and the Netherlands in 1978 and Mexico in 1980.[77] Despite some scientific controversy about whether protein was in fact the central concern in global malnutrition – nutritionist Donald S. McLaren published 'The great protein fiasco: Dogma disputed', in 1974 – soy companies in the United States found ways to turn this worldwide concern to their advantage in selling meat extenders.[78]

Mexico, with its proximity to soybean farms in Texas and throughout the US South, was a prime target. In 1973, the industry's trade organisation, The American Soybean Association, sent Gil Harrison to serve as its Mexico programme director. The soy industry collaborated with the Mexican government to sell a new soy-meat product called PROTEIDA, with 25 per cent texturised soy protein and seventy per cent ground meat, produced in Mexican government packinghouses.[79] This created a major market for the US soy industry, with Mexico 'purchasing 18 ½ million bushels of U.S. soybeans in 1973 up from zero 3 years ago'.[80] Meat extenders thus helped to extend the US soy industry in the 1970s.

But by the 1980s, meat extenders had failed to fulfil the high hopes of their early days. As meat prices began to fall, and concerns about malnutrition turned away from protein, soy extenders were largely relegated to pet foods, rather than human foods.[81] Another significant impediment, in

---

75   Shurtleff and Aoyagi, *Soy Protein Ingredients*, p. 8.

76   Advisory Committee on the Application of Science and Technology to Development, *Feeding the Expanding World Population: International Action to Avert the Impending Protein Crisis: Report to the Economic and Social Council* (United Nations, 1968).

77   Shurtleff and Aoyagi, *Soy Protein Ingredients*, pp. 10–11.

78   D.S. McLaren, 'Dogma disputed: The great protein fiasco', *The Lancet* **304** (13 July 1974): 93–96.

79   Wang et al., 'Soybeans as human food', 36.

80   'Soy-enriched meat products in Mexico', *Soybean Digest*, March 1974: 34.

81   Edwards, 'Soy extenders remain "food of the future"'; Warren James Belasco, *Meals to Come: A History of the Future of Food* (Berkeley: University of California Press, 2006).

both the US and Mexico, was labelling requirements. In California, the 1974 Brigg's Amendment ruled that any hamburger patties that used soy extenders had to be labelled 'imitation hamburger'.[82] Similarly, in Mexico, health authorities disallowed soy products from being labelled as meat, in fear that this would 'open the gates to unscrupulous people who would use large amounts of low-cost soy to extend expensive meats, but mislabel and misadvertise the products'.[83]

By 1980, Robert Boyer, the original innovator of meat analogs, remarked: 'We're at the Model T stage right now with analogs. I'm impatient to get to the Lincoln Continental stage'.[84] He hoped that rising meat prices would drive innovation in soy extender development and regulation to help the product achieve its potential. But the optimism about soy as a meat extender, as a food of the future, did not hold. As we have seen, those rising meat prices did not materialise exactly because soy was used as cheap livestock feed, yielding cheap meat. Instead of being transformed into a meat substitute or meat extender, soy was transformed into meat itself. This negated the *economic* need for soy-extended meat blends, if not the environmental need.

Conclusion

Throughout the last century, soy has experienced a meteoric rise in North America and around the world. Especially in the years after World War Two, soy industry promoters saw the bean as a 'Cinderella crop' whose time had finally come to transform from mistreated housemaid to the belle of the ball.[85] While soy has indeed emerged as an influential and widely-used crop, its true nature has still largely remained in the shadows. So too have the incredibly powerful multinational corporations and trade organisations that control soy globally. The United States Soybean Export Council has offices in more than seventy nations, promoting US soy for animal feed. And the industry is controlled by just four massive companies: Archer Daniels

---

82   C. Boismenue, The market for soy protein isolates, concentrates, textured soy protein products, and soy flour in America today, interview by William Shurtleff, 13 Nov. 1990. As cited in Shurtleff and Aoyagi, *Soy in Mexico*, p. 258.

83   G.R. Harrison and W. Shurtleff, Why haven't soyfoods caught on Latin America?, 17 April 1989, SoyaScan Notes. As cited in Shurtleff and Aoyagi, *Soy in Mexico*, p. 241.

84   Boyer, Development of meatlike products. As cited in Shurtleff and Aoyagi, *Meat Alternatives*, p. 530.

85   B. Ford, *Future Food: Alternate Protein for the Year 2000* (New York: William Morrow and Company, 1978).

Midland (ADM), Cargill, Bunge and Louis Dreyfus Company. These are less familiar names to many, just as soy's global dominance and foundation as livestock feed continues to be largely unknown. They operate behind the scenes, in ways that are not transparent to consumers.

However, soy has come more into the limelight in recent years – with historian Matthew Roth calling 1999 the year of 'peak soy'.[86] Around the turn of the twenty-first century, soy garnered more attention as a human food, whether used in veggie burgers, as tofu or tempeh in increasingly common ethnic foods, in the popular soy milk, or as edamame. But all of these uses were marginal, relative to the quantities of soybean meal and oil directed to livestock feed and processed food ingredients. And in the years since, soy has come under attack due to a wide range of concerns.[87] Soy has been genetically modified to be resistant to the Monsanto Company's star herbicide, Roundup, and has thus been denounced by opponents of GMOs. Soy oil has been blamed for the heart-unhealthy trans fats that it carries when it is partially hydrogenated. Soy has become a punching bag of the alt-right and other parties as they point to its estrogenic effects, which they claim are weakening American men. And soy is rightly condemned for being a resource-intensive crop that drains aquifers in its thirst and encourages deforestation in its hunger for cropland.[88]

Although this transformation of soy has had many ramifications, the most destructive to our natural environment has been soy's role in helping to create and perpetuate a meat-centred global diet, leading to skyrocketing pollution and greenhouse gas emissions in the last half-century. Despite many consumers' image of soy as a meat substitute, it has in fact made possible the low meat prices that prevent the demand and need for meat substitutes and meat extenders.

But ultimately, all these accusations are condemnations not of soy itself so much as of the industrial processes that soy has come to represent and underlie. Its malleability has turned soy into a shapeshifter that permeates much of the twentieth and twenty-first century American food system. Wherever you see industrial food and its environmental effects, soy is there.

---

86   Roth, *Magic Bean*, p. 230.

87   Ibid.

88   Roth, *Magic Bean*, Epilogue; Du Bois, *Story of Soy*.

Anna Zeide

Acknowledgements

The author would like to thank the Virginia Tech Open Access Subvention Fund for supporting the open access publication of this chapter.

CHAPTER 11.

IMAGINING SOY IN GERMANY – CHANGING SCIENTIFIC VISIONS IN THE TWENTIETH CENTURY

Janina Priebe

Introduction

From the turn of the twentieth century, scientists in botany, agriculture and nutrition, as well as economists, saw soy as a 'crop with a promise for the future'.[1] In the wider European perception, in contrast, soy (*Glycine max*) remained largely invisible for most of the century due to its versatility and widespread but hidden applications, for instance as feed grain in industrial livestock production, and its unrecognisable uses in innumerable food and non-food products.[2] Today, too, the central role of soy in European lifestyles remains largely hidden, despite the growing popularity of soy-based foodstuffs. Around ninety per cent of the yearly total soy import to Europe (European Union, Norway and Switzerland), approximately 34.4 million metric tons, is used as feed in intensive animal farming for meat and dairy production.[3]

The remoteness of soy from the perception of the European public gave scientists a key role in producing and mediating knowledge about soy throughout the twentieth century. From 1900, scientists became mediators of knowledge about the environment and natural resources, and they acquired

---

1   I. Prodöhl, "'A miracle bean": How soy conquered the West, 1909–1950', *Bulletin of the GHI Washington* **46** (2010): 111–29.

2   I. Prodöhl 'Versatile and cheap: A global history of soy in the first half of the twentieth century', *Journal of Global History* **8** (2013): 461–82.

3   European Soy Monitor IDH and IUCN, 'European Soy Monitor. Insights on the European Supply Chain and the Use of Responsible and Deforestation-Free Soy in 2017', (IDH (The Sustainable Trade Initiative) and IUCN (International Union for the Conservation of Nature) Netherlands, 2019).

THE AGE OF THE SOYBEAN: 227–246 **doi:** 10.3197/63800040695086.ch11

increasing influence as advisors to those in power.[4] In these positions, scientists were able to recognise and formulate the link between soy and larger visions of societal progress. In twentieth-century Europe, soy was a crucial ingredient for scientists to reimagine the connections between society and nature on the highly industrialised continent.

As the scientific community engaged in more serious attempts to assess the uses and potential of cultivating soy in the central European countries from the early 1900s, science has incorporated the newly gained knowledge about soy, either willingly or unintentionally, into larger visions of power. These visions, their presumptions and objectives, however, changed in different political, cultural, economic and environmental contexts.

These scientific visions created sociotechnical imaginaries of soy throughout the twentieth century. In science and technology studies (STS), the concept of sociotechnical imaginaries seeks to close the gap between scholarly analyses of scientific representations and the influence of science on and its alterations by political power. In this conception, sociotechnical imaginaries are the 'collectively imagined forms of social life and social order reflected in the design and fulfillment of nation-specific scientific and/or technological projects'.[5] According to STS scholar Sheila Jasanoff, multiple, at times conflicting, imaginaries can coexist within a society. Imaginaries are not limited by the nation state but can originate in the visions of individuals, articulated and propagated by organised groups, and are elevated through institutions of power (e.g. legislature, media) into commonly shared notions about the world.[6]

This chapter traces the scientific visions of soy projected by science in the twentieth century, using Germany as a case study to exemplify the ideas that surrounded the import and cultivation of soy in Europe. A central question to be followed in this chapter is: how and why did scientists incorporate soy into larger visions of power? In particular, I focus on visions that got elevated into imaginaries, which manifested in social and environmental changes

4   E. S. Benson. *Surroundings. A History of Environments and Environmentalisms* (Chicago: University of Chicago Press, 2020).

5   S. Jasanoff and S-H. Kim, 'Sociotechnical imaginaries and national energy policies', *Science as Culture* **22** (2013): 189–96.

6   S. Jasanoff and S.-H. Kim, 'Future imperfect: Science, technology, and the imaginations of modernity', *Dreamscapes of Modernity. Sociotechnical Imaginaries and the Fabrication of Power* (Chicago /London: The University of Chicago Press, 2015), pp. 1–33.

that turned soy into a terraforming technology with a global impact.[7] These imaginaries provide an entry point to understanding the role of soy in first preparing for and then accelerating social, economic and environmental developments from the 1950s onwards.[8]

Since the 1900s, Germany was, with interruptions due to the world wars, Europe's largest importer of soy. At the turn of the twentieth century, the uses of soy were a limited but growing field of interest in science and technology in Germany. A major strand of discussions about soy took place in the scientific journal *European Journal of Lipid Science and Technology*, published since 2000 by the European Federation for the Science and Technology of Lipids (Euro Fed Lipid), a federation of scientific associations representing individual scientists and companies, and in the journal's German-language predecessors[9] since the turn to the twentieth century. In this chapter, I use key publications in these journals as the lens through which I identify and contextualise the scientific visions of soy in twentieth-century Germany.

Since their first publication in 1899, these journals have provided a forum for discussing scientific findings about and industrial applications of vegetable and animal oils and fats, often addressing major developments in European research. The developments this chapter addresses are mainly confined to the central and continental European countries, with particularities when it comes to the uses, consumption and perception of soy. To some extent, however, the imaginaries of soy projected by science in twentieth-century Europe have a common ground, despite the various and complex national histories involved. The overarching feature is Europe as a place of soy consumption and limited cultivation. Soy made its entrance to Europe solely as soyfood, initially available only to elite circles of society, mainly in Britain and France, from the seventeenth century. Seasoning with soy sauce, a product of fermented soybean paste, spread widely in the eighteenth century in continental Europe. European travellers had knowledge about the soybean at least from the sixteenth century but, besides experimental cultivation in the late nineteenth century, mainly in Austria, the soy plant remained a

7   S. Sörlin and N. Wormbs, 'Environing technologies: A theory of making environment', *History and Technology* **34** (2018): 101–25.

8   J.R. McNeill and P. Engelke. *The Great Acceleration. An Environmental History of the Anthropocene since 1945* (Cambridge: Harvard University Press, 2014).

9   *Chemische Revue über die Fett- und Harz-Industrie* (1894–1915); *Chemische Umschau auf dem Gebiet der Fette, Oele, Wachse und Harze* (1916–1932); *Fettchemische Umschau* (1933–1935); *Fette und Seifen* (1936–1952); *Fette, Seifen, Anstrichmittel* (1953–1986); *Lipid/Fett* (bilingual) (1987–1999).

curiosity in botanical gardens.[10] While the early years of the 1900s marked a turning point, the end of the Second World War initiated profound changes in scientific and technological visions of soy, introducing the age of soy in Germany. In the second half of the century, the cultivation of soy in Germany was elevated into a vision that offered sustainable ways to escape the degrading environmental impacts of soy cultivation elsewhere, and not least to – once again – secure access to protein. Ever since, the diversity of uses and the dimensions of how soy cultivation and consumption impacts bodies and landscapes globally have increased at an unprecedented pace.

The Turning Point

At the beginning of the 1900s, conjunctions of global environmental and societal developments, and imperialist aspirations and nationalist objectives initiated the incorporation of soy into scientific visions. The intersection of these developments culminated in the years 1908 and 1909. A failed linseed harvest in Argentina and failed harvests of cottonseeds in the United States created a severe shortage of vegetable fats and oils in Europe and North America, where there was a growing demand for seed oils used in applications as various as soap, margarine and dyes, and in numerous industrial purposes.[11]

In 1908, the first large shipments of soybeans to Europe were made from Northeast China, a region comprising the three north-eastern provinces referred to as Manchuria.[12] European traders benefitted from a unique political situation in a specific environmental context, and they linked Asian environments more closely to European consumption, establishing soybean processing plants at large harbours along the continent's coasts.[13] At the same

10   D. Wolffhardt, 'Anbauversuche Mit Sojabohnen in Österreich', *Fette, Seifen, Anstrichmittel* **84** (1982): 92–101; Prodöhl, '"A miracle bean"', 111. See also the extensively annotated bibliography by W. Shurtleff and A. Aoyagi, *History of Soybeans and Soyfoods, 1100 B.C. To the 1980s* (Lafayette, USA: Soyinfo Center, 2007).

11   D. Wolff, 'Bean there: Toward a soy-based history of Northeast Asia', *South Atlantic Quarterly* **99** (2001): 241–52; N. Koning. *The Failure of Agrarian Capitalism. Agrarian Politics in the Uk, Germany, the Netherlands and the USA, 1846–1919* (London, New York: Routledge, 2001).

12   The term Manchuria is problematic in contemporary writing because the term associates the geographical region with Japanese imperialism. 'Manchuria' is, however, commonly used to denote this region for simplicity in historiographical accounts dealing with this time period. H. Mizuno and I. Prodöhl, 'Mitsui Bussan and the Manchurian soybean trade: Geopolitics and economic strategies in China's Northeast, Ca. 1870s–1920s', *Business History* (2019): 1–22.

13   J. Priebe, 'From Siam to Greenland: Danish economic imperialism at the turn to the twentieth century', *Journal of World History* **27** (2016): 619–30; D. Ben-Canaan, F. Grüner and I. Prodöhl, 'Entangled histories: The transcultural past of Northeast China', in D. Ben-Canaan, F. Grüner

11. Imagining Soy in Germany

time, Manchuria's environment changed profoundly and rapidly. Against the backdrop of Russian and Japanese tensions, Chinese infrastructural investments and immigration increased the region's agricultural production and population massively.[14] Manchuria turned into the world's largest global exporter of soy to meet Europe's increasing demand for vegetable oil.

In 1913, amid growing tensions and arms build-up on the European continent, the journal *Chemische Revue über die Fett- und Harz-Industrie* published the protocol of a board meeting of the German Colonial Economic Committee [Vorstand des Kolonial-Wirtschaftlichen Komitees]. The necessity of national colonial production of soybeans was a significant point of discussion during the meeting. As the committee noted, the domestic demand had increased rapidly, mainly because of the growing wealth of the lower segments of society and the resulting change to consumption patterns, which aggravated German dependence on imports.[15] In contrast to other countries, Germany did not receive oil seeds from its colonies and was thus Europe's largest importer of oilseeds by 1914. Soybeans from Manchuria[16] were processed at plants founded within only a few years, mainly in northern harbour cities, such as Hamburg and Stettin (today Szczecin in Poland), encouraged by the removal of import duty on soybeans in 1910.[17] The dependence on oilseed imports, and the political dependence it entailed, created an uneasiness that would shape scientific endeavours in the cultivation and uses of soybeans for the next decades. The imaginary of soy as a resource for a modern lifestyle and its uses in a wide range of industrial applications, enabled through global trade, was challenged because of the national dependence it created.

and I. Prodöhl (eds), *Entangled Histories. Transcultural Research - Heidelberg Studies on Asia and Europe in a Global Context* (Cham: Springer, 2013), pp. 1–11.

14   Wolff, 'Bean there'.

15   *Chemische Revue*, 'Oelrohstoffversorgung Deutschlands' (Germany's supply of crude oil), 1913.

16   Soybeans were also referred to as 'Chinese oilbeans' or 'oilnuts' in contemporary German scientific texts; C. Grimme, 'Oelnüsse aus Singapur', *Chemische Revue über die Fett- und Harz-Industrie* **18** (6) (1911): 125–26.

17   J. Drews, *Die "Nazi-Bohne". Anbau, Verwendung Und Auswirkung Der Sojabohne Im Deutschen Reich Und Südosteuropa (1933–1945)* (Münster: LIT Verlag, 2004); Shurtleff and Aoyagi, *History of Soybeans and Soyfood*.

Janina Priebe

Soy as a Means to Reduce Shortages and Dependence

Soy-based food, such as soy flour and soy milk, had just entered the German market before the First World War when several experiments and patents involving soy derivates came to a halt. In the course of the war, the massive inflation of the German currency, famines and hunger revolts renewed and intensified the search for sources of nutrition for the population and military. Physicians propagated the nutritional value and protein content of the soybean. During and after the war, and likely due to poor food processing, however, many regarded soy products as a low-quality substitute for (or stretching of) meat or coffee, not least because the methods of the proper preparation soybeans and meal were poorly communicated.[18] In many ways, the use of soy emphasised the notion of shortage in the public perception, despite scientists promoting soy as a miracle cure with health and nutritional values.

After the end of the First World War, food shortages and hyperinflation once again made topical the question of dependence on imports of agricultural products in the defeated Central Powers, Germany, Austria-Hungary, the Ottoman Empire and Bulgaria. In 1919, a front-page article of the *Chemische Umschau* recapitulated the scientific debate about soy cultivation in Germany. There had been a fierce debate between individual scientists during the war about whether soy plants could be successfully cultivated on a larger scale in Germany, and how the plant could be 'acclimatised'.[19] A number of scientists were eager to point out that the cultivation of soybeans in Germany was not limited by the climate but by the lack of familiarity with the plant in the wider public and among farmers, as well as by their unwillingness to adapt agricultural practices. They drew parallels, for instance, to the cultivation of maize in Germany. Although maize, too, required sunny and warm conditions, cultivation had proved successful after efforts were taken to adapt the plants and cultivation practices to the German conditions.[20]

The scientific debate about soy focused on agricultural practices, nutritional values and industrial applications. The debate unfolding in the scientific journal cited above, however, also positioned soy in a larger context of the national economy, public health and the wealth of the nation. The idea

---

18   Drews, *Die "Nazibohne"*.

19   *Chemische Umschau*, 'Zur Frage des Anbaues und der Akklimatisation der Soja in Deutschland', 1919.

20   B. Heinze, 'Ueber Den Anbau Der Chinesischen Oelbohne (Soja Hispida) in Unserem Eigenen Lande Und Deren Bedeutung Für Unsere Land- Und Volkswirtschaft Und Für Die Volksgesundheit', *Chemische Umschau auf dem Gebiet der Fette, Oele, Wachse, und Harze* **29** (1922): 361–63.

of cultivating soy within German borders (or, before the First World War, the consideration of cultivating it in German dependencies) was gradually incorporated into scientific imaginaries of decreasing the perceived harmful dependence on imports – an imaginary that was eventually elevated into political objective.

Soy Production as a Matter of Nationalist Interest

Internationally, the 1920s and 1930s were marked by optimism about unlocking new resources through advancing agroindustry. A prominent example with soy as a central ingredient is Henry Ford's development of a car built with various components of soymeal-based plastics and soy-based enamel in the late 1930s.[21] In Germany, meanwhile, the potentiality of soy and the solutions it could hold shaped the main concerns of the National Socialist regime from 1933. The novelty and imagined possibilities that lay in soy as a new versatile resource were a key driver, as in the wider developments of innovation and technology that combined industrial and agricultural spheres during these decades. The acceptance of soy-based foodstuffs and supplements by the wider German public, however, lagged behind the hopes set on soy as a scientifically conceived solution to urgent concerns about dependence on imported oils and fats, and as a solution to substituting scarce protein sources in the diet of workers and soldiers.

In 1933, between fifty and sixty per cent of the total amount of oils and fats used in Germany was imported, and Germany was the world's largest soybean importer. Of the total protein, between twenty and thirty per cent came from imports. One-third of the imported protein was processed in foodstuffs, two-thirds in animal feed.[22] Of all major oilseeds, soybeans had taken centre stage only over the last couple of years. While soybeans only made up a minor proportion of imported oilseeds in 1913, they made up half of all oilseed imports in 1933 (the other half being linseed).[23] The dependence on imports was recognised as a potential threat. Autarchy and independence from international entanglements with global trade and foreign powers was

---

21   *The Times*, 12 June 1937; Prodöhl, 'Versatile and cheap', 477.

22   Wirtschaft und Statistik. Hrsg vom Statistischen Reichsamt, Berlin W 15 Kurfürstendamm 193/94, 13. Jahrgang, nr. 24. 1933; B. Pelzer-Reith and R. Reith, 'Fischkonsum Und „Eiweißlücke" Im Nationalsozialismus', *VSWG: Vierteljahrschrift für Sozial- und Wirtschaftsgeschichte* **96** (2009): 4–26.

23   A. Marcus, 'Bedeutung Der Alten Deutschen Kolonien Als Rohstoffquelle Für Die Deutsche Fettversorgung', *Fettchemische Umschau* **41** (1934): 123–27.

the declared goal of the Nazi regime. Promoting the demand of consumers for domestic produce was a crucial step, giving German agriculture a key position to fulfil the regime's objectives.[24] An obvious step was to increase the share of protein in foodstuffs and not use it up in animal feed. Another strategy was to increase internal production.

Similar considerations circulated in other in other European countries. In the deteriorating transnational and global regime under the British aegis in the 1930s and 1940s, efforts in several European countries were directed toward establishing their own soy cultivation on a more substantial basis. The scientific engagement with breeding and the crossing of cultivars for northern climatic and planting conditions got more determined. After several attempts and breeding experiments, the first successful soy crop was raised in England in 1933; adapted cultivars were introduced in Sweden in the 1940s; and soy plantation was made mandatory in regions considered suitable within the German Reich.[25]

In the German scientific community, however, it was by then widely established that the extensive demand for oilseeds for both nutritional and industrial purposes could not be covered by cultivation inside German borders. Other ways were sought to circumvent the unfavourable climatic conditions for the cultivation in Germany and still build up reliable access on a large scale. IG Farben, a German chemical conglomerate and major contractor of the Nazi government from 1933, took on a key role. Set up under the control of IG Farben but officially independent, a newly established corporation, Soja AG, led soy cultivation in Romania. Through agreements with the Romanian government, exclusive rights to cultivation as well as the exclusive export of soybeans to Germany were secured via the Soja AG, but effectively led by IG Farben.[26] A similar arrangement was pursued and realised in Bulgaria. With these arrangements, the Nazi regime effectively controlled the cultivation of soybeans in South-Eastern European regions close to the German Reich.[27]

Romania became the centre of European soybean cultivation, together

24   Pelzer-Reith and Reith, 'Fischkonsum Und "Eiweißlücke"', 6.

25   F. Fogelberg and J. Recknagel, 'Developing soy production in Central and Northern Europe', in D. Murphy-Bokern, F.L. Stoddard and C.A. Watson (eds), *Legumes in Cropping Systems* (CABI, 2017), pp. 109–24.

26   Drews, *Die "Nazibohne"*, p. 246.

27   C. Freytag, *Deutschlands 'Drang Nach Südosten'. Der Mitteleuropäische Wirtschaftstag Und Der 'Ergänzungsraum Südosteuropa' 1931–1945* (Göttingen: Vienna University Press, 2012).

with Bulgaria, Hungary and Ukraine. As late as August 1939, a German trade deal was also intended to secure the import of 167 000 mt soybeans from the Soviet Union over the next two years in exchange for industrial machines.[28] Overall, the cultivation efforts and new transnational trade connections initiated in several European countries were part of the overarching aim to establish a food regime that was independent of the global regime under British control.[29]

In Nazi Germany, soy-based food and meat extensions were primarily used in collective kitchens, such as in factories, and the Wehrmacht, the German armed forces between 1935 and 1945. The regime's struggle to improve acceptance of the food offered in factory canteens mirrored the wider lack of popularity of soy-based foodstuffs. The goal of this effort was, however, to create maximised physical productivity with the lowest possible nutritional intake. This goal had been explored with both scientific zeal and excruciating brutality in the prison camps.[30] As a result, international observers attributed the good physical condition of the German army to widely-used soy-based food in army canteens and field kitchens.[31]

As was noted by Warren Goss, an observer at a Hamburg oilseed processing plant destroyed in 1944, the European soybeans processed there were almost exclusively used for soy based flour sold to the army. The soy oil was, during the war, the major source of fat in low-quality margarine, the main source of fat for the population.[32] Only about twenty years later, however, margarine made the transition from a cheap wartime substitute for butter to a valued product of advanced technology in processing vegetable oils, a remarkable development that was discussed in the scientific community.[33] The lack of popularity and dislike of soy-based foodstuffs shifted during the first years of the post-war period toward a more positive popular perception. The key was in refined processing technologies and advancing science.

28   C.M. Du Bois, *The Story of Soy* (London: Reaktion Books, 2018).

29   E. Langthaler, 'Gemüse Oder Ölfrucht? Die Weltkarriere Der Sojabohne Im 20. Jahrhundert', *Umkämpftes Essen. Produktion, Handel und Konsum von Lebensmitteln in globalen Kontexten* (2014): 41–66.

30   A.A. Weinreb, 'Matters of taste: The politics of food and hunger in divided Germany 1945–1971' (MA diss., University of Michigan, 2009), p. 205.

31   Drews, *Die "Nazibohne"*, pp. 168–69.

32   W.H. Goss, 'Processing oilseeds and oils in Germany', *Oil & Soap* (Aug. 1946): 241–44.

33   K.-F. Gander, '100 Jahre Margarine - Vom Ersatzprodukt Zum Grundnahrungsmittel', *Fette, Seifen, Anstrichmittel* **72** (1970): 97–103.

Warren H. Goss' investigation of German oilseed industries after the defeat of Nazi Germany was of high interest to the US. It highlights the central role technology and science played in the uses and perceptions of soy in European and American industrialised societies. Goss' work was sponsored by the *Subcommittee of Food and Agriculture of the Technical Industrial Intelligence Committee*.[34] Between 1945 and 1947, the committee organised the securing of German science and technology for improving industrial production.[35] The visions and ideas of the mid-twentieth century paved the way for fundamental changes in European diets and industries that would both propel and be driven by the extensive and varied uses of soy. In research and innovation in Europe and in the US (which were closely tied with their trade relations), soy came to be seen as having great potential as a raw material for industrial production, because it was a cheaper alternative than synthetics.

The Take-off Face for Soy in Europe

The early twentieth century marked a turning point for soy in Germany, but it was only the prelude to the role soy would attain over the century. The central position of soy since the 1950s was a harbinger, and a result of the large-scale impacts industrial societies had on global environments during the Great Acceleration. While developments before the Second World War stood as a turning point, the end of the war marked the take-off phase. The global entanglements of soybean cultivation and processing, the connections of trade with beans, oil, meal and cake, and their various linkages with soy uses in food, feed and industrial manufacturing got ever more complex.[36] Most important, however, was the shift toward soymeal and cake as the main feed in European industrial livestock farming in the second half of the twentieth century.

The profound changes that characterised the decades since the 1950s manifested on several levels, from individual diets to land-use demands in the new centres of soy production in the United States and South American countries. Between 1950 and 2009, the consumption of meat in Germany

---

34   Goss, 'Processing oilseeds and oils in Germany', 241.

35   For the history of soybeans in the US, see Matthew D. Roth, 'Magic Bean: The Quests That Brought Soy into American Farming, Diet and Culture' (Diss., State University of New Jersey, 2013).

36   J.E. McHale, 'The economics of processing soybeans', *Journal of the American Oil Chemists' Society* **36** (1959): a6–a12.

almost doubled, while the consumption of legumes, a primary direct source of protein, decreased over the same time period.[37] The severity of these changes becomes visible in the concept of 'ghost acreage', coined by the Swedish scientist Georg Borgström in the 1950s.[38] The concept of ghost acreage allows calculation of the land area currently used elsewhere – i.e. outside the state's borders – that contributes to the state's consumption. Historically, centres of consumption, for instance in Europe, relied on production in regions considered to be peripheries – in rural areas of the country, in colonies and annexed lands, as well as through trade relations to ensure the import of raw materials from across the world. Borgström pointed out, however, that the conception of future food shortages in the modern globalised society was particularly tied to anticipated shortage of protein in human nutrition. 'Ghost acreage' thus illustrated a science-based notion of the globally connected fate of societies and environments and the urgency of keeping the use of resources within the limits of the planet.

The relevance of this idea is tangible in the central role of soy in enabling the 'meatification' and 'oilification'[39] of European diets after the Second World War, which was the most significant transition related to the role of soy in Europe since the mid-twentieth century. These trends in European diets went hand in hand with the expansion of ghost acreage. The direct and indirect consumption of soy increased, and the impact on land that was used and altered elsewhere increased too. The growing neo-Malthusian fear of future global malnutrition articulated by concerned scientists such as Borgström stood in stark contrast to contemporary excesses when it comes to the European consumption of meat and fats (as compared to the first half of the century). The perceived problem of future malnutrition was growing in its severity because food security seemed to be relying on a hidden, global web created through versatile resources such as soy, whose nutritional value was refined (e.g. into meat) but also diminished (e.g. meat from soy-meal fed animals compared to soy-based foodstuffs).

Although there was a slight increase in the global export of almost all

---

37   T. Dräger de Teran, 'Gut Für Uns, Gut Für Den Planeten: Gesunde Ernährung Und Eine Geringe Lebensmittelverschwendung Können Unseren Ökologischen Fußabdruck in Erheblichem Ausmaß Reduzieren', *Journal für Generationengerechtigkeit* **13** (2013): 11–17, at [11.]

38   The term 'ghost acreage' was first used in the Swedish original of Boström's book *The Hungry Planet: The Modern World at the Edge of Famine* (New York: Macmillan, 1962), which first appeared in 1953.

39   Ernst Langthaler, 'The soy paradox: The Western nutrition transition revisited, 1950–2010', *Global Environment* **11** (2018): 79–104, at 81.

major oilseeds (except cotton, sesame and palm oil, which all decreased) between 1934 and 1958, analyses pointed to soy as the major game-changer in the global oilseed economy in the 1950s. As a scientist from the Max-Planck Institute for Plant Breeding Research in West Germany, Dr Rudorf, highlighted in a presentation in 1959, the global export of soybean oil almost doubled between 1934 and 1956, from 432,000 to 736,000 tons, and reached 1,024,000 tons in 1958.[40] He attributed this growth to the relation between living standards and the consumption of soy oil, which both increased in industrialised and developing countries. The massive increase in use of protein-rich soy and the central role of vegetable oils and fats in food and hygiene products in industrialised societies was a continued development throughout the 1960s and 1970s, and it was a major issue of interest in the scientific community involved in research about oils and fats.[41] Researchers recognised the potential of science and technology surrounding soy uses for the improvement of living standards, quality of life and life expectancy.[42] This insight gave rise to the sociotechnical imaginary of soy as the means to create unprecedented wealth after wartime austerity. In this way, soy became crucial in the years referred to as the 'economic miracle', from the 1950s to the early 1970s, in the Federal Republic of Germany. These years were also witness to the emergence of the German environmental movement, which is here understood as a wide range of perspectives both critiquing the growing wealth and consumption and claiming that this wealth should be used to improve environmental quality.[43] Once again, the scientific visions of soy took on a determining role for broader socio-economic and environmental developments, and the resulting counter-developments and movements. In the following, the role of these visions will be traced with a particular focus on the West German developments.

After the Second World War, the US became the world's leading producer of soybeans, and its massive increase in production, achieved through state subsidies, mechanisation and herbicide use in a modernised agricultural industry, was combined with intensive market development in Europe. In the

---

40   W. Rudorf, 'Weltwirtschaftliche Veränderungen Im Anbau Wichtiger Ölpflanzen', *Fette, Seifen, Anstrichmittel* **62** (1960): 477-83.

41   K.W. Fangauf, 'Sojabohnen Und Sojaeiweiss in Der Menschlichen Ernährung', *Fette, Seifen, Anstrichmittel* **71** (1969): 454–58.

42   B.W. Werdelmann, 'Fettchemie Als Aufgabe', *Fette, Seifen, Anstrichmittel* **76** (1974): 1–8.

43   F. Uekötter. *The Greenest Nation? A New History of German Environmentalism*, History for a Sustainable Future Series (Cambridge, Mass.: The MIT Press, 2014).

US, the production of soybeans increased 313-fold between 1924 and 1973. The cultivated area increased 36-fold for soy, whereas the area used for grains like wheat and oat decreased. The yield by hectare increased 8.7-fold.[44] In West Germany, this development came to shape both industries and diets. The Soybean Council of America, a trade association, began operations in West Germany in 1958, with the purpose of promoting and extending the uses of US soy in foreign markets. Within the General Agreement on Tariffs and Trade (GATT), US exports of feed crops to the European Economic Community (EEC) were exempted from import duties, as agreed in negotiations between 1959 and 1962.[45] The excessive supply, rather than the demand, established soybean meal and cake, initially only considered by-products of oil pressing, as major components of animal feed in Europe's industrial livestock farming.[46] Methods of processing soy in a way that made it suitable as animal feed were discovered in the 1910s.[47] Only in the second half of the century, however, would these scientific advancements intersect with other developments that allowed their global impact to unfold. The developments during these years completed the transition from soy as an Asian 'food crop into a global cash crop'[48] and led to wide-ranging implications for the societies and environments in both the Global South and North.

Dreams and Nightmares of Science

The scientifically conceived visions of soy gradually shifted to enhancing the organism of soy itself. Soy attained a pivotal role in realising dreams but also nightmares of biotechnology in the second half of the twentieth century. In the 1960s and 1970s, the science and technology of soy was gradually absorbed by a global discourse of malnutrition and science-led solutions. Until the mid-1970s, protein malnutrition and the 'protein gap' in developing countries were a top priority of the United Nations' global

---

44   E. Langthaler, 'Ausweitung Und Vertiefung. Sojaexpansionen Als Regionale Schauplätze Der Globalisierung', *Austrian Journal of Historical Studies* **30** (2019): 115–47, at 129.

45   Animal feed from the US that contained soy was exempted from import duties in Western European countries from 1947; ibid. For details on GATT, see Sara Dillon. *International Trade and Economic Law and the European Union*: Bloomsbury, 2002).

46   Langthaler, 'The soy paradox', 101–03

47   G.L. Hartman, E.D. West and T.K. Herman, 'Crops that feed the world 2. Soybean – worldwide production, use, and constraints caused by pathogens and pests', *Food Security* **3** (2011): 5–17.

48   Langthaler, 'Ausweitung Und Vertiefung', 116.

Janina Priebe

agenda.[49] The fear of the 'population bomb',[50] too, put food technology in the limelight. As a reaction to these broader global developments, German scientists framed soy cultivation as a tool to solve global problems, optimised through science and technology.[51]

The fear of the protein gap revived research about the cultivation of soy in Europe that had already advanced in the 1940s. A scientist who published about renewed experiments with soy cultivation in Austria in the early 1980s referred to the year 1973 as the year of the 'protein crisis [die Eiweisskrise]'.[52] The protein crisis was connected to the oil crisis of the 1970s in the public consciousness because they both laid bare the fragile interconnections of global trade and supply. The protein crisis was, moreover, aggravated by the conjunction with poor harvests over two consecutive years, the prior intentional reduction of national stocks of grains and the increased buying-up of stocks by the Soviet Union.[53] In Germany, the potential role of soy was increasingly tied to securing the national supply in such unpredictable times. The 1980s saw active efforts to promote soy cultivation. The renewed interest is tangible in a scientific study that investigated the cultivation of soybeans in two locations in Western Germany with different soil and climatic conditions.[54] In the same year, West German farmers founded the German Soy Support Association [Der Deutsche Sojaförderring e.V.], which provided information, training and educational resources to soy cultivators and the interested public.[55]

The same period saw the emergence of the German environmental movement in the 1960s and its manifestation in the founding of the Green Party in 1980. Cracks in the Western industrialised consumer societies made

49   R.D. Semba, 'The rise and fall of protein malnutrition in global health', *Annals of Nutrition and Metabolism* **69** (2016): 79–88.

50   P.R. Ehrlich. *The Population Bomb. Population Control or Race to Oblivion?* (Sierra Club/Ballantine Books, 1968).

51   G. Lehmann and S. Möller, 'Möglichkeiten Zur Verbesserung Der Welternährungslage Unter Berücksichtigung Von Maniok, Soja, Milcheiweiss Und Blutplasma', *Fette, Seifen, Anstrichmittel* **83** (1981): 453–61.

52   Wolffhardt, 'Anbauversuche', p. 92.

53   E. Zahn, 'Konsumtheorie, Konsumforschung Und Die Wandlung Der Konsumgesellschaft', in W. Rippe and H-P. Haarland (eds), *Wirtschaftstheorie Als Verhaltenstheorie.* (Berlin: Duncker & Humblot, 1980), pp. 63–85.

54   R. Marquard, W. Schuster and R.J. Honarnejad, 'Produktivität, Öl- Und Eiweissqualität Von Sechs Sojabohnensorten in Anbauversuchen Auf Zwei Deutschen Standorten', *Fette, Seifen, Anstrichmittel* **82** (1980): 89–93.

55   Sojaförderring in Deutschland, https://www.sojafoerderring.de/aktuell/sojafoerderring/ (accessed 13 Aug. 2020).

new types of austerity visible that had little to do with the outright lack of food and clothing during the world wars, but they nevertheless left deep marks in the public perception of what level of lifestyle could be sustained. The emerging environmental movement, largely originating in consumer protection against toxic residues in agricultural products, and the environmental justice movement against discriminating distribution of industrial pollution, brought the agroindustry under scrutiny.[56]

During the early years of the formation of environmental interest groups in the Federal Republic of Germany, the focus of action lay on the protection of local environments from the hazards of industrial production.[57] Other threats created by human technology were more abstract but were undeniably global and irreversible. The atomic bombs that destroyed Hiroshima and Nagasaki in 1945 showed the existential threat to humanity that could be unleashed through science and technology. The same insights, however, were seen to hold the promise of preventing all human suffering in the future, not least through using inexhaustible nuclear energy to meeting basic human demands. Within decades of the Second World War, the discourses about nuclear science, the atomic age and the perils of the atomic bomb would be reflected in (and intersect with) the discourse on biotechnology and genetically modified (GM) crops.[58] German environmentalism, fuelled by the anti-nuclear movement in its early years, mirrored the German resistance to GM crops as a large-scale technology to solve humanity's problems.

Soy became a crop particularly central to the application of biotechnology in the later part of the twentieth century. The conjuncture of economic, political and environmental developments changed the core of scientific and technological visions of soy in Europe in the early 1990s, just as they had been changed profoundly at the beginning of the century. While the environmental history of soy had become truly global and accelerated since the 1950s, it now became an all-encompassing and entangled history of large-scale environmental transformations and impacts on human and non-human bodies and livelihoods.

The first GM food products, among others a growth hormone for increased milk production of cows, were commercially used from the 1980s in the US.

56   Benson, *Surroundings*.

57   Uekötter, *The Greenest Nation?*

58   J. Radkau, 'Hiroshima Und Asilomar: Die Inszenierung Des Diskurses Über Die Gentechnik Vor Dem Hintergrund Der Kernenergie-Kontroverse', *Geschichte und Gesellschaft* **14** (1988): 329–63.

The first shipments of GM soy from the US to Europe arrived in 1996 to be used as animal fodder, in mixed shipments with non-GM beans and not labelled as GM products.[59] Monsanto, a US-based agrochemical company, had recently created a genetically engineered soy plant that would resist the company's own herbicide RoundUp. The uses of RoundUp were limited before, but it could thereafter be widely and generously applied to GM plants. After the broad introduction of GM soy to US farmers, RoundUp sales in the US increased vastly.[60]

The case of European resistance to GM soy in the 1990s is an example of scientific and technological visions, despite being either already realised or at least realisable, not being elevated into a common imaginary. The cultivation of GM soy is today prohibited in, among others, all EU countries. An example of how scientific visions and policy interact to change socio-economic and environmental landscapes is, as in the mid-twentieth century, the cultivation of soybeans in Romania. The production of soy in Romania increased until 2007, but admission to the EU in the same year demanded that the use of GM crops be discontinued.[61] Nevertheless, recent industry estimates speak of a global GM-soy production on approximately 94 million hectares (in 2017), which amounts to 76 per cent of the global soy cultivation area, mainly in the United States, Argentina, Brazil and Paraguay.[62] There is, after all, no restriction on importing and using GM soy in the EU to be used as fodder in the meat- and dairy-industry. This contradiction between the harsh anti-GM stand in domestic cultivation and neglect of the impacts of large-scale GM cultivation for EU uses in the South-American production centre has been decried on several occasions by political parties and non-governmental interest groups.[63] The diverging attitudes towards GM crops since the late 1990s have been described as a 'global rift'[64] exemplify

59   R. Schurman, 'Fighting "frankenfoods": Industry opportunity structures and the efficacy of the anti-biotech movement in Western Europe', *Social Problems* **51** (2004): 243–68.

60   Du Bois, *The Story of Soy*.

61   I. McFarlane and E. O'Connor, 'World soybean trade: Growth and sustainability', *Modern Economy* **5** (2014): 580–88.

62   IDH and IUCN, 'European Soy Monitor. Insights on the European Supply Chain and the Use of Responsible and Deforestation-Free Soy in 2017', p. 26

63   J. Leroux, 'Meat, animal feed and the EU's unbearable hypocrisy on GMOs'. The Greens /Efa in the European Parliament.

64   Y. Tiberghien, 'Europe: Turning against agricultural biotechnology in the late 1990s', in S. Fukuda-Parr (ed.), *The Gene Revolution. GM Crops and Unequal Development* (Oxon: Earthscan from Routledge, 2007), pp. 51–68, at p.2007 52.

the range of implications that technological and scientific visions can have on regulatory frameworks and cultural perceptions.

Pointing to the strong links between food and national identities in European countries, the development of regulatory and societal frameworks opposing the cultivation of GM soybeans in Europe has been referred to as largely the result of populist appeals that go against scientific knowledge and clearance through public expert agencies.[65] The emergence of the German resistance against GM soy, however, has to be seen both against the historical backdrop of the intersection between anti-nuclear and environmental movements and in a wider context of scientifically mediated knowledge about food and feed products at the time (for instance, the scandal of bovine spongiform encephalopathy, BSE). Moreover, concerns were raised about the growing influence of large multinational agro-industrial businesses monopolising GM patents.[66] The many trajectories intersecting and merging into German resistance to GM originate, among other developments, in the entangled environmental histories and scientific imaginaries of soy cultivation and uses at the turn of the twenty-first century.

Future Outlooks and Conclusion

What are the directions of recent scientific visions surrounding soy in Germany that could become crucial for imaginaries to be realised in the twenty-first century? Will there, once again, be a divergence of interests and attitudes that leads to a growing chasm between global regions,[67] or will there be a great escalation in which visions and socio-economic realities clash?

In 2016, the European Commission approved, among other GM soy varieties, the import of Monsanto's Ready 2 Xtend GM soy.[68] Given the regulations for labelling and tracing of GM crops, these GM soy varieties are allowed for both food and feed use in the EU but are not allowed to be cultivated. In 2019, the US, the major producer of GM soy, had a 72 per

65   See, for example, S. Markowitz, 'The global rise of populism as a socio-material phenomenon: Matter, discourse and Genetically Modified Organisms in the European Union', in F.A. Stengel, D.B. MacDonald and D.Nabers (eds), *Populism and World Politics* (Hampshire: Palgrave Macmillan, 2019), pp. 305–35.

66   S. Fukuda-Parr, 'Introduction: Genetically Modified Crops and development priorities', in S. Fukuda-Parr (ed.), *The Gene Revolution*, pp.2007 3–14.

67   K. Pomeranz. *The Great Divergence: China, Europe, and the Making of the Modern World Economy*, The Princeton Economic History of the Western World: Princeton University Press, 2001).

68   K. Plume, 'EU approves Monsanto, Bayer Genetically Modified soybeans' (Reuters, 2016).

cent share of soybean imports to the EU. The EU is, in turn, the largest importer from the US, with a share of 22 per cent, followed by China with 18 per cent as of April 2019.[69] Against the background of the recent decades, however, the direct marketing of GM crops to European consumers remains complicated and much debated.[70]

Scientific visions once again connect the role of soy to global food security, now in the context of climate change and a growing world population expected to result in unprecedented environmental impacts within the next decades. While approximately 17 million metric tons were produced globally in 1960, the World Health Organization estimates global soy demand will reach 500 million tons by 2050, primarily for food and feed products.[71] In Germany and other European countries, there are recently increasing efforts to meet the demand for non-GM soy by scaling up own cultivation in colder climates but with less environmental impact. In Germany, for instance, companies recognise the growing demand for sustainable and non-GM soy-based products in the face of consumer concerns about deforestation and herbicide use on GM-soy plants in the major producer countries in North and South America.[72] Within the EU, soybean acreage has increased since 2014, from 2 million hectares in 2013, to 3.2 million hectares in 2015. The largest potential for expanding soybean cultivation is found in Central and Eastern Europe, in the Danube region.[73]

Another realm of scientific visions concerns the role of soy in the transport sector. Studies published in the *European Journal of Lipid Science and Technology* are witness to the central role oilseeds, such as soybeans, have acquired for the transport sector as the basis for biomass-based fuels.[74] However, the increased use of bio-based fuels is seen to accelerate land use changes and

69   European Commission, press release 16 April 2019. EU-U.S. Joint Statement: the United States is Europe's main soya beans supplier with imports up by 121 %: https://ec.europa.eu/commission/presscorner/detail/en/IP_19_2154 (accessed 4 Aug. 2020)

70   J. Scholderer, 'The GM Foods debate in Europe: History, regulatory solutions, and consumer response research', *Journal of Public Affairs* 5 (2005): 263–74.

71   F. Austen, 'Germany, land of blooming soy fields?', *Deutsche Welle* (25 Oct. 2017): https://www.dw.com/en/germany-land-of-blooming-soy-fields/a-41099867.

72   Ibid.

73   This is also due to the new member countries that have joined the EU since then. D. Costin Dima, 'Soybean demonstration platforms: the bond between breeding, tecnology and farming in Central and Eastern Europe'. *5ᵗʰ International Conference 'Agriculture for Life, Life for Agriculture'* 10 (2016): 10–17.

74   See, for example, M. Mittelbach, 'Fuels from oils and fats: Recent developments and perspectives', *European Journal of Lipid Science and Technology* 117 (2015): 1832–46.

deforestation with a negative impact on climate change mitigation.[75] Another strand of international scientific visions includes the further variation of soy-based products and there is, for instance, a growing focus on uses of micro bacteria (e.g. from algae) in the publications of the *European Journal of Lipid Science and Technology* since the 2000s. Recent studies highlight the potential of soybean oil as a basis for bio-based plastics that promise a variety of uses in biomedical applications because of their biocompatibility.[76] This stands, once again, for hopes being placed on agro-industrial innovation, reminiscent of the optimism of the early 1900s.

When it comes to future prospects of soy in Germany in the twenty-first century, the discourse of food security at a national level stands out. The German Federal Ministry of Food and Agriculture has developed the Plant-based Protein Strategy [*Eiweisspflanzenstrategie*],[77] which unites several national and European objectives to increase the domestic cultivation of high-protein plants. The joint European effort aims to improve and increase certified cultivation, processing and commercialisation of plant-based protein generally, and soybeans in particular.[78] The prime goal in supporting domestic cultivation and production is the improvement of ecological services, such as soil quality, climate change mitigation and biodiversity, through the cultivation of protein plants. Other arguments are regional development and, not least, increasing the supply of domestic, non-GM protein. The weight given to the argument of improving ecological services and environmental quality is new to the twenty-first-century visions of soy.

While the onset of the Great Acceleration witnessed soy as an accelerator and enabler of meat- and oil-based diets and lifestyles, the twenty-first century seems to revitalise soy as a solution for global problems in the context of climate change and ecological crises. In contrast to the similar framing during the 1960s and 1970s, however, the European stance toward soy cultivation and import is today complicated not only by growing domestic demand but also by the environmental impacts of large-scale cultivation

---

75   S. Majer et al., 'Implications of biodiesel production and utilisation on global climate – a literature review', *European Journal of Lipid Science and Technology* 111 (2009): 747–62.

76   See, for example, S. Miao et al., 'Soybean oil-based polyurethane networks as candidate biomaterials: Synthesis and biocompatibility', *European Journal of Lipid Science and Technology* 114 (2012): 1165–74.

77   Eiweißplanzenstrategie des Bundesministreriums für Ernährung und Landwirtschaft. https://www.bmel.de/DE/themen/landwirtschaft/pflanzenbau/ackerbau/eiweisspflanzenstrategie.html

78   Ibid.

of GM soy weighed against the impacts of climate change on global food supplies.

The changing scientific visions of soy in Germany throughout the twentieth century exemplify how the imaginaries of soy shifted between and permeated national and global interests, as well as societal and environmental concerns. The case of Germany also shows how the elevation of scientific visions into imaginaries manifested in society depended on a wider context of cultural and historical perceptions of soy. Today, the increasing (and increasingly various) use of soy as a transforming technology of nature, including human nature, is a symptom as well as the motor of the profound global changes in society and environment initiated during the great acceleration. The consumption of soy in Europe and the visions and fears surrounding soy imports and cultivation are part of a global environmental history that involves the extension of agriculture into forested land, changed compositions of nutrients in the soil, and the diet of humans and farm animals worldwide. On a global scale, soy cultivation and global trade in soy products have transformed the physical appearance and the functions, outside and inside, of environments, societies, human and non-human bodies – and they will continue to do so in the future.

CHAPTER 12.

BETWEEN BRAZIL AND PARAGUAY: AN ENVIROTECH HISTORY OF GLOBAL SOYFARMING

Jó Klanovicz

Anyone who travels across the Brazil-Argentina-Paraguay border experiences the exuberance of the Iguazu Falls, the giant nature of Itaipu's Hydroelectric Power Plant and the surviving fragments of the Atlantic Forest. This border offers a synthesis of the Great Acceleration, where the primitive force of the waters coexists with the brute force of technology and its attacks on biodiversity.

By intermingling shades of green and gold with the waters and the concrete, another face has brought together environment and technology in the region since the end of the twentieth century: the agrolandscape of soy.[1] The plantations *inscribe* stories of economic relevance, are politically disputed, biologically and ethically controversial, and continuously reconstruct the interaction between humans and non-humans. Between the arrival of soybean culture in the region in the 1970s and the transgenic triumph of crops at the beginning of the twenty-first century, soy has an *envirotech history* characterised by what Dolly Jorgensen calls 'frontier work', where human and non-human agents have been transiting at the limits between the biological, the political and the biotechnological, with implications for the very notion of agriculture in contemporary times, to the extent that we can think of soy, metaphorically, as animal protein.[2]

---

1   This chapter owes much to the academic dialogues that are being established in the areas of Environmental History and the History of Science and Technology, and to the exchange of information with Biological Sciences and Agronomy. In this sense, I would like to thank Paulo Rogério de Oliveira and Cacilda Rios Faria (both from the *Universidade Estadual do Centro-Oeste do Paraná*, Unicentro).

2   By using these notions of frontier work and of Envirotech History, the chapter is supported by discussions such as D. Jørgensen, F.A. Jørgensen and S.B. Pritchard (eds), *New Natures: Joining Environmental History with Science and Technology Studies.* (Pittsburgh: University of Pittsburgh Press, 2013); and M. Reus and S.H. Cutcliffe (eds.), *The Illusory Boundary: Environment and Technology in History* (University of Virginia Press, 2010).

THE AGE OF THE SOYBEAN: 247–264 **doi:** 10.3197/63800040695086.ch12
© Jó Klanovicz

What is different in soy farming landscapes?

With the timely incorporation of the notion of Anthropocene in historical studies, the first distinct dimension is the scale of soy farming. In 2018, more than six per cent of the agricultural area of the planet was covered by the crop. Of the 346 million tons of soy that the planet harvested in 2018, more than fifty million each were produced by Brazil (112 million tons), Argentina (54 million tons) and Paraguay (9.2 million tons).[3]

These countries had no significant soy plantations in the first half of the twentieth century. In 1950, Brazil planted just over 24,000 hectares of soy, which was concentrated in the state of Rio Grande do Sul, in rotation with wheat.[4] Paraguay began to have soy plantations in the late 1960s. Argentina, a traditional wheat producer, started to intensify soybean culture at the same time. Only a few decades separate an incipient and secondary activity from a reality that created an agroecosystem of intertwined stories of seeds, pathogens, agricultural policies, science, work and migration, and global trade that transformed the Southern Cone into the world centre of soy in the twenty-first century.

The second difference is the radical bio-technologisation of soy farming, which seems to have extrapolated the modernising logic characteristic of the second half of the twentieth century. Soy farming is more than a kind of Green Revolution mechanised or chemicalised crop: its main trace is its genetic improvement in a globalised world. In this way, soy cannot be compared to the super-productive grains of the Green Revolution. Soy is a leguminous plant that produces pods, it is extremely leafy and its productivity does not reside in the improvement of the seeds but rather in the broad planting area. Transgenic soy is the symbol of this dimension. When a high-tech plant is associated with the globality of its expansion, it is not exaggerated to consider it a distinct historical phenomenon.[5]

Even though it is planted, does soy continue to be agriculture? Unlike coffee, orange, cotton, apple, wheat or rice, we know or discover that we eat soy only in the small print of food labels and, for the time being, we have not yet experienced a moment in which we would happily buy a kilogram of transgenic soy at the supermarket.[6]

3 USDA, Soybeans, Data & Analysis, 2018: fas.usda.gov/commodities/soybeans

4 O.A.F. Conceição, *A expansão da soja no Rio Grande do Sul, 1950–75* (Porto Alegre: FEE, 1986).

5 B. Glaeser (ed.), *The Green Revolution Revisited: Critiques and Alternatives* (London: Routledge, 2011); N. Uphoff, *Envisioning 'Post-Modern Agriculture': A Thematic Research Paper* (Wassan, 2011).

6 Regarding these monocultures that are consumed on a daily basis, see S. McCook, *Coffee is Not Forever: A Global History of the Coffee Leaf Rust* (Athens: Ohio University Press, 2019); F. Uekötter

12. Between Brazil and Paraguay

The soybean *kainós* between Brazil and Paraguay has implications that exceed the chemical-machinery-capital triad. In this context, I understand the Soyacene as the origin of new work, land, capital and technology relations in a techno-vegetable capitalism, which has become an emerging historical force with geomorphic impact. In this scenario, soy farming would no longer be agriculture but rather the meeting point of environmental, economic, social, scientific and political events of the twenty-first century.[7]

This chapter aims to discuss soy's agro-landscape in the frontier between Brazil and Paraguay as the current centre of the Soyacene. There are symmetrical agencies between ecology and technology when discussing soy farming in the region, with significant implications for soybean's enviro-tech history.

Becoming soybean

Nowadays, soybeans present biological-historical transformations that can be thought of through the notion of domestication as a historical process that began to promote the ecological interaction between humans and plants in a distant past and the gradual and continuous writing of a botanical dimension of culture.[8]

For Ruth Mendum, 'seeds of domesticated plants are a literature, a hard drive and a coded record of past information. Historically speaking, breeding is a special kind of alteration that has made domesticated plants among the most ancient of technical products.'[9] In this sense, molecular studies suggest that the cut between a wild soybean (*G. soja*) and domesticated germplasm (*G. max*) occurred between 6,000 and 9,000 years ago, between the northeast and the central region of China – two regions that have been historically considered as the place of origin for soy crops by agronomists such Nikolai Vavilov and Theodore Hymowitz.[10] In turn, studies of protein, cytogenetic

(ed.), *Comparing Apples, Oranges, and Cotton: Environmental Histories of Global Plantation* (Hamburg: Campus, 2014).

7   Regarding the Great Acceleration, see J.R. McNeill, and P. Engelke, *The Great Acceleration: An Environmental History of the Anthropocene since 1945* (Cambridge: Belknap Press, 2014). For a Brazilian discussion on the Great Acceleration, see also *Varia Historia* **34** (65) (2018), special issue on Brazil's Great Acceleration (ed. by A. Acker and G. Fischer).

8   P. Gepts, 'Domestication of plants', in N.V. Alfen (ed.), *Encyclopedia of Agriculture and Food Systems* 2 (San Diego: Elsevier; 2014) pp. 474–86, at p. 478.

9   R. Mendum, "Subjectivity and plant domestication: decoding the agency of vegetable food crops', *Subjectivity* **28** (2009): 316–33.

10   See N.I. Vavilov, *Five Continents*, (Rome: IPGRI, 1997); T. Hymowitz, 'On the domestication of the soybean', *Econ. Bot.* **24** (1970): 408–21.

and morphological variation position the origin of the plant in several regions: the encounter of agricultural sciences with archaeology suggests that the plant first appeared between Russia, China, Korea and Japan.

In Asia, in addition to the natural emergences of wild soy in regions ranging from Russia to Japan, varieties of domesticated soybean circulated intensively through the establishment of terrestrial and maritime trade routes. For instance, soy trade already appears in classic Japanese texts like the *Kojiki* (Records of Ancient Matters), but also figures in other Asian areas, such as the south of India or the Moluccas Islands.[11]

In Europe, since the advent of the navigation era, the arrival of the seed was a matter of time. Soy appeared in the late eighteenth century, brought by travellers and missionaries. Some of the first European records were made by the Dutch physician Engelbert Kaempfer, who lived in Japan between 1690 and 1692 while working for the Dutch East India Company. Kaempfer registered soy, soy sauce and miso in *Amoenitatum Exoticarum Politico-Physico-Medicarum* (1712).[12] Subsequently, Friedrich Haberlandt, *Die Sojabohne: Ergebnisse der Studien und Versuche über die Anbauwürdigkeit dieser neu einzuführenden Culturpflanze* (1878), reports various experiments carried out with the grain in Europe between 1875 and 1878, covering the spectra of human feeding, oil production and nutritional potential of the legume. Haberlandt compiled registers of more than a hundred tests performed on the continent, with seeds acquired at the 1873 Vienna World's Fair from China, Japan, Tunis and Transcaucasia.[13]

At the same time, Haberlandt noted there were at least twenty varieties being investigated in Europe and that the plant already had a few different names: *Dolichos Soja* (Jaquin, 1781), *Glycine soja* (Lineu, von Siebold and Zuccarini), *Soja japonica* and *Soja hispida*. The author carefully cited other Europeans who had already written about the plant, such as Franchet and Savatier, for Japan, and Maximonicz, for the Amur River; and it was already known that it could flourish in all seasons, with a vegetation period in colder times.[14] The research also discussed the high nutrient content of soybeans,

11   M. Hiraga and S. Hisano, *The First Food Regime in Asian Context? Japan's Capitalist Development and the Making of Soybean as a Global Commodity in the 1890s–1930s*, (Kyoto: Kyoto University, 2017).

12   F. Haberlandt, *Die Sojabohne. Ergebnisse der Studien und Versuche über die Anbauwürdigkeit dieser new einzuführenden Culturpflanze*, (Wien: Carl Gerold's Sohn, 1878), p. 8.

13   Ibid.

14   Ibid., p. 9.

analysed in Germany with Japanese seeds, which showed that soy had a higher protein concentration than beans, lupine, peas and lentils.

Europe's interest in soy was related to a commodities and technology circulation regime, typical of the age of empires, as Ernst Langthaler has already pointed out.[15] On the continent, there was a growing demand for protein and lubricants at the end of the nineteenth century, and expansionist interests, in terms of territories, future markets, ports and railways, were increasingly bringing Europe closer to locations such as the Chinese northeast.

An encounter of imperialistic interests happened in the Chinese northeast at the end of the nineteenth century, positioning China, Japan, Russia, Great Britain, France and the United States in the arena. The territory began to be the target of securitisation policies and several interactions among the international actors started to be built from the juxtaposition of geographical interests; colonising the region meant a demonstration of strength, and the region's soy would soon be incorporated into plant imperialism.

In 1861 the British forced the opening of the port of Yingkou, which led to the installation of a Japanese consulate in 1876. The main exported products were soy and its by-products (especially flour and oil). Japan became the main soy flour destination in 1890 and the only international trader from 1892. The Meiji method of agricultural production promoted the replacement of fish flour as a fertiliser in Japanese agriculture with soy flour all over the empire at the same time.[16]

Japan found in soy the possibility to redesign the use of a locally grown plant, introducing it to the global economy. For Midori Higara and Shuji Hisano, the consolidation of the Japanese Empire between 1880 and 1930 was closely linked to soy, feeding back national capitalism – especially after the Russo-Japanese War (1904–1905). In the 1930s, Manchuria became soy's global source until World War II cut off supplies and enabled the United States to become a producer.[17]

In the process of expansion of the Japanese military and imperial power, the main key was international trade. Japan promoted trade companies as a national strategy; Nippon Yusen (Mitsubishi Group) started cabotaging services from Yingkou in 1890, and Mitsui started investing in the region

15   E. Langthaler, 'Broadening and deepening: Soy expansions in a world-historical perspective', *Historia Ambiental Latinoamericana y Caribeña (HALAC) Revista de la Solcha* **10** (2) (2020): 244–77.

16   Hiraga and Hisano, *The First Food Regime.*

17   Ibid.

in 1892. During the post Russo-Japanese War period, Japan expanded its regional presence, gaining the lease rights to Port Arthur and part of the Chinese Eastern Railway. After this conflict, Mitsui & Co. exported the first shipment of soy oil to Liverpool, purchased by Lever Brothers, which started to use it as a raw material for margarine production.

Japanese soy capitalism acquired worldwide importance after the Russo-Japanese War and was reinforced during World War I when soy oil export to the United States took off. After the war, it began to export the product more intensively to Europe, with Germany as the main buyer. At this point, four companies held global soy exports: Mitsui & Co. and Mitsubishi Corporation (which traded the product) and Honen and Nisshin (which produced oil). Together, they were responsible for 83 per cent of Manchuria's soy export volume.[18]

In the United States, Japanese soy had attracted attention since the beginning of the century, not only for its cheaper oil but also due to its capacity to provide forage. Cotton companies discovered that they could press soybeans to produce oil in the off-season of cotton. The use of soy flour for animal feed in Europe also caught the attention of the US Department of Agriculture (USDA). In addition, the circulation of scientific papers on the nutritional, fat and protein supply potential gained strength when, in 1904, George Washington Carver reinforced the protein potential of the legume, starting to promote its cultivation throughout the country.

After World War I, producers and extension workers created the American Soybean Association (ASA), which would sponsor the Oriental Agricultural Exploration Expedition in 1929. Midori and Hisano note that the US government and ASA sought to promote the soybean crushing industry in the United States, in addition to building fat and oil reserves for war. Politically, they lobbied for margarine to be produced with soy planted in the USA.[19] As a result of an intense political and scientific increase in soy, the USA became the leading soy producer, while Japan became its main importer after World War II. The US census of agriculture shows that, in 1934, just over 23 million bushels soybeans were grown on almost 150,000 properties.[20] In the 1930s, twelve states cooperated with the federal government to create a regional industrial soybean products laboratory in Urbana, Illinois. As with

---

18   Ibid., p. 6.

19   Ibid.

20   A.W. von Struve, 'The soybean crop in The United States', *Science* **93** 2404 (1941): 86–87.

many other symbols of American know-how and modern science, soy was propagandised at state fairs, including a special train that displayed exhibits and represented two national associations of producers and processors; US successes attracted scientists and agronomists from other countries, including South Africa and Brazil.

As a result, soy stopped being 'oriental' and increasingly occupied larger areas in the Americas, with the US becoming the centre of global soybean production until the 1970s. Ernst Langthaler has divided soy expansion into three periods:

> a) The Chinese soy expansion (1900s–1930s), that shows a predominant shift of the external frontier, associated with the peasant mode of farming; b) The US soy expansion (1930s–1970s) that represents a predominant shift of the internal frontier, connected to the entrepreneurial mode of farming, and c) The Brazilian soy expansion (1970s–2010s) that reveals a flexible combination of extensive and intensive frontier shifts, corresponding with the capitalist mode of farming.[21]

After World War II, soy was better known, controlled and genetically programmed to the extent of scientific and technological evolution. Moreover, the plant assumed different signatures, and constructions – in the form of certified seeds and better management based on primers and technological packages – which, however, continued to be dependent on short days or on latitudinal position in planting areas.

In the same period, studies on soy's photoperiod managed to define the characteristic of industrialised soy as a C_3 plant in terms of photosynthesis. In other words, soy came to be understood as a very photosensitive plant, which means that the transition between stages of plant development until flowering could happen in direct response to day length – and, especially, night length. The key to soybeans' flowering was longer nights – and, in this way, the latitudinal position of planting became important for a good harvest. As a result, the US began to establish a map with twelve maturation areas, so as to better adapt plants to a wider range of latitudes. This attempted to optimise the production of soy varieties and to use them both as forage and protein reserve.

Another important challenge in the construction of the modern plant was nitrogen fixation. From the beginning of the twentieth century, soybeans were known as skilled fixers of underground nitrogen, a characteristic that made them suitable for rotation agriculture, especially with intensive nitrogen consumers such as corn (*Zea mays*). By the 1950s, seeking an alternative to

21   Langthaler, 'Broadening and deepening'.

traditional mineral nitrogen fertilisation, nitrogen fertilisation via biological fixation from bacteria began to be intensively researched by the microbiology community in the attempt to produce greater nodulation in the plant's roots (productivity improvement).

From the 1960s, American industries met emerging breeding programmes outside the United States, reaching new environments in the global north and south, becoming a major coloniser of the Southern American agricultural frontier.

The Hegemonic Republic of Soybeans

'Soy has no frontiers' (*'A soja não tem fronteiras'*), declared a 2003 advertising campaign by the multinational company Syngenta. The campaign showed a map of South America with a green cloak that covered Brazil, Paraguay, Uruguay, Argentina, and Bolivia, which, together, formed the 'United Republic of Soy' (*'República Unida da Soja'*). Along the borders between Paraguay and Brazil, soybeans became a landscape that can be understood as a physical measure of technological change. Since the 1960s, the Paraguayan region of Alto Paraná, and western region of the Brazilian state of Paraná shared a frantic deforestation process closely associated with the widespread diffusion of soy farming. Thinking of this region in the context of the United Republic of Soy, one sees an example of brute force monoculture characterised by the link between large-scale projects and cross-border pressures over Atlantic Rainforest fragments.

Between the 1960s and 1980s, Brazil and Paraguay were two authoritarian states, deeply marked by the persecution by military governments of political opponents, ethnic minorities and small farmers, all considered risks to progress and national threats. Moreover, both countries turned out to be authoritarian regarding the natural world, especially along their borders (the Brazilian state of Paraná, and Paraguayan *departamento* of Alto Paraná).

In Paraguay, the long dictatorship of General Alfredo Stroessner (1954–1989) promoted a 'march to the east' from the 1970s, seeking to resettle farmers from the central part of the country on the *frontera oriental* (the eastern border), a policy implemented by the Rural Welfare Institute (*Instituto de Bienestar Rural*, IBR).[22] The creation of smallholding agrarian colonies on the country's border – IBR's first goal – was not successful, due to slow land titling. However,

---

22   A. Menezes. *A herança de Stroessner: Brasil – Paraguai, 1955–1980.* (São Paulo: Papirus, 1987); C. Moraes, 'Interesse e colabroação do Brasil e dos Estados Unidos com a ditadura de Stroessner (1954–63)', *Diálogos* **11** (1–2) (2001): 55–80.

255

12. Between Brazil and Paraguay

IBR's goal to consolidate the agricultural border ended up being promoted by the entry of Brazilian farmers who were already familiar with both agrarian techniques and crops of state interest, such as wheat and soy.[23]

After the military coup of 1964, Brazil redefined its policy of agricultural modernisation, reinforcing a technocratic stance supported by a developmentalism that shifted the focus from the national agrarian problem of agrarian reform to the technique.[24]

This process accommodated at least two concerns of the Brazilian civil-military government: it assured political stability in border areas during a regime in which territories located up to 150 kilometres from foreign borders were considered areas of national security interest (*Áreas de Interesse de Segurança Nacional*). In this context, promoting large crops in regions such as Marechal Cândido Rondon, in the west of the Paraná (one of the southern Brazilian states), would halt the organisation of resistance movements against the military regime, such as rural workers' unions or 'communist cells'.

During this period, agricultural cooperatives were strongly encouraged, and part of the know-how of cooperative associates was directly implemented – or, better, transferred – a few kilometres to the west, into Paraguay, along with seeds, technical assistance, scientific publications and capital. Disguised in scientific garments under the cover of neutrality, soybean farming quickly expanded into forests along the border. In this context, small farmers were sacrificed by the politicisation of regional soy farming, mainly since the region had been recently occupied (between the 1930s and 1940s) under the claim of it being a 'demographic void'.[25] Once planted, soy became part and parcel of the territorial border.

Still in the 1970s, in western Paraná, technical assistance and rural extension agencies fostered agricultural cooperatives with agendas that promoted technical meetings to discuss soybean culture in its most diverse aspects. These included: seeds, varieties, inoculation, fungicide treatment, fertilisation, liming, soil conservation against pests and diseases.[26]

23   H.M. da Silva, 'O problema agrário e a colonização da fronteira oriental do Paraguai: "la marcha para el este"', *Revista Percurso* **7** (2) (2015): 47–61.

24   W. Dean, *A ferro e fogo: a história e a devastação da Mata Atlântica brasileira* (São Paulo: Cia das Letras, 1996).

25   A.M. Myskiw, *Colonos, posseiros e grileiros: conflitos de terra noroeste paranaense (1961/66)* (Niterói: Universidade Federal Fluminense, 2002), p. 13.

26   *Rondon Comunicação, Caravanas técnicas* (Marechal Cândido Rondon, 1974); Curitiba: *Biblioteca Pública do Estado do Paraná* (Paraná State Public Library).

Palotina, another municipality on the border, became the largest Brazilian soy producer from 1976 to 1979, in an agro-landscape, newspapers defined as 'super-technical' and deforested.[27] Throughout the 1970s, Paraná's regional press, not only in Palotina and Marechal Cândido Rondon, but also in Guarapuava (*Jornal Esquema Oeste*) or Londrina (*Folha de Londrina*), would work to convince farmers from several regions to plant soybeans, considering the economic benefits of the crop and also the possibility of taking advantage of the technological apparatus already used for other crops.[28] The state, which produced just over 8,000 tons of soy in 1961, reached 4.15 million tons in the early 1980s, and 6 million tons in 1991, with a concentrated production in the regions of Campo Mourão, Ponta Grossa and areas closer to the border with Paraguay, such as Toledo and Cascavel.[29]

During the 1970s, farmers also began to plant soybeans on a commercial scale rather than in family plots, partly in response to the collapse of the massive Peruvian anchovy fishery, a leading worldwide source of protein supplements in livestock and poultry feed. The anchovy collapse created a protein shortage that drove soybean prices skyward. These steep price rises, combined with a US soybean export embargo in 1973 to curb domestic food and feed inflation, pushed Brazil into the soybean market.

As part of broader agricultural growth, the Brazilian government heavily invested in a comprehensive soybean research programme, including breeding varieties adapted to local soils and growing conditions at low latitudes. As a result, Brazil's soybean production increased from a million tons in 1969 to 15 million tons in 1980. Initially, production growth was concentrated in the traditional farming regions in the south – namely in the states of Rio Grande do Sul, Santa Catarina, Paraná and São Paulo – although, after 1990, it rapidly spread into the tropical savanna of the Cerrado.

The establishment and the solidification of soy as a viable crop owed

---

27   *Jornal Esquema Oeste*, 'Soja terá grande expansão de produção' (Guarapuava: Cedoc-Unicentro, 1979); J. Klanovicz, and L. Mores, 'A sojização da agricultura moderna no Paraná, Brasil: uma questão de história ambiental', *Fronteiras: Journal of Social, Technological and Environmental Science* 6 (2) (2017): 240–63.

28   *Jornal Esquema Oeste*, 'Agricultura ganha esperança no sul' (Guarapuava: Cedoc-Unicentro, 1972); 'Aprenda a plantar soja: vale a pena' (Guarapuava: Cedoc-Unicentro, 1972); 'A produção de soja cresce no Paraná' (Guarapuava: Cedoc-Unicentro, 1973); 'Os novos rumos da alimentação' (Guarapuava: Cedoc-Unicentro, 1976); 'Soja terá grande expansão de produção' (Guarapuava: Cedoc-Unicentro, 1979); *Folha de Londrina*, 'O boom do soja: questão social à vista' (Londrina: FL, 1973); 'Soja: tudo correu bem' (Londrina: FL, 1976).

29   M.G.Moreira, *Soja*, 2014–15: http://www.agricultura.pr.gov.br/arquivos/File/_deral/Prognosticos/ Soja_2014_15.pdf.

much to the proper performance of cultivars introduced from the south of the United States. Moreover, the first soybean breeders wisely prioritised the introduction of cultivars and then elite lines.[30]

The varieties tested in the south of Brazil (123 between 1940 and 2000) were mainly derived from US seeds. These included the Acadian (between the 1940s and the 1960s), Bienville (from 1969 to 1977), Bossier (from 1973 to 1993), Davis (from 1968 to 1998), CNS (in the 1960s), Hampton (from 1968 to 1980), Hardee (from 1968 to 1987) and Hill (from 1967 to 1974). In genetic terms, such varieties resulted from the selection of pioneer seeds and were experimented with consideration to latitude and biological nitrogen fixation.[31] Part of this agro-environmental production would also reach Paraguay.

In the process of construction of research units focused on soy, microbiology played a significant role in Brazil, especially thanks to researchers such as Johanna Döbereiner (1924–2000), who, from the 1950s, promoted an investigation on the biological nitrogen fixation in soy.[32] Döbereiner also convinced producers to prioritise soybeans that depended less on artificial fertilisers. While this issue was being globally solved, soybeans advanced northwards and westwards towards Paraguay and the Brazilian Cerrado.

In the 1970s, technological packages and soy cultivation systems expanded to several country regions, published by the Brazilian Agricultural Research Corporation (*Empresa Brasileira de Pesquisa Agropecuária*, Embrapa), in partnership with regional agencies of agricultural research.[33]

30   E.R. Bonato and A.L.V. Bonato, *Cultivares que fizeram a história da soja no Rio Grande do Sul* (Passo Fundo: Embrapa Trigo, 2002).

31   Ibid.

32   Embrapa. 'A cientista que poucos brasileiros conheceram', 2020: https://sistemas.sede.embrapa.br/40anos/index.php/personagens/detalhes/1.

33   Regarding soy in Amazonia, see F.C. de Camargo, *Sugestões para o soerguimento econômico do vale amazônico* (Belém: Instituto Agronômico do Norte, 30 maio 1948). About midwest and southeast regions, which include the Cerrado, see W.C.de Menezes, and W.A. de Araújo, *Contribuição para melhoramento dos solos ácidos e pobres da estação experimental de Sete Lagoas - Minas Gerais - para a cultura do algodoeiro', I Reunião Brasileira do Cerrado* (Sete Lagoas: Instituto de Pesquisas e Experimentação Agropecuárias – IPEACO, 1961); and *Embrapa, Sistemas de produção para a Soja – Goiás. Itumbiara*, 1975 (parallels 4 to 19, latitude S), *Embrapa, Pacotes Tecnológicos para a Soja: Triângulo Mineiro e Alto Paranaíba, Ituiutaba*, 1974. About soy farming in southern Brazil. see *Ministério da Agricultura, 'Levantamento de reconhecimento dos solos do estado do Rio Grande do Sul. Primeira etapa, planalto rio-grandense', Pesquisa Agropecuária Brasileira* 2 (1967): pp. 71–209; CTAA, 'Contribuição ao estudo da Soja no Brasil', *Boletim Técnico* n. 10; *Embrapa, Pacotes Tecnológicos para a Soja (sul do Rio Grande do Sul)*, 1974; *Embrapa, Pacotes Tecnológicos para o Trigo e a Soja*, Florianópolis, 1975; 'Embrapa, Aptidão agrícola dos solos do nordeste do Estado do Paraná' (interpretation of soils),

Since the mid-twentieth century, the growth of Brazilian soybeans and by-products production and export has ended US dominance over the world soybean market. From less than one per cent in the early 1960s, Brazilian soybean output grew to over fifteen per cent of global production. By 1977, Brazil exported more soybean meal and almost as much oil as the United States.[34] While Brazil is undoubtedly projected to overtake the US during the twenty-first century, climate change might negatively impact soybean crops due to the intensification of droughts. For example, the southern states of Brazil, which are responsible for forty per cent of national soybean production, lost more than 25 per cent of their yield in the 2003/2004 and 2004/2005 growing seasons. In this context, biotechnology – with Genetically Modified Organisms (GMOs) – might play an increasingly important role.

Since the 1970s expansion process of the agricultural border in Paraguay, Brazilians with capital and experience in mechanised agriculture who have ventured in search of cheap land outside of Brazil due to necessity, economic pressure or simply betting on the search for a better life, have found new horizons of opportunities.

Approximately 400,000 Brazilians emigrated to Paraguay between the 1970s and the 1980s, looking for a way to purchase large portions of land from the more than six million hectares offered on the eastern border. Thousands of these hectares were sold to Brazilians in this period by members of the *Partido Colorado* and Paraguayan military.[35] In this context, the departments of Alto Paraná, Itapúa, Canindeyú, Caazapá were also reached by soy.[36]

Paraguay opened arable areas through deforestation and pressure on traditional and indigenous populations on the eastern border. Soon, soy would dominate the landscape, replacing other crops like wheat or corn. In 1995, the legume occupied 28 per cent of the national summer cultivated area and

Curitiba, 1975. For soy packages focused on animal production, *Embrapa Zona da Mata, Pacotes tecnológicos para gado de leite* (Minas Gerais: Coronel Pacheco, 1975). The volume of publications on soy allowed Embrapa to historicise its own process of specialised development since the late 1980s (see E.R. Bonato, *A soja no Brasil: história e estatística* (Londrina: Embrapa-Soja, 1987)).

34   A.J. Cattelan and A. Dall'Agnol, 'The rapid soybean growth in Brazil', *Oilseeds & Fats Crops and Lipids* **25** (1) (2018): 12.

35   B.S. Bassi, *É ele o maior latifundiário brasileiro no Paraguai: Tranquilo Favero*. De Olho Nos Ruralistas, 2020: https://deolhonosruralistas.com.br/deolhonoparaguai/2018/08/16/o-rei-da-soja-tranquilo-favero-protagoniza-conflitos-no-paraguai/#:~:text=Com%20isso%20ele%20lidera%20a,de%20hectares%20sob%20seu%20controle.

36   INBIO - Instituto de Biotecnologia Agrícola, 2016: http://www.inbio.org.py/

accounted for twelve per cent of Paraguayan exports.[37] By 2018, it occupied seventy per cent of arable lands, representing forty per cent of the country's exports, and consolidating the territory as the fourth largest soy exporter in the world, and the sixth-largest producer, according to USDA, in 2019.

From just over 600,000 hectares of arable land and none cultivated with soy in the 1960s, Paraguay's soybeans and agricultural border expansion reached more than 4,500,000 arable hectares in the early 2010s (with more than fourteen per cent of these lands in the hands of Brazilians) – of which 2,500,000 were occupied by soy.[38] The US, Brazil, Argentina and Paraguay, throughout this period, displayed significant, continuous and accelerated soy farming expansion.

Brazilian landowners in Paraguay started grouping into agricultural cooperatives, following the example of southern Brazil. As the departments of Alto Paraná, Canindeyú and Itapúa began to be 'soyised', Brazilian landowners formed the *Central Nacional de Cooperativas* (National Cooperative Central), Unicoop.

Hegemonic Seeds and GMOs as the Epithet of Soyacene

Part of the radical soy expansion in the Brazil-Paraguay border region (Paraná, and Alto Paraná) has been characterised by what Ignácio Narbondo denominates as the *sojización* – literally 'soyisation' – of agriculture. This is a fundamental point of intellectual construction regarding the consequences of *Braziguayan* soybeans. Soybean farming in the region has altered surface interactions between humans and the natural world and constituted new natures beyond technological packages. It has been materialised through porous entanglements between the natural and the synthetic, often dictated by trademarks and patents.

The process of soy expansion in the Southern Cone started to structure itself in the tensions between business and state regulations, and seed transgressions and circulations, which have already been qualified in this chapter as having been politically contested and ethically controversial since the 1990s. The structure of science and technology – an external market that consumed soy in an increasingly accelerated regime – and the impasses posed by the new scenario of GMOs form the hegemony of soybeans in South

37   R Fogel and M. Riquelme, *Enclave soyero merma de soberanía y pobreza* (Assunción: Ceris, 2005).

38   FAOSTAT, *Organización de las naciones unidas para la alimentación y agricultura dirección de estadística,* 2016: http://faostat3.fao.org/browse/area/169/S (Accessed 21 June 2016).

America. They are found in the vigorous institutionalisation of a history of biological traits, hybridisations, breeding and germplasm preservation. Seeds and borders have become linked to international and national norms related to seed circulation and the clashes of soy agricultural biotechnologies.

Between Brazil and Paraguay, the history of soy must be written considering the modernisation of seeds. In this sense, soybean regional trajectories evolve in the context of the institutionalisation of seed control. Soyisation is also parallel to the emergence of the International Seed Testing Association (ISTA), and the International Union for the Protection of New Varieties of Plants (UPOV) since the 1960s.

Between 1969 and 2000, Bonato and Bonato detected several genetically modified cultivars (in order of relevance): Bragg (from the late 1960s to 2000), Bossier (from the 1970s to the 1990s), Paraná (1974–1996), BR-1 (1976–1993), Davis (1970–1995), Santa Rosa (1969–1996), Hardee (1969–1984), Ocepar 4-Iguaçu and Ocepar-14 (1990–2000), Hale-7 (1969--1975), Bienville (1969–1977), M-Soy 6101 (1998–2000), Década (1982–1989), among others.[39] The seeds with the most significant contribution to the production process, according to the authors, were Bragg, Santa Rosa and BR-4, and, on a smaller scale, Hill, Bienville, Bossier, Davis, Paraná and Ocepar 4-Iguaçu.

The search for certified cultivars, registered or in the public domain depended on the productivity observed in experimental stations. Furthermore, the process of constitution of an international monoculture in the Southern Cone faced the paradox of the significant genetic variability stored in the germplasm banks, which clashed with the reduced diversity of commercially used cultivars. Seed selection and cultivation are related to plantation productivity and resistance to diseases that accompany monocultures. Until the beginning of the 1970s, in the Southern Cone, there were few soy diseases; the main concern was the bacterial pustule (*Xanthomonas axonopodis* pv. glycines). However, other problems would still arise, such as the frogeye leaf spot (*Cercospora sojina*).

In particular, the research promoted by the Southern Institute of Agricultural and Livestock Farming Experimentation (*Instituto de Pesquisa e Experimentação Agropecuária do Sul*, IPEAS), at latitude 28ºS, took advantage of the investigation records developed at the *Instituto Agronômico de Campinas* (IAC), São Paulo, carried out since the 1930s, at latitude 23º S. When compared to the IAC's data, the varieties acquired by IPEAS (Bragg, Davis,

---

39   FAOSTAT, *Organización*.

12. Between Brazil and Paraguay

Hardee, Hill and Hood) allowed researchers to overcome the limitations of day length by breeding selected cultivars in the 1940s with American varieties of late maturation in the groups VII and VIII (Acadian, Pelican), resulting in the cultivar Santa Rosa (in the 1960s–1970s).[40] The success with research on new varieties, such as Santa Rosa, allowed Rio Grande do Sul to create, in 1969, the State Commission for Soy Seeds (*Comissão Estadual de Sementes de Soja*, CESOJA-RS), which started to elaborate dissemination plans for varieties more suited to the regional soil, weather and latitude, and more resistant to soybean diseases.[41]

Within the expansion scope of scientific and technological institutional-isation, studies produced year after year began to display the strengths and weaknesses of imported seeds and new domestic varieties. IPEAS found that the Bossier seed, which circulated in Rio Grande do Sul, Santa Catarina and Paraná (southern Brazil), and arrived in Paraguay, was much more susceptible to stem canker and to frogeye leaf spot (*Cercospora sojina*). The Bienville variety, one of the most cultivated seeds outside of Rio Grande do Sul, was susceptible to the bacterial pustule (*Xanthomonas axonopodis* pv. glycines). In turn, Davis soybeans were resistant to frogeye leaf spot, and Bragg was successfully disseminated in lower latitudes.

In Paraguay, the seed dissemination strategy was structured by companies such as Sementes Iruña, AgroSanta Rosa, Sementes Tupi, Sementes Colonias Unidas and Agrotec, in partnership with Embrapa, which was especially interested in receiving royalties from BRS-133. In this sense, Paraguay saw the circulation of many Brazilian seeds, as it is the case of Ocepar-14, BR-16, FT Estrela and FT Maracaju. It would not be long before Paraguay started cultivating the first glyphosate-resistant soy of Argentinian origin and without an official name, which was called *branquinha*.[42] While soybean planting processes were intensified in Paraguay, Embrapa Soja started to exercise agricultural importance in the country, becoming its technological reference (Table 1).

---

40   Bonato and Bonato, *Cultivares*.

41   Ibid.

42   C. Rodrigues, *Avaliação técnica e comercial da empresa Sementes Iruña – Paraguai* (Pelotas: Universidade Federal de Pelotas, 2009), p. 77.

Jó Klanovicz

1983	1984	1985	1986	1987	1988	1989
Paraná	Paraná	Paraná	Paraná	Paraná	Paraná	Paraná
Davis	Davis	Davis	Davis	FT-1	FT-1	FT-1
Bragg	Bragg	Bragg	Bragg	Década	Década	Davis
	Halesoy-71	Halesoy-71	Halesoy-71	Davis	Davis	BR-4
	Lancer	Lancer	Lancer	BR-4	BR-1	Bragg
	Bienville	Bienville	Bienville	Bragg	Bragg	Bico Preto
	Década	Década	Década	Bossier	Santa Rosa	Cristalina
		Bossier	Bossier	Cristalina		
		BR-1	BR-1			
		Cristalina	Cristalina			
			Santa Rosa			
			FT-1			

Table 1.

Varieties of soybean seeds in Paraguay – Sementes Iruña. Source: Rodrigues, *Avaliação técnica e comercial da empresa Sementes Iruña – Paraguai*, p. 77.

This cross-border dynamic of soy meets issues related to the phytogenetic breeding and to the protection of cultivars, a process that led to the establishment of a protection programme for plant breeders' rights in 1994 (Law n. 385/94). A new professional and agricultural class, pertaining to both plantations and politics, has been constituted over time, merging public and private interests, nationally replicating the international controls over seeds. In Paraguay, they appear in the *Boletín Nacional de Cultivares de la* DISE-SENAVE, which gathers more than 300 protected cultivars and 700 cultivars in the commercial register[43].

According to Dolia Garcete, the Paraguayan Association of Plant Breeders (*Asociación Paraguaya de Obtentores Vegetales*, PARPOV) is organised of 21 seed breeders: Agroseed Criadero y Semillas; Algodonera Guaraní SA; Bras Max Genetica Ltda; Coodetec; Asociados Don Mario; EMBRAPA; FT SEMENTES SA; Instituto Agronômico do Paraná (IAPAR); Igra Semllas;

---

43   D.M. Garcete González, *Industria Semillera en el Paraguay: en el contexto de la tasa tecnológica*, (Pelotas: Universidade Federal de Pelotas, 2013).

12. Between Brazil and Paraguay

Monsanto Paraguay SA; Nidera Semillas; OR Melhoramento Sementes; Pioneer Semillas; Pure Circle South America SA; Relmo Paraguay SA; Syngenta Crop Protection; TMG (Tropical mejoramiento & Genética); Dirección de Investigación Agrícola (DIA); Advanta Semillas; Cadec SA Productores de Semillas.

Dolia Garcete notes that, on 27 November 1990, the Cooperativa Colonias Unidas Agroindustrial y Comercial Limitada was the first to enrol in Paraguay's national seed system – which had, in 2012, 276 registered companies (141 active ones).[44] The country also started to structure, throughout this period, the Agricultural Biotechnology Institute (*Instituto de Biotecnología Agrícola* – INBIO), which emerged to promote proper access to products derived from agricultural biotechnology, incorporating them in national production and promoting research in Paraguay. It is essential to highlight that Brazilian public and private companies play a leading role in the list of seed breeders for Paraguayan crops. This is the case of Embrapa Soja, FT Sementes, IAPAR (all from the Brazilian state of Paraná), and OR (from Rio Grande do Sul). However, the circle of Brazilian companies is broader in Paraguayan soy, involving more companies from Paraná, such as Cooperativa Lar (from Cascavel) or, COPAGRIL (from Marechal Cândido Rondon). COPAGRIL, for instance, has played a prominent role in issuing reports for seeds tested by Paraguayan producers since the early 1980s.

A more recent stage in soy's natural-cultural entanglement in the Southern Cone emerged in the late 1990s, with the RR (Roundup Ready) seed. At the turn of the 1990s to 2000, the debate over GMOs moved Brazilian public opinion, especially since, in 1997, the country approved its Cultivar Protection Law (Law 9.456, April 25). Other similar laws also started to be published in South America to provide legal guarantees related to the circulation of seeds and to what in biology is called 'escape of genes'.

Veja magazine, the most-read magazine in Brazil at that time, stated that the next step after the emergence of transgenic seeds would be to improve the nutritional quality of food. *Veja* stated that transgenic soybeans were crucial, since the seed had a 'poor nature'. Public opinion approved biotechnology's role in the sense of producing seeds that were resistant to pests and pesticides and plants that would adapt well to poor soils and dry climates.[45]

In Paraná's soy producing centre, the regional media appropriated the

44   Ibid.
45   *Veja* magazine, 8 July 1998.

debate, as in the case of the *Jornal Esquema Oeste*, which welcomed GMOs, based on the idea that the country was currently prevented from collecting US$ 17 million per year in royalties on grain distribution. In 1999, pressure from producers was already intense, propelling the creation of new laws that would allow transgenics in food, especially since Embrapa, the largest producer of soybean cultivars in the country, was very interested in the Cultivar Protection Law, in addition to being Monsanto's partner. As a result, the Cultivar Protection Law was approved according to UPOV/1978 regulations, assuring breeders' rights. *Veja* magazine, one of the main propagators of neoliberal policies, abstained from the debate until 1998, then asserting transgenic soy as an essential element to escape Brazil's economic backwardness from the multinational private initiative.

The debate in the magazine began when the first seeds of transgenic soybeans were illegally planted in Rio Grande do Sul in 1998, and then, after the approval of the Cultivar Protection Law and the formation of agricultural conglomerates in Brazil, by companies such as the North Americans Monsanto and DuPont, the Swiss Novartis, and the German AgrEvo.[46]

In the late 1990s, Monsanto started selling its herbicide Roundup, with its GM seed Roundup Ready in Brazil. It launched a new phase of the control or editing process regarding a plant, and marked soy's leap to another position – not that of agriculture conventionally called modern, but of a product created from an agricultural platform. While corn crops are super resistant to weeds – but not to insects – soy is more susceptible to weeds.

These new challenges for soybean farming were intensively debated in the public sphere at the end of the 1990s and beginning of the 2000s, triggering legal battles around GM seeds. Beyond transgenics, the horizon that biotechnologised soybeans have unveiled for a product offered from an agricultural platform represents both new and old patterns of agricultural modernisation between Brazil and Paraguay. Rapid expansion of soybean has been followed by multicentric emergence of social and environmental conflicts, especially at that frontier; biotech seeds increase the dependency of farmers on just a few industries and corporations.

---

46 J.C Araújo and M. Mercadante, *Produção transgênica na agricultura* (Brasília, Diretoria Legislativa, 1999).

CHAPTER 13.

CRISIS NARRATIVES FROM THE DUTCH SOYACENE: REGIONAL SUSTAINABILITY HI/STORIES AT SITES OF SOY CONSUMPTION

Erik van der Vleuten and Evelien de Hoop

Introduction

Since May 2019, the Netherlands have been caught in a peculiar crisis – and soy is crucially involved. This so-called Nitrogen Crisis was triggered by a Dutch Council of State ruling. The ruling invalidated government-issued nitrogen emission permits, because procedures to issue these permits had not complied with European Union rules for protecting designated nature conservation areas. Thousands of ongoing housing and infrastructure construction projects (which require permits for their nitrogen emissions) came to a sudden standstill. To resume construction, which was a policy priority due to housing shortages, nitrogen emissions in other sectors needed to be cut drastically. In order to do so, many actors focused on intensive animal farming, responsible for over half of all Dutch nitrogen emissions. An emergency Government Commission proposed reducing the sector by half. Others proposed reducing the nitrogen content in animal feed, noting that agriculture's nitrogen emissions originated overwhelmingly from imported soymeal, the protein basis of the sector. Animals absorb part of soy's protein and thus the nitrogen in their bodies and emit the rest through urine and manure, which harms biodiversity through acidification and eutrophication. Large and radicalising farmer groups protested fiercely against such 'nature protection measures', and warned of starving animals and the demise of their sector – which operates with extremely low profit margins. The parties have been at loggerheads ever since.[1]

---

1 J.W. Remkes et al., *Niet alles kan overal. Eindadvies over structurele aanpak op lange termijn* (Amersfoort: Adviescollege Stikstofproblematiek, 2020); J. Schollaardt, *Factsheet Emissies en Depositie van Stikstof in Nederland* (The Hague: TNO 2019); J.W. Erisman, 'Setting ambitious goals for agriculture to meet environmental targets', *One Earth* 4 (1) (2021): 15–18.

THE AGE OF THE SOYBEAN: 265–288 **doi:** 10.3197/63800040695086.ch13
© Erik van der Vleuten and Evelien de Hoop

Erik van der Vleuten and Evelien de Hoop

The Nitrogen Crisis suggests that not only soy production regions, but also soy consumption regions deserve attention when considering histories of the Soyacene. Most current research focuses on local or global (e.g. greenhouse gas emissions) ecological changes and social conflicts at sites of soy production, predominantly in the Americas. In order to better understand production-related socioecological and international relations, that historiography studies massive deforestation, land use conflicts, pesticide pollution, child labour and more, and the complicity of agribusiness, science and innovation, international trade, politics and markets, and much more.[2] Conversely, historical studies of soy *consumption* have not focused on broader and intertwined social, environmental and economic changes at sites of consumption, but on the ambivalent roles of soy in human diets – as a health food and meat alternative as well as a core ingredient in processed foods (soy oil) and the meat industries (soybean meal) that undergird modern industrialised diets.[3] To our knowledge, the historiography of broader regional change at sites of soy consumption, on a par with and in relation to histories at sites of production, is still in its infancy.[4]

This chapter explores such broader histories at sites of soy consumption. We speak of soy's 'sustainability histories' to denote interrelated economic, social and environmental histories regardless of whether or not historical actors use the term 'sustainability'.[5] As the Nitrogen Crisis illustrates, soy consumption might particularly manifest in regional sustainability histories of areas with intensive animal farming. We focus on such areas in the Netherlands, which have come to host some of the most intensive animal farming in Europe and the world (witness the staggering manure emissions per hectare, see Figure 1). Dutch agricultural history tells us that cheap imported soy, processed into compound feed, was pivotal to this development – imported soy became as important to intensive animal farming as artificial

2   For a review, see C.M. da Silva and C. de Majo, 'Towards the soyacene: Narratives for an environmental history of soy in Latin America's Southern Cone', *Historia Ambiental Latinoamericana y Caribeña* **11** (1) (2021): 329–56.

3   For further references, see E. Langthaler, 'The Soy paradox: The Western nutrition transition revisited, 1950–2010', *Global Environment* **11** (1) (2018): 79–104.

4   F. Haalboom, 'Oceans and landless farms: Linking Southern and Northern shadow places of industrial livestock (1954–1975)', *Environment and History* (Online First 2020); E. de Hoop and E. van der Vleuten, 'Sustainability knowledge politics: Southeast Asia, Europe and the transregional history of palm oil sustainability research', *Global Environment* **15** (2) (2022): 209–45.

5   J.L. Caradonna (ed.), *Routledge Handbook of the History of Sustainability* (Routledge, 2018).

fertilisers were to modern arable farming.[6] That literature also observes the remarkable rise of intensive animal farming in this region coinciding with spiking soy imports at nearby Rotterdam Port, 'the hub of soybean and bean product [soymeal, soy oil] trade for all of Europe and the surrounding areas', according to the *Soybean Update* in 1983.[7] More recently published *Soy Barometers* tell us that, by the early 2010s, Dutch soy imports ranked second only after Chinese imports (which, however, were of a different order of magnitude). By then, Dutch soy imports embodied a foreign land use of some 2.6 million hectares, roughly corresponding to no less than eighty per cent (!) of Dutch domestic land territory, and dwarfing the country's own few hundred hectares of domestic soy production.[8]

This chapter not only highlights the Soyacene's global sustainability history in agricultural soy consumption regions but also unpacks the diversity of relevant sustainability history narratives within such regions, thus rejecting notions of regions as monolithic entities. As we shall see, the past five decades have birthed very different, and politically conflicting, stories about the past and future of soy, animal farming, and sustainability challenges in the area under study. We here focus on four such hi/stories (i.e. narrations of the past in relation to the present and the future, by historical and contemporary actors including professional historians), which we tentatively identify as an 'agricultural miracle' narrative, an 'environmental pollution' narrative, an 'animal suffering' narrative and a 'global footprint of soy consumption' narrative.[9] These four narratives highlight important yet distinct features of the Dutch Soyacene. Each hinges crucially on massive soymeal imports for animal feed consumption, even though focus and attention on the role(s) of soy may vary greatly.

6   J. Bieleman, *Five Centuries of Farming: A Short History of Dutch Agriculture 1500–2000* (Wageningen: Wageningen Academic Publishers, 2010); J. Bieleman, 'Landbouw en milieu—een eeuwig spanningsveld?', in G. Castryck and M. Decaluwe (eds), *De relatie tussen economie en ecologie gisteren, vandaag en morgen* (Verloren, 1999), pp. 25–36.

7   As quoted in W. Shurtleff and A. Aoyagi, *History of Soybeans and Soyfoods in the Netherlands, Belgium and Luxembourg (1647–2015). Extensively Annotated Bibliography and Sourcebook* (Lafayette, CA: Soyinfo Center, 2015), source nr. 1414.

8   J.W. van Gelder, B. Kuepper, M. Vrins, *Soy Barometer 2014. A Research Report for the Dutch Soy Coalition* (Amsterdam: Profundo, 2014), pp. 11, 15, 27.

9   On hi/stories: E.M. Cheung, 'The hi/stories of Hong Kong', *Cultural Studies* **15** (2001): 564–90. Compare: William Cronon, 'A place for stories: Nature, history, and narrative', *The Journal of American History* **78** (4) (1992): 1347–76.

Erik van der Vleuten and Evelien de Hoop

A Statue for Pigs: The Agricultural Miracle of Intensive Animal Farming

Dutch agricultural history has a long tradition of research on agricultural innovation in relation to the socio-economic fortunes of rural communities.[10] As such, it interpreted the rise of soy-enabled large-scale intensive animal farming as an innovative and successful response to the postwar crisis among smallholding communities in the impoverished Southern and Eastern sandy-soil regions of the Netherlands. This narrative was especially dominant in the field of agricultural history during the 1990s and early 2000s. It focused on mass-pig and poultry farming, where soy became the dominant protein basis, on the sandy soil regions connecting the South-eastern provinces of Noord-Brabant and Limburg. The spectacular rise of intensive animal farming in this area constituted, in the words of Prime Minister Wim Kok in 1996, a veritable agricultural miracle.[11]

We highlight four key features of this agricultural miracle narrative. First, the narrative considers the agricultural crisis of the late 1940s as the trigger for the spectacular rise of intensive animal farming. Although this post-war agricultural crisis was of course international, Dutch historiography presented the economic and social prospects in the South-eastern provinces of the country as particularly gloomy.[12] Local and provincial farmer organisations and state-employed agronomical experts spoke of a 'Small Farmers Question', which referred to the many unprofitable mixed-agriculture smallholders on the region's poor sandy soils – so very different from their affluent, export-minded and internationally reputed colleagues elsewhere in the Netherlands. The numerous children on these poor family farms had no prospects of starting a farm of their own (smallholder plots could not be further subdivided) and a thorough dislike for jobs in urban industries.

---

10 Bieleman, *Five Centuries*, pp. 16–17; P. Kooij et al., *De Actualiteit van de Agrarische Geschiedenis. Historia Agriculturae Vol. 30* (Groningen/Wageningen: Nederlands Agronomisch Historisch Instituut, 2000), p. 2.

11 Kok used the German term *Landwirtschaftswunder*. A.H. Crijns, *Van overgang naar omwenteling in de Brabantse land-en tuinbouw 1950-1985. Schaalvergroting en specialisatie* (Tilburg: Stichting Zuidelijk Historisch Contact Tilburg, 1998), p. xiii

12 Ibid. and T. Duffhues, *Voor een betere toekomst: Het werk van de Noordbrabantse Christelijke Boerenbond voor bedrijf en gezin 1896-1996* (Nijmegen: Valkhof Pers, 1996); J. Korsten, *Standhouden door veranderingen. De Limburgse Tuinbouwbond als behartiger van agrarische belangen 1896–1996* (Nijmegen: Valkhof pers, 1996).

Mass unemployment and impoverishment of rural communities loomed.[13]

Second, the narrative credits the crisis response measures of regional farmer organisations, provincial authorities and state-employed agricultural consultants for the rise of affluent large-scale animal husbandry. Early measures aimed at dissolving the perceived cleavage between rural agriculture and urban industry by attracting industry to the rural area and prepping the young rural generation for new educational, employment and emigration possibilities; industrial entrepreneurship soon entered the countryside and unemployment vastly decreased.[14] Other measures boosted farm productivity, profitability and scale increase through e.g. research and innovation, education and consultation, financing schemes for farmers, common agricultural sales and exports facilities, and cooperative feed, fuel and tools purchase. Meanwhile, national policies under Minister of Agriculture Sicco Mansholt strengthened Dutch exports – and so did the new European Economic Community (with Mansholt as agricultural commissioner) by establishing a common market through tariff policies, production subsidies, and a cold chain for transporting perishable products. These efforts combined, so the narrative continues, fostered an entrepreneurial and innovative attitude among sandy soil livestock farmers; a new generation of agricultural entrepreneurs established industry-scale pig and poultry farms and associated agricultural industries, astounding the nation by 1960 and EEC competitors in the 1970s and 1980s.

This narrative typically illustrates these changes with spectacular numbers on the rise in large-scale pig and poultry farming. In the province of Noord-Brabant, for example, the human population less than doubled from 1.2 to 2.1 million between 1950 and 1985, while pig numbers rose from under 300,000 to almost 5 million and poultry from 3.6 million to over 25 million. Average farm sizes increased from under ten to over 500 pigs, and from under 200 to over 18,000 chickens.[15] Pig farming became iconic for the financial success of 'non-land based' agriculture: 'the pig had drawn the small-scale sandy soil farmer out of his misery' and had rightfully 'gained itself a statue' – referring to the bronze statue in front of the

13   Also: A. Maris et al., *Het kleine-boeren vraagstuk op de zandgronden. Een economisch-sociografisch onderzoek van het landbouw-economisch instituut* (Assen: van Gorcum, 1951); A. Maris and R. Rijneveld (eds), *Het kleine-boerenvraagstuk op de zandgronden. Ontwikkeling in de periode 1949–1958. Rapport 347* (The Hague: LEI, 1960).

14   Also: *Noord-Brabant welvaartsbalans. Ontwikkelingsplan 1965*, 2 vols (Den Bosch: Provincie Noord-Brabant, 1965).

15   Duffhues, *Voor een betere toekomst*, p.14; Crijns, *Van overgang naar omwenteling*, pp.107 and 113–38.

provincial government building in Noord Brabant, donated in 1979 by the regional Pig Farmers Association to 'celebrate the economic development of pig farming'.[16]

A third feature of this narrative concerns the role of soy and agricultural feed companies in this agricultural transition. As noted, agricultural historians stated that high-protein feed was to intensive animal farming what artificial fertilisers were to arable farming, spiking productivity per hectare. Soy became the dominant protein source from the 1970s, ultimately constituting some ninety per cent of the protein base in pig and poultry farming. Our narrative elaborates that feed constituted the largest variable cost in pig and poultry farming; that agronomists researched the most cost-effective feed nutrients at experimental farms; that agricultural consultants constructed feed schemes tailored to individual farms; and that policymakers supported feed imports – under Mansholt, both the Dutch Government and EEC policies exempted feed from import tariffs. Soy (the cheapest protein source) and tapioca (a cheap carbohydrate source, replacing wheat) were massively imported by commercial and cooperative trade companies, which emerged as crucial historical agents: they negotiated, purchased, imported, processed and distributed the cheapest possible compound feed. For example, the cooperative trade firm Cooperatieve Handelsvereniging's company history commemorated the art of negotiating deals with soy producers in Argentina, Brazil and Paraguay (and tapioca producers in Thailand). Domestically it offered, like its competitors did, financing schemes to farmers in return for feed contracts. By doing so, the firm incited farmers to invest in soy-based large-scale animal farming and invited the veritable 'invasion of pigs and chickens' in sandy soil agriculture, while simultaneously growing into one of the largest EEC players.[17]

Fourth and finally, socio-technical transitions are rarely straightforward and unproblematic, and the conventional agricultural history narrative typically ends with an observation of several backlashes visible to all by the 1980s. For example, mixed-agriculture smallholding, which post-war crisis measures had sought to preserve as the traditional cornerstone of regional

16   Duffhues, *Voor een betere toekomst*, p.282; Crijns, *Van overgang naar omwenteling*, pp. xii–xiii.

17   H. Siemens et al., *Terug naar de Kern. 100 jaar Cehave Landbouwbelang* (Apeldoorn: Agrifirm, 2011), p.15. Also: H. Veldman, E. van Royen and F. Veraart, *De geschiedenis van Cebeco-Handelsraad 1899–1999* (Eindhoven: SHT/ Cebeco, 1999); S.F. Van der Laan, *Een varken voor iedereen: De modernisering van de Nederlandse varkensfokkerij in de twintigste eeuw* (Utrecht: Utrecht University, 2017), pp. 69–70.

agriculture, was unintentionally ousted by large-scale specialised pig, poultry, cattle or dairy farms. Older generations of farmers lamented the risk taking, money loans and loss of core farming values ('true farmers' should have land *and* animals) of the new generation of agricultural entrepreneurs and their industry-scale farms and agricultural industries. Individual farmers are quoted as saying, for example, that 'we should leave space for smaller farms' and that 'our gamble with pig farming paid off, but gambling can be addictive, and some people cannot stop'.[18] Farmer organisations, agricultural experts and provincial authorities agreed to halt unchecked growth of the sector, but found themselves unable to do so.

The second backlash was environmental: intensive animal farming's environmental pressures grew as the sector grew, and newspapers and policymakers fiercely debated a national 'manure problem' by the 1980s.[19] The conventional agricultural history narrative typically casts this problem as 'the next challenge', coming to the farming community in the form of 'public perceptions', 'social critique', and 'insensitive new environmental policies', now calling for a severe *reduction* of the sector. The narrative documents farmers' protests against these 'external pressures', and suggests that the long-term solution is not reducing the number of animals, but empowering the farming community to tap into its proven capacity of 'innovating to survive': organisational and technological innovation would turn agrarian entrepreneurship into 'agrarian stewardship' while transitioning toward a more sustainable future.[20] This interpretation, however, was fiercely challenged by a second narrative to which we now turn.

18   Cited in Crijns, *Van overgang naar omwenteling*, p.104 (also pp. 90–91).

19   From 0–2 newspaper articles per year before 1982 to >100 per year in 1985–1994 according to the national publication database. www.delpher.nl, keyword search on 'mestprobleem OF mestproblematiek' (consulted 29 Nov. 2021).

20   Crijns, *Van overgang naar omwenteling*, p. 220; Diffhues, *Voor een betere toekomst*, pp. 323–26; Korsten, *Standhouden door veranderingen*; Bieleman, 'Landbouw en milieu'. Also L.G. Horlings, *Duurzaam boeren met beleid: innovatiegroepen in de Nederlandse landbouw* (Nijmegen: Katholieke Universiteit Nijmegen,1996), p. 18.

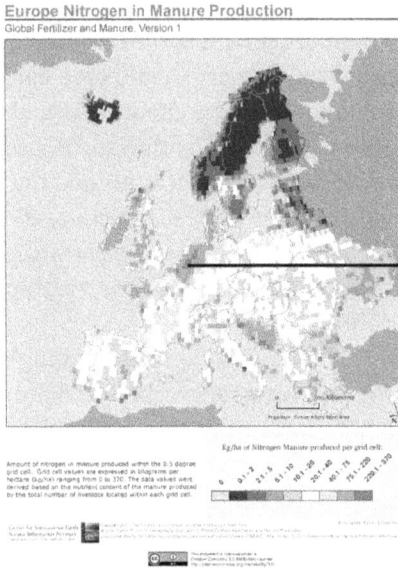

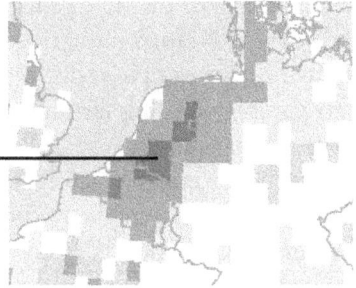

Erik van der Vleuten and Evelien de Hoop

Figure 1.

Nitrogen emissions through manure in kg/ha around 2010. The Southern and Eastern Netherlands top the list in Europe and worldwide. CC BY 3.0. [21]

Europe's Dung Heap: Environmental Pollution and Policy

Whereas the agricultural history narrative chiefly sought to describe and explain the economic miracle of intensive animal farming in Dutch sandy soil regions, a second narrative by investigative journalists and other critics (within and beyond academia) focused on this miracle's detrimental effects on domestic land, water and air quality.[22] Imported soy was again crucial to the narrative – but now as transcontinental carrier of the most debated

21   Source: P. Potter et al., *Global Fertilizer and Manure, Version 1: Nitrogen in Manure Production* (Palisades, NY: NASA Socioeconomic Data and Applications Center, 2012). https://doi.org/10.7927/H4KH0K81 (Accessed 27 November 2021). CC BY 3.0.

22   J. Frouws, *Mest en Macht: een politiek-sociologische studie naar belangenbehartiging en beleidsvorming inzake de mestproblematiek in Nederland vanaf 1970* (Wageningen: Landbouwuniversiteit Wageningen, 1994); F. Bloemendaal, *Het mestmoeras* (The Hague: SDU, 1995); L. Lamers, *De kool, de geit, en het Nederlandse mestbeleid* (Wageningen: Landwerk, 2016); B. Hermans, *De Mestmarathon. Kroniek van Ruim 42 Jaar Nederlands Mestbeleid* (Utrecht: Natuur & Milieu, 2016).

pollutants, notably nitrogen (which as protein was the very reason for importing soy), but also phosphorous and potassium. This narrative is highly concerned with exposing and explaining persistent policy failures to curb domestic agricultural pollution – from the first in-house agricultural research and policy debates thereon in the 1960s to the present-day Nitrogen Crisis. Thus, whereas agricultural histories especially studied and voiced the perspectives of agricultural organisations and policy, this second narrative focused on inconvenient truths and irresponsible agricultural policies and practices that – in the prosaic words of investigative journalists of Follow the Money – ultimately turned the country into the 'dung heap of Europe'.[23] This narrative also spilled over into the emerging fields of Dutch environmental and sustainability history, which identified intensive animal farming as a key contributor to landscape, environmental and ecosystem degradation.[24] Here, we again elaborate on four aspects of this environmental pollution and policy narrative.

First, the narrative prominently mentions scientific knowledge about intensive animal farming's environmental implications for land, water and air. It mobilises such knowledge claims to emphasise the scale and urgency of the problem, and also to criticise and dismiss recurring policymaker and farmer arguments that the problem had been unknown to them, thus preventing them from taking action earlier (a typical *story of awakening*[25] that obscures and depoliticises the long history of environmental problems). Thus, the narrative spotlights how agricultural research institutes investigated the use of animal manure in arable farming since the 1960s. Researchers appreciated the advantages, such as reduced purchasing of artificial fertiliser; but, from the second half of the 1960s, they also warned of future excessive distribution of manure slurry on fields and dumping of surplus slurry in ditches. Calculations and experiments predicted excessive concentrations of nitrogen, phosphorous and potassium in both soil and water. By 1970 it was known how this caused eutrophication, algal growth, oxygen lack, rotting organic materials and species reduction in surface waters, as well as species

23   H. Ariëns and E. Meelker, 'De stinkende achterkant van vleesfabriek Nederland', 8 May 2021. Available at ftm.nl (consulted 28 July 2021).

24   J.L. van Zanden and S.W. Verstegen, *Groene geschiedenis van Nederland* (Utrecht: Spectrum, 1993), pp. 63–92; H. Lintsen et al., *Well-being, Sustainability and Social Development: The Netherlands 1850–2050* (Cham: Springer Open, 2018), pp. 397–416.

25   C. Bonneuil and J.B. Fressoz, *The Shock of the Anthropocene: The Earth, History and Us* (London/ NY: Verso Books, 2016), p. xiii.

reduction in ecosystems that require nutrient-poor soils.[26] When animal holdings boomed in the 1970s and 1980s, aerial ammonia emissions caused a penetrating stench that became a familiar fact of life to rural citizens – and was allegedly recognised by foreign visitors as a characteristic smell of the Dutch countryside.[27] In the early 1980s, agricultural researchers also con-nected aerial ammonia emissions to acid rain and the local acidification of soils and waters by nitric acid: agriculture was held responsible for almost half of the acid rain problem.[28] Meanwhile drinking water companies worried about nitrates threatening ground water quality.

A second observation on this narrative concerns the role of soy in envi-ronmental pollution. Here, we find a paradox. On one hand, soy is central to the narrative as a transcontinental pollutant carrier – especially of nitrogen, the most debated pollutant from intensive animal farming. Soy also carried phosphorous and potassium, but it shared those honours with large quantities of imported carbohydrate sources such as tapioca and maize. On the other hand, however, contributions to this narrative rarely mentioned and elabo-rated the role of soy explicitly. Instead, our narrative typically black-boxes the foreign sources of domestic pollution. The oft-heard phrase was that 'we import the feed, export the pigs, and are stuck with the mess';[29] the Dutch environmental pollution narrative focused on the how and why of this 'mess', but made remarkably little effort to unpack the imported feed sources of domestic pollution (or, for that matter, the domestic and foreign consumption of pig meat). As a result, it also remained oblivious to associated social and environmental conditions at foreign sites of feed production. As such, it is the direct opposite of the fourth narrative that we discuss below.

Third, concerning the domestic 'mess', the environmental pollution narrative spares no effort to detail and document the persistent failure of policymakers and farmers to address domestic pollution problems. It in-terprets this failure as ill-will, fraud and policy system failure. Agricultural journalist Frits Bloemendaal's 1995 book already presented the preceding decades of Dutch agricultural pollution policies as a history of deception.

---

26   S. Algra, 'De invloed van de landbouw op het natuurlijk milieu', *Landbouwkundig Tijdsschrift* **84** (4) (1970): 155–64.

27   Van Zanden and Verstegen, *Groene geschiedenis*, pp. 84–85.

28   N. van Breemen et al., 'Soil acidification from atmospheric ammonium sulphate in forest canopy throughfall', *Nature* **299** (1982): 548–50. Quantitative contribution to acidification: Horlings, *Duurzaam boeren*, p. 18 table 1.1.

29   Minister of Agriculture Cees Veerman (2003) cited in Lintsen, *Well-being*, p. 403.

13. Crisis Narratives from the Dutch Soyacene

Two decades later, prominent newspaper columnist Tom-Jan Meeus (who had published on the manure crisis since 1990) called it a 'scandal that had been allowed to persist for 50 years', characterised by a repetitive pattern of scandals, political response, further animal farming sector growth, further scandals and so on: 'I would not know a similarly dark policy history with a similar lack of progress.'[30]

This history of scandals starts with the 'denial', 'silencing and neutralizing problem signals', and 'delaying tactics' by the farmer-policymaker alliance throughout the 1970s.[31] The narrative highlights the experiences of prominent agricultural researcher Chris Henkes. In interviews, he repeatedly commemorated how his early findings had consistently been suppressed: 'those who claim that the Ministry of Agriculture lacked knowledge of the effects of manure surpluses until the 1980s, denies reality'.[32] It also highlights faulty claims by the Ministry of Agriculture, e.g. that the growth of intensive animal farming had already stopped, or that detergents in household sewage and industrial waste, not agriculture, were to blame for environmental eutrophication.[33] A third example is that the Commission appointed by the Ministry to examine manure issues insisted on basing its calculations on national manure production averages, deliberately obscuring local or regional manure surpluses in pig farming regions (a practice for which it was reprimanded later).[34]

In the mid-1980s, the first law to constrain pig and chicken farming expansion was issued; this feat is generally interpreted as a breakdown of the alliance between the Ministry of Agriculture and farmer organisations. With this breakdown, a second phase commenced in the 'history of scandals', featuring repeated cycles of fraud, policy response, new frauds and so on. Already in the late 1980s, the Ministry of Agriculture tacitly allowed farmers to use legal loopholes to expand their animal holdings (the Minister

---

30 Bloemendaal, *Het mestmoeras*, pp. 7 and 235 ff.; T.J. Meeus, 'Het schandaal dat 50 jaar kon voortbestaan', *NRC* 15 Nov. 2018, p. 2.

31 Bloemendaal, *Het mestmoeras*, pp. 9–18; Frouws, *Mest en Macht*, pp. 75–82; Hermans, *De Mest-marathon*, pp. 10–11.

32 Henkes, cited in Ariëns and Meelker, 'De stinkende achterkant'. Also 'De geschiedenis van het mestprobleem', Argos, VPRO Radio 15. Jan. 1993. Available at https://www.vpro.nl/argos (accessed 29 Nov. 2021).

33 Bloemendaal, *Het mestmoeras*, pp. 14–15.

34 Ibid. p. 13.

was forced to step down in 1990 for this and other scandals).[35] The policy deception story continues to this day, and includes recent exposures (winning a Dutch award for 'best investigative journalism' in 2017) of systemic farmer fraud in the Southeastern provinces, illegally and massively distributing excess manure on fields instead of paying for delivery to a manure processing plant. The most recent example is the invalidation of the government accounting policy based on which nitrogen permits were issued, which triggered the current Nitrogen Crisis.[36]

Fourth and finally, this environmental history narrative typically ends on a half-hearted note of hope regarding prospects for a more sustainable future. On one hand, from the 1980s until today, authors observe how public outrage over environmental degradation and political failure creates initiatives looking to transcend past habits and initiate solutions for more sustainable futures. On the other hand, they simultaneously note how hopeful developments are already countered and watered down by farmer interests before they are even realised. Considering the long track record of avoiding measures addressing root problems (notably: reducing absurdly high concentrations of animals) and of favouring temporary, fraud-sensitive, end-of-pipe administrative and technological fixes (e.g. manure accounting systems, stables emission filters, or manure processing plants), developments towards more sustainable futures remain uncertain at best.[37]

Pigs in Despair: Animal Welfare and Animal Rights

A third important narrative on the sustainability history of the Dutch Soya-cene focuses on its implications for animals and changing animal-human relations. It is well-represented by historian Dirk-Jan Verdonk's impressive *Vegetarian History of the Netherlands* (2009), which we use as the basis to discuss this narrative.[38] Inscribed in the historiographical tradition of animal history and multi-species history, vegetarian history is a research strategy

35   Ibid., p. 15.

36   J.P. Dohmen and E. Rosenberg, 'Het mestcomplot', *NRC.NEXT* 11 Nov. 2017; J.W. Erisman et al., *Stikstof: de sluipende effecten op natuur en gezondheid* (Uitgeverij Lias, 2021).

37   Bloemendaal, *Het mestmoeras*, pp. 231–36; Hermans, *De Mestmarathon*, pp. 40–43. On ammonia filter fraud: G. Jansen and H. de Jonge, 'Namaak luchtwassers bij varkensboeren zorgen juist voor meer stankoverlast', *NOS Nieuwsuur* (14 July 2018, 17:14), available on www.nos.nl (Accessed 5 Dec. 2021).

38   D.J. Verdonk, *Het Dierloze Gerecht. Een vegetarische geschiedenis van Nederland* (Amsterdam: Boom, 2009). Another example of multispecies history: A.F. Haalboom, Negotiating Zoonoses: Dealings

to problematise human-centric agricultural and food histories that ignore animals or see them as mere resources for human lives.[39] Indeed, the 'agricultural miracle' and 'environmental pollution' narratives centred on human lives and environmental concerns external to the agri-food sector. The third narrative, by contrast, brings to the fore changing farm animal experiences (notably animal suffering) and relations between 'humans and other animals'.[40] Here, we shall discuss animals' changing living conditions; human actors' diverse understandings of the problem of animal suffering, of who speaks for the animals and of envisaged futures; and the role played by soy in all this.

With regard to animals' living conditions in intensive animal farming, this narrative unpacks how animals were turned into increasingly effective machines for producing meat, milk and eggs. Verdonk colourfully illustrated what this looked like for the animals concerned. For example, chicken, pigs and cows were now confined to indoor spaces that often barely exceeded the (combined) size of their residents; their bodies had been scientifically bred to grow and mature exceedingly fast, and their feed content and feeding schemes were designed solely for maximised weight increase, regardless of associated cardiovascular and skeletal dysfunctions. Similarly, daylight management schemes stimulated growth or milk and egg production, causing, for example, ocular dysfunction. Animals were slaughtered at an increasingly young age thanks to faster growth; body-parts such as beaks and tails were clipped to prevent animals from mutilating others in response to overcrowded stables; and they were deprived of having sex, as artificial insemination allowed the farmer to control the reproduction cycle. Intriguingly, artificial insemination also necessitated that pig farmers learn the skill of sexually arousing sows manually: multispecies history indeed.

A second feature of this narrative concerns diverse human interpretations of the problem of animal suffering, of solutions and better futures, and of who could speak for the animals in the first place. The animal suffering narrative, like the environmental pollution narrative, observes that agricultural authorities and farmers did not wholeheartedly voice and address animal suffering: as in the case of environmental pollution, they ignored or actively suppressed knowledge about animal suffering in the 1960s and

with Infectious Diseases Shared by Humans and Livestock in the Netherlands (1898–2001) (dissertation, Utrecht University, 2017).

39   Historiographical embedding: Verdonk, *Het Dierloze Gerecht*, pp. 15–29 and 409–10; F. Dieteren, 'Review of Het Dierloze Gerecht', *Low Countries Historical Review* **126** (3) (2011): 118–20.

40   Verdonk, *Het Dierloze Gerecht*, p. 19.

278

Erik van der Vleuten and Evelien de Hoop

1970s – unless productivity was directly threatened by animal suffering. Farmers were locked into the methods of intensive animal farming if they were to stay in business, and staff of the Ministry of Agriculture allegedly argued that '[intensive animal farming] was an inevitable development, a necessity, an economic necessity ...'[41] Neither did researchers raise their voice on animal suffering. Agricultural scientists were employed and funded by the Ministry, and Verdonk noted that scientists' norms of positivist science made studying animal inner wellbeing difficult. If, by exception, scientists did speak up, the Ministry tried to intervene, just as it had in the pollution case. For example, ethologist Gerrit van Putten studied and filmed pigs in stables and on transport from the late 1960s, and observed that the animals suffered severely. Van Putten would later be nationally and internationally lauded, but working for a Ministry research institute in the 1970s, he was issued repeated gag orders, and reports and film material were locked away. This practice was exposed when a public television broadcaster retrieved such film material by court order, broadcasting it with the warning that 'those who have seen this video will no longer enjoy their steak'.[42]

From the early 1970s, and in liaison with such scientists, activists increasingly voiced their concerns over animal suffering. Their trigger was the new *Flevohof* educational theme park, opened in 1972 and displaying agricultural innovation. Appalled by industrial farming's treatment of animals, a band of youngsters established the action group Tasty Animal [*Lekker Dier*] and organised playful actions to make animal suffering visible and political. Their initial concerns resonated with Ruth Harrison's *Animal Machines* (1964) on animal suffering in the UK; soon, however, the Australian ethicist Peter Singer's *Animal Liberation* (1975) became the main source of inspiration: as suffering beings, animals should not be submitted to abusive human will. Henceforward the action group worked for animal rights, as opposed to merely improving animal welfare under industrial conditions. Animal welfare and animal rights discourses co-existed and collided, but did not gain widespread prominence in public debates until the outbreak of swine-fever in 1997, which involved the enforced killing of over eleven million pigs and

41   Expressed by E.H. Ketelaar, in documentary: C. Tromp, Y. Nijland, C. Samson and M. Euwe, 'Episode 593: Lekker Dier', *Andere Tijden* (11 July 2013). Available at: https://anderetijden.nl/aflevering/593/Lekker-Dier (accessed 25 July 2021).
42   Verdonk, *Het Dierloze Gerecht*, p. 311. Gerrit van Putten was interviewed for the documentary 'Lekker Dier'; for Verdonk, *Het Dierloze Gerecht*; and for Crijns, *Van Overgang naar Omwenteling*.

three million piglets.[43] After this major blow to agricultural productivity, agricultural policymakers and farmer organisations agreed that intensive animal farming needed to change – if only to prevent similar catastrophes in the future. The animal welfare problem definition, not the animal rights perspective, dominated: research and legislation sought to improve animal welfare in various ways – but always in line with the production-oriented definition of animal welfare that had emerged in science and policy, namely that 'welfare is understood as living in reasonable harmony with the environment from a physical and ethological perspective … The environment therefore needs to be such that it meets the adaptive capacity of the animal.'[44]

The animal rights perspective did not disappear, however. By 2002, continued public outrage had birthed a new political party, the Party for the Animals [*Partij voor de Dieren*]. The party soon gained seats in Parliament, providing it with a prominent stage to voice animal rights issues. The party sought – and still seeks – to defend animals' rights, de-centre human interests and work for systemic transformation towards nature-centred, not human-centred, sustainability that stretches far beyond how animals are treated within the Netherlands.[45] In addition, vegan animal rights movements mushroomed (sometimes as local chapters of international NGOs), including Animal Rights, People for the Ethical Treatment of Animals (PETA), Anonymous for the Voiceless, Bite Back and Proveg.[46] These organisations also approach animals as sentient beings that ought to have rights. Unlike the dominant discourse within government and among meat-eating publics, they argue for a future in which animals are no longer part of the food production system.

Third and finally, we observe that the role of imported soy in this animal suffering narrative is – similar to the pollution narrative – simultaneously

---

43   B. Elzen, F.W. Geels, C. Leeuwis and B. van Mierlo. 'Normative contestations in transitions "in the making": animal welfare concerns and system innovation in pig husbandry', *Research Policy* **40** (2) (2011): 263–75.

44   Citation from Verdonk, *Het Vleesloze Gerecht*, p. 306, based on NRLO, *Raport van de Commissie* (Den Haag, 1975).

45   R.L. Langeveld, Het leven op aarde gaat niet alleen over mensen: Een kritische vergelijking van de ecocentrische belangenbehartiging van de Partij voor de Dieren en Greenpeace Nederland in de Nederlandse parlementaire democratie, 1992–2018. (MA Thesis, Utrecht University, 2020).

46   Animal Rights, 'Animal Rights'. www.animalrights.nl (accessed 12 Dec. 2021); PETA, 'Dieren zijn niet van ons om op te experimenteren, te eten, te dragen, te gebruiken voor amusement of te mishandelen op welke manier dan ook'. www.peta.nl (accessed 12 Dec. 2021).; Anonymous for the Voiceless, 'Become an animal rights activist'. www.anonymousforthevoiceless.org/join (accessed 12 Dec. 2021); Bite Back, 'Dierenrechtenorganisatie'. www.biteback.org (accessed 12 Dec. 2021); ProVeg, 'ProVeg Nederland'. www.proveg.com/nl (accessed 12 Dec. 2021).

crucial and often black-boxed. Verdonk's *Vegetarian History* is a case in point. It acknowledges the global footprint of Dutch animal farming, citing how feed production for Dutch agriculture requires no less than five times Dutch land territory (according to the 2014 Soy Barometer discussed earlier, soy would account for just under a fifth of this),[47] and lamenting global deforestation from an animal perspective: 'we eat them, as meat eaters, and lay claim to their habitat, mostly for feed production'.[48] Subsequently, he regrettably limits the study to Dutch developments, explicitly eschewing a transcontinental 'entangled history' and subscribing to methodological nationalism.[49] The resulting invisibility of foreign soy cultivation and trade seems to have carried over to domestic soy uses; even the soy-based *bloated chicken* [*plofkip*] – broiler chicken on an excessive protein diet to grow (and suffer) excessively, a prominent campaigning symbol of activists – is absent from his work. Looking beyond Verdonk's pivotal work, we observe that soy barely features in the future visions of those who advocate animal wellbeing without substantial changes in intensive animal farming, but that it does feature in the future visions of those who wish for a major overhaul or eradication of animal farming in the form of drastic reductions in soy imports and consumption. For example, soy has featured explicitly in the Party for the Animals narrative since the party's inception, arguing for abolishing the import of soy for animal feed as a stepping-stone toward a future in which the sustainability challenges associated with its cultivation and the problems associated with intensive animal farming are simultaneously addressed.[50]

Pigs as Embedded Soy: Dutch Agriculture and American Ecosystems

A fourth narrative on sustainability in the Dutch Soyacene adds to the previous narratives in two important ways. First, it is explicitly and primarily concerned with Dutch agriculture's environmental and social footprint at sites of soy cultivation, notably in Latin and North America. Second, as

47   van Gelder et al., *Soy Barometer 2014*, p. 40.

48   Verdonk, *Het Vleesloze Gerecht*, pp. 13, 404 (n. 10).

49   Ibid., p. 23.

50   Partij voor de Dieren, *Verkiezingsprogramma 2006*. https://www.partijvoordedieren.nl/downloads/2014/08/1408630865_Verkiezingsprogramma_2006.pdf (accessed 11 Dec. 2021); Partij voor de Dieren, 'Duurzame sojateelt'. https://www.partijvoordedieren.nl/standpunten/sojateelt (accessed 1 Dec. 2021).

such, this narrative explicitly places soy centre stage. This shows in the names that its makers, developmental and environmental NGOs, chose for their collaboration between 2003 and 2018 – the *Dutch Soy Coalition* – and key publications such as the 'soy barometers'. The latter intriguingly presented Dutch intensive agriculture and its transnational supply and product lines in explicit soy-terms. For example, exported pig meat or eggs were represented as quantitative equivalents of 'embedded soy'. Below, we first trace the historical origins of this 'global footprint of soy consumption' narrative, and then take a closer look at the perspectives, problem definitions and solutions presented in the soy barometers. We end with a reflection on current public debates on soy triggered by two documentaries on deforestation in the Amazon and the Cerrado aired on Dutch national television.

First, with regard to the historical origins of this narrative, the Dutch Soy Coalition itself situated its roots in 1981 when it looked back on its work in 2018.[51] In 1981, developmental NGO Solidaridad and environmental NGO Friends of the Earth Netherlands [*Milieudefensie*] co-published a report entitled 'Soy-yes soy-no: large-scale production: the consequences for poor farmers in Brazil and for ourselves'.[52] This report dismissed big business and government claims that soy was 'the answer' to the world's food problem, as a protein- source for both animals and humans; instead, it emphasised a wide diversity of socio-ecological implications of soy production, and that the amount of protein available for human consumption already exceeded global needs. As the title suggests, the report traced the soy supply chain from Brazilian cultivation to Dutch consumption, warning Dutch consumers 'not to forget the interests of 3rd world inhabitants'.[53] Among the follow-up reports, a 1994 Friends of the Earth Netherlands report argued that Dutch per capita meat consumption should decline by fifty per cent to halt soy- and tapioca-induced soil exhaustion in Thailand, the US and South America, as well as to avert the domestic manure crisis.[54] Over the years, environmental NGOs' campaigning on soy intensified. Greenpeace protested against the import of soy produced with harmful socio-ecological effects at sites of

51   De Nederlandse Sojacoalitie, *Na 15 jaar eind aan Nederlandse sojacoalitie* (2018). https://www.bothends.org/uploaded_files/document/Sojacoalitie.pdf (Accessed 9 Dec. 2021).

52   Solidaridad and Milieudefensie, *Soja sonee, produktie op grote schaal: de gevolgen voor arme boeren in Brazilië en voor ons* (1981).

53   Ibid., back cover.

54   Milieudefensie, *Vlees op de korrel: pleidooi voor een duurzame produktie en consumptie van vlees en zuivel.* (Milieudefensie, 1984).

production.[55] Friends of the Earth demanded abolition of animal suffering and environmental damage in the Netherlands and at feed production sites, advocating for fair, local and circular food systems.[56] Both Ends joined the Latin American *Rios Vivos Coalition* that protested the canalisation of the Paraguay-Paraná river system for soy transport. In 2003, these NGOs, and others, established the Dutch Soy Coalition to jointly address the negative effects of soy production and the role of the Netherlands therein. Activities of this coalition included multi-stakeholder seminars, public campaigning, finding international allies, negotiating measures with social, corporate and policy partners, and, of course, researching and exposing the soy supply chain. The Coalition was disbanded in 2018, allegedly because its members felt that 'soy should no longer be seen as a single issue' and should be integrated in broader ongoing debates on deforestation, intensive agriculture, human rights, protein transition and more.[57] Individual members continued to collaborate on specific soy-related activities and publications.

Second, this transcontinental perspective on Dutch soy, including its interpretations of the main problems and solutions involved, was elaborated, deepened and represented in the Coalition's key publications – the Dutch 'soy barometers' of 2009, 2012 and 2014. These were followed up by European 'soy monitors' of 2017, 2018 and 2019, which represent a continuation of the same narrative although they were published by one of the Coalition's members, IUCN-NL, together with the Sustainable Trade Initiative (IDH). Most research and writing of the first four reports was done by a small group of people of the Amsterdam-based research bureau *Profundo*.[58] As noted, an intriguing feature of this research was its thorough quantitative mapping of

55   Greenpeace, 'Sporen van criminele soja'. https://www.greenpeace.org/nl/natuur/4324/sporen-van-criminele-soja/ (Accessed 12 Dec. 2021).

56   Milieudefensie, 'Archief: Burgerinitiatief 'Stop fout vlees'. https://milieudefensie.nl/archief/burgerinitiatief-stop-fout-vlees (Accessed 12 Dec. 2021); Milieudefensie, 'Onderwerpen: Voedsel'. https://milieudefensie.nl/onderwerp/voedsel (Accessed 12 Dec. 2021). Noteworthy is Milieudefensie's 'travellog of a soy bean', which beautifully illustrates socio-ecological challenges at sites of soy production in relation to the consumption of soy by animals for meat and dairy in the Netherlands: Milieudefensie, 'Actueel: Reisverslag van een sojaboon', https://milieudefensie.nl/actueel/reisverslag-van-een-sojaboon (Accessed 12 Dec. 2021)

57   De Nederlandse Sojacoalitie, *Na 15 jaar eind aan Nederlandse sojacoalitie*, p. 1 and 3–10.

58   A. Herder (Profundo), T. Mohr (Both Ends), G. van der Bijl (Solidaridad), E. van Wijk and E. Herman (Fairfood International), *Sojabarometer 2009: Soja die je niet ziet* (Amsterdam: Nederlandse sojacoalitie, 2009); IDH and IUCN-NL, *European Soy Monitor: Insights on the European Supply Chain and the Use of Responsible and Deforestation-free Soy in 2017.* (Amsterdam, 2019). Researched by B. Kuepper and M. Riemersma of Profundo. Coordinated by N. Sleurink of IDH and H. van den Hombergh of IUCN-NL.

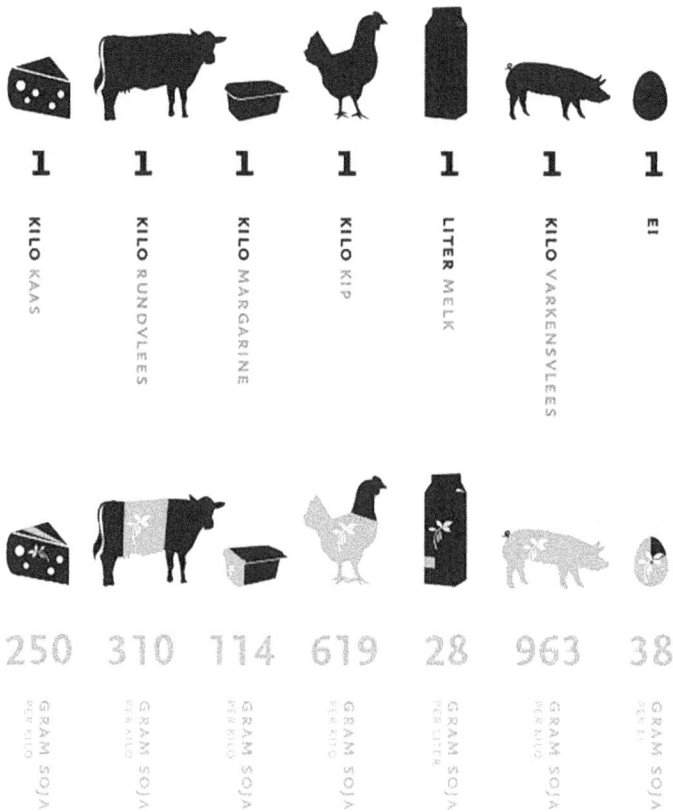

Figure 2.

Visualisation (*avant la lettre*) of the concept of 'embedded soy'. Here, one kilo of pork equals 963 grams of embedded soy.[59]

soy supply chains. The reports amply discuss the databases, difficulties and methodologies involved to produce the 'best available data'. In a nutshell: in 2013 this data involved some 276 million metric tons of global soybean production, some 174 million tons of global exports, 8 million tons of

59   Herder et al., *Sojabarometer 2009*, p. 2.

soybean and soymeal imports (of which 6.5 million tons originated from Latin America), making the Netherlands the second largest importer after China. Some 3 million tons of imported soy was domestically crushed in large plants into soymeal (and oil) and added to the 4.7 million tons of soymeal imports. Most soymeal was exported as feed on EU markets, under 3 million tons were processed or consumed domestically – mostly by animals. Finally, soy was consumed by humans or exported in the form of processed products (meat, eggs, dairy products and so on). These processed products, too, were quantified in soy equivalents: for example, some 1.4 million tons of domestically consumed meat had a 'soybean equivalent' of 0.5 million tons, corresponding to 174 thousand hectares of foreign land use.[60] Later European soy monitors called this 'embedded soy'– the amount of soy needed to produce products (Figure 2).[61]

Concerning the main problem at stake, these soy barometers and monitors are univocal: Dutch agriculture's use of soy wreaks havoc in the Americas. The 2009 barometer highlights 'negative consequences of soy production' such as large-scale deforestation and soil degradation, (often violent) land use conflicts, food insecurity, slavery, and the use of GM crops and pesticides.[62] The 2014 barometer elaborates on, and quantifies whenever possible, threats to ecosystems and social justice in specific key cultivation hotspots: the South American Amazon rainforest; the Cerrado wooded grasslands and Gran Chaco woodlands; a variety of wetlands in such as the Pantanal (Brazil, Paraguay and Bolivia) and the Parana Delta (Argentina); and the North American Great Plains.[63] Remarkably, and in stark contrast to the previous narratives, the implications of soy consumption in the Netherlands (pollution, animal suffering, and also health effects of processed foods that are particularly consumed by low-income groups) are absent here. This is remarkable given the narrative's claim of highlighting 'sustainability issues in the soy value chain'[64] and the elaborate mapping of 'embedded soy's' trade- and consumption trajectories.

This absence of sites of consumption features similarly in the envisioned solutions. Like for the previous narratives, the problem definition entails a

---

60   van Gelder et al, *Soy Barometer 2014*, p. 40.
61   IDH and IUCN-NL, *European Soy Monitor 2017*, p. 15.
62   Ibid. p. 5.
63   Van Gelder et al., *Sojabarometer 2014*, pp. 17–26.
64   IDH and IUCN-NL, *European Soy Monitor 2017*.

specific set of envisioned solutions. The overall strategy of the Soy Coalition and its individual members included diverse solutions: responsible cultivation of soy, replacing soy in animal feed with alternative protein sources and reducing the consumption of meat. However, the soy barometer and soy monitor reports focus almost exclusively on the solution of making soy cultivation 'responsible' through certification. The 2012 Barometer explains this selective focus: most stakeholders in the Netherlands – also agricultural stakeholders – could agree that imported soy should be cultivated 'responsibly', so that Dutch agriculture and consumers would no longer contribute to social and ecological damage at sites of soy production. As such, the 2014 Barometer spends 25 pages on diverse soy production certification schemes and the introduction of certified soy in the Netherlands, versus only three on soy replacement options, and none on reducing meat consumption (the latter solution, which threatened meat farming, was explicitly excluded from these reports, that were clearly consensus-oriented).[65] Notably, after the Dutch Soy Coalition disbanded in 2018, and authorship of the European Soy Monitors changed (Dutch) hands (to the Sustainable Trade Initiative IDH and research bureau Schuttelaar and partners), the discussion of governing responsible soy cultivation broadened: it now included national legislation and sustainability initiatives (such as the Amazon Soy Moratorium, an agreement between diverse partners in Brazil) at sites of production, and initiatives that seek to influence production practices from the demand-side, predominantly through certification.[66]

Third, we end our discussion of this narrative with a recent challenge to the dominant solution of certification of responsible soy cultivation. After Dutch importers switched to certified soy, farmers and retailers assured their customers that milk, cheese and meat did not contribute to deforestation and social exploitation. However, two documentaries aired on Dutch television in December 2019 and in November 2021 shattered that illusion.[67] The latter

65   Herder et al., *Sojabarometer 20*, pp.16–17; Van Gelder et al., *Sojabarometer 2014*, pp. 1 and 42–67.

66   IDH and Schuttelaars and Partner,s *European Soy Monitor: Insights on European Responsible and Deforestation-Free Soy Consumption in 2018* (Amsterdam, 2020). Contributions by R. Hiel, V. Geling, and T. de
Vries (Schuttelaar & Partners) and C. Lan and N. Sleurink (IDH); IDH, *European Soy Monitor: Insights on European Responsible and Deforestation-free Soy Consumption in 2019.* (Amsterdam, 2021). Prepared for IDH by Schuttelaar & Partners.

67   R. Rietveld, D. van der Wilde and F. Glissenaar, 'Bord vol Ontbossing', *Zembla* (25 Nov. 2021). Available at: https://www.bnnvara.nl/zembla/artikelen/bord-vol-ontbossing (accessed 12 Dec. 2021); R. Rietveld and F. Glissenaar. 'Ramp in het regenwoud – deel 2', *Zembla* (12 Dec. 2019).

traced the origins of imported certified, 'deforestation-free' soy, revealing that most imported soy had involved large-scale deforestation, land grabbing, violence and more. A complex system of trading certificates resulted in a situation in which the actual soy imported in the Netherlands could not be traced back to production sites that meet the certification criteria. Besides, the earlier 2019 documentary had already pointed out that, even if Dutch imports had stemmed from certified plots, certification would not stop large-scale deforestation to provide room for soy cultivation for export elsewhere; the Dutch reliance on demand-side measures to 'change the system' hence seems misplaced, and at the very best clears Dutch consumers' conscience regarding their own footprint.[68] Despite this latter observation, the Dutch animal feed industry – applauded by the WWF – has responded to the revelations by committing to direct supply chains, so that only deforestation-free soy reaches the Netherlands.[69] More radical (and not necessarily more effective) threats expressed by European supermarket chains in May 2021 to boycott all agricultural produce (including soy) from Brazil in response to a bill that would legalise private occupation of publicly owned lands in Brazil, have not, at the time of writing, been turned into action.[70]

Conclusions

This chapter argued that the sustainability history of the Soyacene must not only take global and local developments produced at sites of soy cultivation and trade into account, but also local and global sustainability developments produced at sites of (agricultural) soy consumption. To do so, it studied areas in the Netherlands with some of the world's most intensive animal farming, which is highly dependent on large quantities of cheap imported soy. Besides arguing that regions of soy production and consumption harbour diverse narratives of the Soyacene, this chapter also argues that similar diversity exists *within* such regions. As such, we identified and presented four (hi)

Available at: https://www.bnnvara.nl/zembla/artikelen/ramp-in-het-regenwoud-deel-2 (accessed 12 Dec. 2021).

68   Rietveld, 'bord vol ontbossing'.

69   J. Lamers, 'FrieslandCampina en Agrifirm gaan voor ontbossingsvrije soja', *NieuweOogst*, 23 Nov. 2021. https://www.nieuweoogst.nl/nieuws/2021/11/23/frieslandcampina-en-agrifirm-gaan-voor-ontbossingsvrije-soja (Accessed 12 Dec. 2021).

70   RetailDetail, 'Europese voedingsretailers dreigen met boycot Braziliaanse producten', *RetailDetail*, 6 May 2021. https://www.retaildetail.nl/nl/news/food/europese-voedingsretailers-dreigen-met-boycot-braziliaanse-producten (Accessed 12 Nov. 2021).

stories that emerged in the past five decades or so about the past and future of soy, animal farming and sustainability challenges in the region under study.

We interpret these narratives as simultaneously contradictory and complementary: while they imply very different politics of problem definition and proposed solutions, jointly they bring into view diverse aspects of a broader regional Soyacene history. The 'agricultural miracle' narrative highlights how massive imports of cheap soy through the nearby Rotterdam harbour provided a protein basis for the emergence of internationally competitive industrial animal farming in sandy soil regions, sparking a veritable socio-economic transformation that brought affluence to impoverished rural communities. An 'environmental pollution' narrative and an 'animal suffering' narrative regarded this transformation as generative of intense pollution and intensive animal suffering, with an important role for soy – i.e. carrying nitrogen across the Atlantic that became involved in eutrophication and acidification and in the protein-intense lives of farm animals. A 'global footprint' narrative mapped how Dutch soy use was implicated in ecological and social problems at overseas sites of soy production in the Americas. Given time and space, we could have added other narratives. For example, an environmental sciences research narrative on methane emitted in intensive animal farming has recently gained prominence. Methane constitutes a significant share of Dutch greenhouse gas emissions, and this narrative therefore provides an important addition to research on the greenhouse gas contributions of soy, which usually draws its system boundaries around soy cultivation, transport and processing, excluding consumption.[71]

To end, we argue that the diverse narratives presented in this chapter, and those that feature in other chapters of this book, jointly highlight the need to open up ways of thinking about the future to address the manifold challenges in which soy plays the role of critical enabler in an equitable manner. These diverse soy challenges, we argue, are all part and parcel of the shaping of the Soyacene, which stretches across the globe from sites of production to sites of consumption. The historical narratives presented in this chapter typically implied future visions based on a specific soy-enabled socio-ecological challenge in isolation from other challenges, and most highlight technological and organisational innovation (from emission-free and animal-friendly stables to agricultural stewardship and soy certifica-

---

71   N. Escobar et al., 'Spatially-explicit footprints of agricultural commodities: Mapping carbon emissions embodied in Brazil's soy exports', *Global Environmental Change* **62** (2020): 102067.

tion) as a key solution, although innovation has only provided partial and temporary solutions so far, in practice. This includes Dutch' actors reliance on certification as the means to soy sustainability, including the recent initiative in the Netherlands of setting up direct value chains of certified soy. Here, 'sustainability' means that Dutch consumers can trust that their milk and meat has not contributed to deforestation (as WWF activist Natasja Oerlemans recently put it)[72] while Dutch agricultural industries proceed their business more-or-less as usual, and large-scale deforestation continues in the Amazon, the Cerrado and other sites of production. The dominant focus on such single-issue solutions crowds out space for alternative, more equitable and more sustainable modes of living across the globe. Imagining plural, more inclusive and more sustainable futures requires us to engage anew with the diverse histories of soy presented throughout this book, by bringing into mutual conversation these diverse historical narratives within and between distant regions.[73]

Acknowledgements

The authors thank Valentijn van 't Riet and Victor van den Belt (Friends of the Earth Netherlands) and Jan Korsten (Foundation SHT) and Frank Veraart (Eindhoven University of Technology) for informal discussion and/ or access to sources.

---

72   Rietveld et al., 'Bord vol Ontbossing'.
73   De Hoop and van der Vleuten, 'Sustainability knowledge politics'.

Section V

Soybean and the Agrarian Question: A Regional History

CHAPTER 14.

FROM SELF-RELIANCE TO DEEPENING DISTRESS: THE AMBIVALENCE OF THE YELLOW REVOLUTION IN INDIA

Richa Kumar

Introduction

In this paper, I trace a multi-layered history of the Soyacene[1] in India, bringing together various human and non-human actors including scientists, farmers, processors, government agencies, geological forces, ecological factors and the properties of the beans themselves. Drawing upon my ethnographic research from 2006 to 2012 in the central Indian region of Malwa in the state of Madhya Pradesh, which was the heartland of soybean cultivation, supplemented with additional secondary literature on the oilseed complex in India, I describe how soybean farming is emblematic of the Great Acceleration, where natural resources and human labour have been used in ways that devalues them in the long run, in search of short-term profits.[2]

Soy began to be grown commercially in India in the mid-1970s for conversion into soybean meal that was exported as part of the global cattle-feed complex, with the oil being sold domestically. Although it has never produced more than five per cent of the world's total crop, Indian soy is non-genetically modified, and has found niche markets in the Middle East and southeast Asia. Driven by a domestic agro-processing industry and independent family farmers, soy was hailed as having brought about a yellow revolution in India in the 1980s and 1990s. Transnational agribusinesses entered the Indian edible oil complex in the late 1990s and their economic consolidation over

---

1   C.M. da Silva and C. de Majo, 'Towards the Soyacene: Narratives for an environmental history of soy in Latin America's Southern Cone', *Historia Ambiental Latinoamericana y Caribeña (HALAC)* **11** (1) (2021): 329–56.

2   J.R. McNeill and P. Engelke, *The Great Acceleration: An Environmental History of the Anthropocene since 1945* (Cambridge, MA: Harvard University Press, 2014).

THE AGE OF THE SOYBEAN: 291–311 **doi:** 10.3197/63800040695086.ch14

the domestic supply chain led to an unprecedented expansion in soybean farming across the dryland regions of central and western India in the last two decades.

Despite its initial economic contribution to the nation by supporting the growth of a domestic processing industry, adding to the coffers as an export crop and improving farm incomes, the advent of soy, known as the yellow revolution, brought with it ecological crisis amidst deepening farm distress. As the ultimate flex crop, the soybean complex, with its attendant transnational linkages, successfully substituted water, agrobiodiversity, diets and even farmers, across India's drylands.[3] Although soy doesn't need irrigation since it is grown in the rainy season, it transformed the water harvesting systems of the drylands in conjunction with a broader shift in cropping patterns that has undercut the ability of the drylands to sustain farming in the long run. Farmers are running on a technological treadmill that is threatening their very existence. Moreover, monoculture soybean farming replaced mixed cropping patterns of cereals-pulses-oilseeds, reducing access to nutritive foods for rural communities, while soy oil has replaced traditional, healthier oils in people's diets, leading to growing nutritional concerns across rural and urban India. By documenting these broader relationships that create the Soyacene in India, this paper hopes to contribute towards a re-evaluation of the grandiose narratives of progress that embody historical tropes like the yellow revolution.

Soybean the Saviour

Known as *kali tuur* or *bhatmaas*, soy was a marginal crop grown in hill tracts in various parts of India for more than a thousand years and used to make a healthy black soup until its renaissance sixty years ago.[4] As part of a trajectory of global aid and knowledge flows in agriculture and food systems, a group of Indian and American scientists in the Indian agricultural research system sought to introduce soy in India in the 1960s, as a solution for the high levels of protein malnutrition in the country.[5]

---

3   G. de Oliveira and M. Schneider, 'The politics of flexing soybeans: China, Brazil and global agroindustrial restructuring', *The Journal of Peasant Studies* **43** (1) (2015): 167–94.

4   S.P. Tiwari, O.P. Joshi and A.N. Sharma, *The Advent and Renaissance of Soyabean: A Landmark in Indian Agriculture* (Indore: National Research Centre for Soyabean, 1999); G.S. Chauhan and O.P. Joshi, 'Soybean (Glycine max) – the 21st century crop', *Indian Journal of Agricultural Sciences* **75** (8) (2005): 461–69.

5   Interviews with Dr P.S. Bhatnagar, ex-Director of the National Research Centre for Soybean, Indore, on 31 Jan. 2006 and 28 Feb. 2006.

14. From Self-Reliance to Deepening Distress

Yellow seeded varieties from the US – Bragg, Clarke, Lee and others – were successfully adapted and grown under the All India Coordinated Research Project on Soybean (AICRPS), starting in 1967. Much effort was also put into trying to popularise soy as a food by experimenting with products like soy milk, soy paneer (tofu), soy yogurt, soy nuggets (texturised vegetable protein), soy baby foods, etc., and several recipe books were published. However, all these efforts to create demand failed and the dream of alleviating protein malnutrition in India remained unfulfilled.[6]

Instead, soy began to be grown commercially in India in the mid-1970s for entirely different reasons – to be converted into soy meal that was exported as part of the global cattle-feed complex. In the 1960s the family of Mahadev Shahara had a ginning and oil pressing mill in Malwa in western Madhya Pradesh (MP). Through a series of coincidences, his son, Kailash Shahara, learnt about the use of soybean de-oiled cake as cattle feed and realised it could be processed in their existing groundnut mill and exported for a profit.[7] His firm encouraged farmers in Malwa to grow the local black variety, and started processing it in 1972.

Malwa, in central India, was a groundnut processing region and other oil millers jumped on the soybean bandwagon, just as growing incomes of consumers in places like the Middle East and southeast Asia, the growth of the beef-cattle industry in these regions and the geographical advantage of India to service these markets compared to American producers, were creating demand. Between 1970 and 1980, the growth of soybean acreage in Malwa was rapid.

In 1979, the Madhya Pradesh Cooperative Oilseed Growers' Federation (Oilfed) was established with a goal to help make India self-sufficient in edible oils. As of 1976, India was importing almost thirty per cent of its edible oil requirements.[8] The government wanted to duplicate the success of dairy cooperatives that had boosted milk production. To increase oil production, it would have been logical for the Oilfed to encourage farmers in MP to grow groundnut, which was, historically, an important crop of Malwa and

6   Prof. Bhatnagar, the head of the AICRPS, had a hard time convincing his own wife to use it – it had a beany flavour and an unpleasant aftertaste. Farmers tried to feed it to cattle but the animals suffered from diarrhoea because it cannot be fed to livestock without processing.

7   Interview with Kailash Shahara, Indore, Madhya Pradesh, 2 March 2006. He was head of the largest soybean processing company in India – Ruchi Soya (as of 2006) and the Director of the National Board of Trade, the only open outcry soybean oil futures exchange in the country.

8   C.S.C. Sekhar, 'Agricultural price volatility in international and Indian markets', *Economic and Political Weekly* **39** (43) (2004): 4729–36, at 4733.

gave 32–35 per cent oil upon extraction. In contrast, soybean only yielded 17–18 per cent oil. But, by the late 1970s, soybean oil was already available as a by-product of soy meal extraction in Malwa. It was becoming the blending agent of choice in refined cooking oil and the raw material for making vanaspati (refined, hydrogenated vegetable shortening).

Thus, choosing to promote soy in Malwa, the Oilfed became the 'Soya Sangh' or SoyaFed in MP.[9] It set up large processing factories in the region with solvent extraction technology (using hexane to extract the oil and process the leftover deoiled cake), along with providing extension advice and purchasing soy from farmers directly through village level cooperative societies. The 1980s saw soy cultivation increase by leaps and bounds, reaching 2.5 million hectares (or a quarter of the total cultivable area) in ten years in MP.

Along with the creation of demand, soybean also became popular in the Malwa region because it came as a filler in rainy season *(kharif)* fallows. Alternative crops (corn, groundnut, sesame, black gram and green gram) faced high rates of complete failure in the monsoons due to the frequency of waterlogging in the plateau's deep black soil, and were grown on limited, well-drained soils. Sorghum and pigeon pea were of long duration and could not be harvested in time for planting wheat and gram in winter. They were grown in a limited area where the main winter crops would not be planted.[10]

But, with soy, farmers found a crop that was of just the right duration for the rainy season, and which could tolerate heavy rain, waterlogging and drought well enough to produce some 'average' yield – about eight to twelve quintals a hectare. Soy's tap root system goes 1–1.5 feet below from where it can take water. Temporarily, it can develop aerial roots in water-logged conditions.[11] Its leaves will turn yellow, but it will recover with application of urea.[12] Farmer Omkar Singh Chauhan emphasised its ability to withstand drought – crops like corn and lentils like black gram and green gram will die if there is no rain for 20–25 days but soy will survive, he noted.

The biggest advantage of soybean varieties that were created to suit Indian

9 Interview with S. Lakshminarayan, Secretary to the Government of India, Ministry of Home Affairs, New Delhi and ex-Managing Director, Madhya Pradesh Oil Federation, Bhopal. Delhi, 23 Feb. 2006.

10 Interview with Dr O.P. Singh, Vice President, Research and Development, Dhanuka Group of Companies, New Delhi and ex-Senior Scientist, Sehore Agricultural College, Sehore, Madhya Pradesh, 23 Feb. 2006.

11 Ibid.

12 Interview with Dr K.K. Nema, Scientist, Sehore Agricultural College, Sehore, Madhya Pradesh, 25 Jan. 2006.

conditions was that they grew in 100–120 days and fit into the time of the rainy season. Since it did not affect the winter crop, the commodity balance was not disturbed. For farmers, this was a bonanza, bringing double cropping in the region. As a new crop, it was free from pests, insects and diseases and the soil was rich in nutrients.[13]

Double cropping increased incomes of farmers and financed purchase of tractors and mechanised implements for farming, motors, pumps and tubewells, which facilitated the spread of irrigated high-yielding varieties of winter wheat that had been promoted as part of the green revolution. After liberalisation of the Indian economy in the 1990s, other consumer goods found their way to villages in central India. Traders, farmers, processors alike reaped these rewards with many calling soybean the miracle bean, the meat of the fields and the gold of Malwa. Others called Malwa a prosperous garden. Some upper caste farmers were able to purchase more land from this income while processing company owners built bungalows on the posh M.G. Road of Indore, the largest city in Madhya Pradesh.

This was hailed as the yellow revolution and it became a saviour for the nation.[14] From 1994 to 2006, total soy production in the country almost doubled from 3.9 million tons to 7.3 million tons.[15] Dollars from soy meal exports contributed towards financing imports for industrialisation and creating a growing processing industry in the state. Soybean oil also contributed to making the country self-sufficient in edible oil (at least till 1994).[16]

Liberalisation of the Indian economy in the early 1990s and deregulation of the import and export of oilseed products encouraged private industry to get into a lucrative edible oil and soy meal market.[17] Along with the growing demand for non-GM soy meal in countries like Vietnam, Japan,

13   Interview with Dr G.S. Kaushal, ex-Director, Department of Agriculture, Madhya Pradesh, Bhopal, 7 Feb. 2006.

14   R. Kumar, *Rethinking Revolutions: Soyabean, Choupals and the Changing Countryside in Central India* (New Delhi: Oxford University Press, 2016).

15   SOPA, 'India's Soyabean Production 1981–82 to 2007–08', *Soyabean Processors Association Statistics* (2008): http://www.sopa.org/st3.htm.

16   However, by the early 1990s, edible oil imports started rising again. Despite the growth of soybean acreage, India did not achieve long term self-sufficiency in edible oils. As of 2001, India imported almost 60% of the edible oil consumed annually, which rose to over 70% in the last decade (see R. Chand et al., 'WTO and Oilseeds Sector: Challenges of Trade Liberalisation', *Economic and Political Weekly* **39** (6) (2004): 533–37).

17   Deoiled cake and edible oil moved into the open category list of exports in 1994, no longer subject to quotas or licensing. At the same time, imports of edible oil were subjected to tariffs, which were set by the government in consultation with the local industry.

Indonesia, South Korea and Thailand, this created an impetus for many more processing plants to be built, with large domestic conglomerates like ITC-IBD entering the fray.[18] The public-sector Oilfed in Madhya Pradesh slowly shut up shop, due to economic unviability and corruption (its factories were taken over by private players), but private processing capacity began to rise again. With growing demand, soy acreage expanded to all possible areas in the central plateau (Malwa, eastern Rajasthan and Gujarat) and then the peninsular Deccan plateau (Maharashtra and parts of Karnataka). By 2006, there were several hundreds of small soybean processors across these western and central Indian states.

Transnational agribusinesses also emerged as important players, initially through the joint venture route and then, with the approval of 100 per cent foreign direct investment, direct acquisitions became possible. Cargill and Louis Dreyfus began procuring soy in 2006–07 and the latter became one of the top five soybean crushers in 2007 through the leasing of processing plants, surpassing even domestic giants like ITC-IBD.[19] Bunge purchased the famous Dalda vanaspati brand from Hindustan Lever Limited in 2003, and also acquired a large soy processing plant of Prestige Foods company in Madhya Pradesh.[20] Archer Daniels Midland (ADM) acquired soy processing plants in Maharashtra and Rajasthan in 2011, although it had been in a joint venture with a local company since 1998.[21] Cargill, through the purchase of several edible oil brands, including soybean oil, and the purchase of a port-based oil refinery, became one of the leading edible oil players in the Indian

18   Interview with Rajesh Agrawal, Chairman, Soybean Processors Association, Indore, Madhya Pradesh, 4 March 2006. Also see S. Persaud and M.R. Landes, 'The role of policy and industry structure in India's oilseed markets', *Economic Research Report No. ERR-17*. Washington DC: Economic Research Service, US Department of Agriculture, April 2006: http://www.ers.usda.gov/media/862821/err17_002.pdf.

19   N.N. Srinivas 'Oilseed sector gets hot with entry of Louis Dreyfus'. *Economic Times*, 27 Sept. 2007: http://articles.economictimes.indiatimes.com/2007-09-20/news/27667643_1_commodity-trading-ruchi-soya-oilseed.

20   G. Chandrashekhar, 'US agri major buys Prestige Foods' assets', *The Hindu Business Line* 26 Sept. 2003: http://www.prnewswire.com/news-releases/bunge-announces-acquisition-of-the-soybean-crushing-business-of-india-based-prestige-foods-limited-71122247.html; P. Chakraborty, 'Strategising growth through brand acquisition', *Modern Food Processing*, July 2013: http://www.mrssindia.com/media/data/strategy-mncs-in-edible-oil-industry-july-2013.pdf.

21   ADM Media Relations, 'ADM expands oilseeds processing capacity in India', *News Release. Archer Daniels Midland Company*, 6 Sept. 2011: http://origin.adm.com/news/_layouts/PressReleaseDetail.aspx?ID=351; R. Sahgal and M. Vyas, 'US farm giant ADM buys soyabean plant in India', *Economic Times* 13 April 2011: http://articles.economictimes.indiatimes.com/2011-04-13/news/29413773_1_oil-brand-edible-oil-indian-brands.

India: Soybean Production

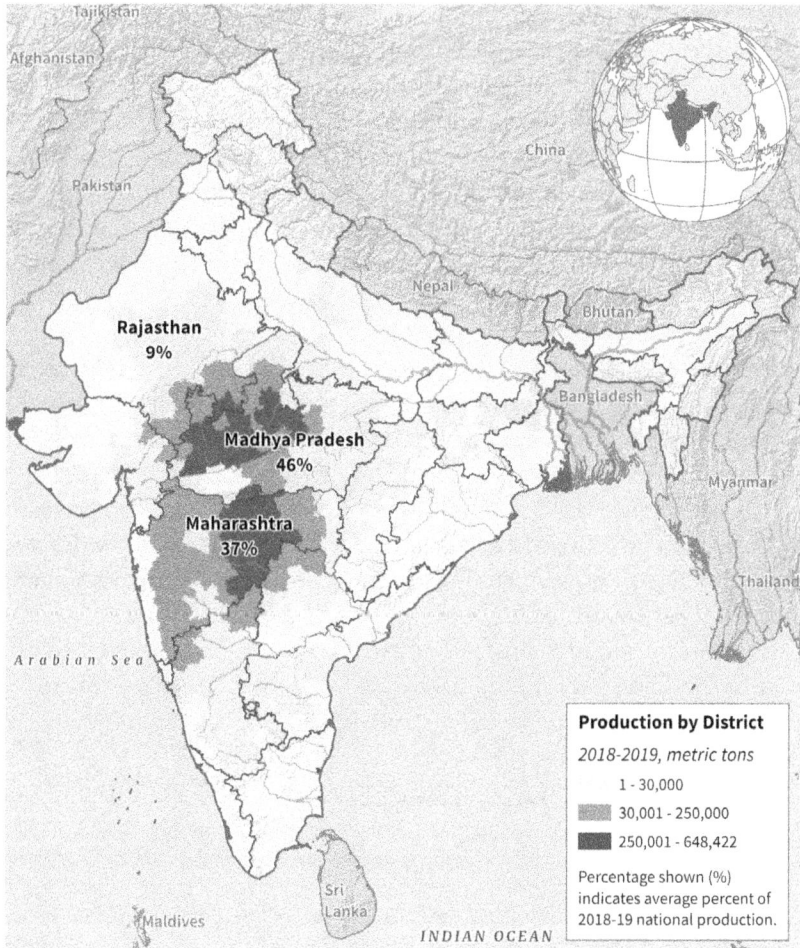

Rajasthan 9%

Madhya Pradesh 46%

Maharashtra 37%

Production by District
2018-2019, metric tons
- 1 - 30,000
- 30,001 - 250,000
- 250,001 - 648,422

Percentage shown (%) indicates average percent of 2018-19 national production.

USDA **Foreign Agricultural Service**
U.S. DEPARTMENT OF AGRICULTURE

Source: India Ministry of Agriculture, Directorate of Economics and Statistics, Market Year 2018/19 data by districts

Figure 1.

India – Soybean Production.

Source: https://ipad.fas.usda.gov/rssiws/al/crop_production_maps/sasia/IND_Soybean.png

market in the 2000s.[22] ConAgra Foods was already in the Indian market since 1997 when it acquired a stake in the ITC promoted company, ITC Agrotech, the producer of Sundrop cooking oil. The company was renamed as Agro Tech Foods Ltd. in 2000, and effectively became a subsidiary of ConAgra once the latter had obtained a majority stake.[23]

The opening of online commodity futures markets in 2003 in soy, soy oil and soy meal, allowed TNCs and domestic conglomerates to hedge their risks, finance local purchases and gain a foothold in the market. They were able to hold large open positions on these exchanges using their physical stock as a backup. But this raised concerns that these legitimate hedging transactions might be leading to abnormal price shifts in domestic soybean futures markets and impacting prices in market yards.[24]

Beginning in the 2000s, the domestic poultry and egg industry also became a major consumer of deoiled cake. In 2007, about 65 per cent of soybean deoiled cake was exported and the rest was sold domestically.[25] By 2013, exports of meal had come down to 37 per cent, and this fell drastically to 7 per cent in 2015 due to falling production, caused by poor weather conditions, and growing domestic demand.[26] By 2017, the poultry, dairy and aquaculture sectors consumed 63 per cent of total deoiled cake production rising to 79 per cent in 2020.[27]

Even though national production went up to 11.2 million tons in 2020 across 12.2 million hectares, domestic demand spiralled out of control.[28] In

22   Srinivas 'Oilseed sector'; Chakraborty, 'Strategising growth'.

23   I.A. Dutt, 'ITC reduces stake in Agro Tech Foods', *Business Standard* 19 July 2010: http://www.business-standard.com/article/companies/itc-reduces-stake-in-agro-tech-foods-110071900059_1.html.

24   R. Kumar, '*Mandi* traders and the *"Dabba"*: Online commodity markets in India', *Economic and Political Weekly* **45** (31) (2010): 63–70.

25   R. Kotian, 'Soybean industry in India', *Presentation* (2008): http://www.slideshare.net/ranjankotian/soybean-industry-in-indiappt-1703901.

26   Calculated using data from SOPA available from Commoditiescontrol Bureau, 'SOPA Revises India's 2020–21 Soybean Import Estimate Upwards To 4 Lakh Tonnes', 11 Aug. 2021: http://www.commoditiescontrol.com/eagritrader/common/newsdetail.php?type=MKN&itemid=601453&comid=,1,&cid1=,3,&varietyid=,1,2,3,&varid.

27   R. Mohan, 'Future of Indian crushing industry and prospects for export of Indian oilmeals from India', *The Solvent Extractors' Association of India–46th AGM*, Mumbai 13 Sept. 2017; Commoditiescontrol Bureau, 'SOPA'.

28   Department Of Agriculture, Cooperation and Farmers Welfare, Ministry of Agriculture and Farmers Welfare, Government of India, Lok Sabha Starred Question No. 35, 20 July 2021: http://164.100.24.220/loksabhaquestions/annex/176/AS35.pdf.

2021, the government was compelled to allow 1.2 million tons of soy meal imports (all genetically modified) in August to help meet a total demand of 5 million tons.[29]

The growth of a large domestic processing industry buoyed by exports and domestic demand, and the dramatic increase in soybean production (it accounts for 45 per cent of oilseed production in India) over the last forty years has firmly established the economic contribution of the yellow revolution to India's development. However, the stirrings of discontent, which were somewhat muted in the early 2000s, have now exploded in full force. Farmers are struggling with environmental degradation and a reproduction squeeze, indigenous industry has been plagued with excess capacity and is facing closures, mergers and acquisitions by large players, and the growing nutrition cost to edible oil consumers is now coming to the fore.

Deepening Distress: Environmental Degradation

Monoculture soybean farming replacing mixed cropping patterns of cereals-pulses-oilseeds across central India, causing decline in soil fertility, biodiversity loss, increasing pest attacks, rising problems of weeds and transforming older systems of groundwater recharge through rainwater, by requiring the draining out of soils in the monsoon. This has caused untold environmental distress, especially in terms of water availability in the dryland regions of central India.

Although soy sheds its leaves at harvest, which dry out and mix with the soil, adding to its fertility for the winter crop, farmers did not replace all the micronutrients that soy monocultures drew out of the soil. They mainly provided nitrogen and, at times, phosphorus because these were the most widely available chemical fertilisers and the only subsidised ones. Deficiency

29   This surge in demand was blamed by SOPA and the poultry industry on the abnormal price rise of soy and soy meal due to excessive speculation and hoarding, in the face of higher exports, which amounted to nearly 27% of total production. The poultry industry called for a ban on futures trading of soybeans and soy meal and called for imports of soy meal to fulfil its needs. See Commoditiescontrol Bureau 'SOPA'; S. Reidy, 'India could allow soybean meal imports', *World Grain*, 17 Aug. 2021: https://www.world-grain.com/articles/15710-india-could-allow-soybean-meal-imports; P. Biswas, 'After processors, poultry industry wants ban on future trade of soybean', *Indian Express*, 3 April 2021: https://indianexpress.com/article/india/after-processors-poultry-industry-wants-ban-on-future-trade-of-soybean-7257608/; V. Fernandes, 'India allows import of GM soy meal; will GM soybean cultivation follow?' *The Federal*, 2 Sept. 2021: https://thefederal.com/analysis/india-allows-import-of-gm-soy-meal-will-gm-soybean-cultivation-follow/.

Richa Kumar

of sulphur and zinc had become common by the 2000s.[30] In the past, animal dung and green manure were used to enhance soil fertility, but the advent of mechanisation and reduction of grazing lands had reduced animal herds.

Farmers rued that in the 1980s, the Oilfed and other government organisations promoted soybean monocultures by providing free seeds and chemical fertilisers; but in the 2000s, the same government was telling them that the land has been spoilt. 'Even we know that', said an upper caste farmer who was one of the first to adopt soy in his village. Scientists like Dr O.P. Joshi of the National Research Centre for Soybean concurred, saying it was a 'bad mistake to tell farmers in the green revolution to forget organic fertilizer and use inorganic only'.[31]

Planting the same crop made it more prone to pests and diseases, pushing soybean farmers into an endless struggle against nature. In 1967, there were hardly four or five insect pests, when soybean was a new crop. By 2006, there were too many to count – from the girdle beetle and the yellow mosaic virus, to the tobacco caterpillar and the gram pest. That year, seventy to eighty per cent per cent of the crop sown in the rainy season in Madhya Pradesh was soy and ninety per cent of that (across India it was 85 per cent) was JS 335 – only one variety of soybean.[32] The tobacco caterpillar had become one of the most dreaded soybean pests. In rocky areas at the edge of the Malwa plateau, adivasi farmers reported an epidemic of *kamliya keet*, an insect that destroyed any broad-leafed plant (soybean and weeds) within a week. It did not attack grassy crops like maize, sorghum, cotton and groundnut, which were exactly the crops that had been replaced by soy in the rainy season![33]

This pushed farmers onto a cycle of relying on more potent and ever-more-expensive pesticides. In the mid-2000s, Avaunt (Indoxacarb), a new molecule from Dupont came on the market. It became the new last resort for farmers, leaving far behind the 'Cyper, Endo, Mono, Trizo, Qui-

---

30   Interview with Dr G.S. Kaushal, 7 Feb. 2006. Farmers added DAP and Urea (phosphorus and nitrogen) to the soils but only manure gave these other nutrients, albeit in smaller quantities.

31   Interview with Dr O.P. Joshi, Scientist, National Research Centre for Soyabean (NRCS), Indore, Madhya Pradesh, 2 March 2006.

32   Interview with Entomologist Dr K.J. Singh of Sehore Agricultural University, 8 Feb. 2006. The popular varieties in the 1970s and 80s were Punjab-1 and JS-72-44, the latter known locally as Gaurav. In 1994, a new higher yielding variety, JS-335, with better germination and shorter duration (95–100 days), was introduced, which replaced almost all other varieties across India. Interview with Dr Saxena, ex-Joint Director of the MP Agriculture Department.

33   Interview with Ratanlal Maru, Ranipura village, Dhar district, Madhya Pradesh, 7 Sept. 2006.

nal'[34] chemicals in its wake. In 2010, a new pesticide known as Coragen had replaced Avaunt – this was also developed by Dupont. Avaunt was not seen to be effective any more.[35]

By 2006, soy had been grown in the same fields for over 20–25 years, reducing its productivity and causing an explosion of diseases and pests.[36] Farmers saw themselves running on a technological treadmill – in a race against insects. But to win, they were dependent upon scientists to create pest-resistant varieties. The ICAR system created several new varieties in the 1990s and 2000s, but most of them failed. Finally, in 2007, a new variety JS 95-60 was released, which was of very short duration (82–88 days) and higher germination. It was also resistant to stem fly and root rot and soon began to replace JS-335. By 2020, each of these varieties were grown on forty per cent of the total area under soybean.[37] But pest attacks continued unabated. In 2020, an attack of the gram pod borer, the tobacco caterpillar, and the green semi-looper along with the spread of the yellow mosaic virus led to significant crop losses across the soybean belt.[38]

Over the last 100 years or so, scientists in private research firms and in public labs have been constantly playing catch-up with newly evolving pests and weeds – the trend towards Bt crops and herbicide resistant crops emerged from this same backdrop. As one fails, the next one is ready. But there is no respite. In the past, farmers were able to use inter-cropping or mixed cropping to prevent spread of diseases and pests to epidemic proportions and to counter the

34   The list refers to older pesticide formulations such as endosulphan, monochrotophos, trizophos, alpha cypermethrin, and quinalphos.

35   Coragen is the brand name of the molecule Rynaxypyr, for which Dupont filed a patent in August 2002. H. Damodaran, 'MNCs going all out to push new-gen pesticides', *The Hindu Business Line*, 27 June 2013: http://www.thehindubusinessline.com/industry-and-economy/agri-biz/mncs-going-all-out-to-push-newgenpesticides/article4857294.ece.

36   Interview with Dr K.K. Nema, Sehore Agricultural College, Sehore, Madhya Pradesh, 25 Jan. 2006.

37   See Soybean Processors Association website: https://sopa.org/MajorSoybeanvarietiesinIndia.pdf; and Directorate of Oilseeds Development, Ministry of Agriculture & Farmers Welfare, Government of India: http://oilseeds.dac.gov.in/Variety/Soybean.aspx.

38   P. Biswas, 'Reports of crop damage in MP, Maharashtra push up soyabean price', *The Indian Express* 28 Oct. 2020: https://indianexpress.com/article/india/reports-of-crop-damage-in-mp-maharashtra-push-up-soyabean-price-6903625/; D.K.Jha, 'Mosaic virus, pest attacks in three states to hit soybean output this year', *Business Standard* 23 July 2020: https://www.business-standard.com/article/companies/mosaic-virus-pest-attacks-in-three-states-to-hit-soybean-output-this-year-120072301291_1.html.

302

Richa Kumar

properties of one plant against another.[39] Friendly pests and natural predators were utilised. But the arsenal of the farmer drew upon a multi-crop system, which has been entirely set aside. Instead, farmers are completely dependent on private companies to develop ever more potent pesticides.

On a monoculture field, everything else is designated as weeds – unwanted plants that must be 'killed' to give the monoculture variety exclusive access to soil nutrients. However, many 'weeds' were naturally occurring plants in the area, which were often eaten by animals. The use of herbicides not only killed traditional weeds, which were suitable to the climate and could be eaten by cattle, but also led to the explosion of minority weeds such as *dudhi* which were inedible.[40] The advent of herbicides further militated against inter-cropping as the herbicide for soy (a broad leaf plant) was fatal to maize and lentils (narrow leaf plants).

The biggest problem resulting from soy farming in the dryland region of Malwa was its impact on the availability of groundwater. Soy farming replaced important farm practices related to water management in an agrarian system that took into account low water availability. By draining water instead of replenishing it and by financing farm investment in water-extraction technologies, soy played a crucial role in transforming Malwa ecologically.

Along with displacing sorghum, maize, lentils and groundnuts, soy spread in areas which were left fallow in the kharif season in Malwa. Land was historically left fallow because the deep black soil of the Malwa plateau became waterlogged in the rain. Food crops were grown only on high, well-drained ground.

But it was this waterlogging which was crucial to sustaining the water harvesting systems of this dryland region where the only source of water was the rain that had been stored in ponds or in the ground (thus enabling groundwater recharge). Nearly sixty per cent of the land remained uncultivated in this way. In the winter season, unirrigated wheat varieties were grown primarily on residual soil moisture in deeper and water retentive black soils.[41]

39   As Dr K.J. Singh explained, 'Mix soybean and maize and broadcast the seed. The corn will go visibly tall and you can sell the cobs. Birds will sit on the maize and eat insects on the soybean.' He also recommended that early, mid and late varieties were planted by farmers so that insect problems did not linger. But farmers were constrained by soil fertility and water availability in choosing these varieties. Interview with Dr K.J. Singh, Entomologist, Sehore Agriculture College, Sehore, Madhya Pradesh, 8 Feb. 2006.

40   Interview with Karan Singh Pawar, the ex-MLA from Dhar district in Madhya Pradesh and a large landowner, 24 Aug. 2006.

41   P.S. Vijay Shankar, *Funders' Report on Agrarian Programs* (Bagli, Dewas District: Samaj Pragati Sahyog, 2006).

14. From Self-Reliance to Deepening Distress

Soy, however, requires well-drained fields. Although it grows in water-logged conditions, the yield is two to three times higher in well-drained soils. So farmers then shifted from storing/saving water during the rains to draining out all the water from the field. Moreover, after taking a crop of soy, the residual soil moisture was inadequate to support an unirrigated wheat crop. As soy acreage increased, the survival of unirrigated wheat became a question mark.

In the 1970s and 1980s, irrigated varieties of wheat popularised under the green revolution came to Malwa. Around the same time, rural electrification was gaining momentum and technologies for more efficient water extraction from wells, such as diesel and electric motors, followed suit. Soon, the technology for digging tubewells, known as cutter machines, arrived. Lacking any other source of surface irrigation and with well-irrigation being insufficient, farmers turned wholeheartedly to adopt tubewell technology to grow irrigated wheat. Tubewells replaced wells and ponds as a source of water. The new varieties of water-intensive wheat became popular, which increased wheat output by four times. 'While only 13% of wheat area was under irrigation in 1970–71, nearly 90% of the wheat area was under irrigation by 1993-94.'[42]

The increased income from soy gave farmers the resources to invest in technologies to extract groundwater in winter and, consequently, encouraged the planting of high yielding varieties of wheat and water-intensive chickpeas over a larger and expanding area of cultivation. The growing remuneration from soybean and wheat (the latter thanks to increasing Minimum Support Price and purchase by the state), further promoted agricultural intensification in Malwa by facilitating the purchase of newly available mechanised technologies such as tractors and allied implements for tilling, sowing, weeding, spraying and threshing. Farmers even began growing short duration potatoes, peas and onions before planting irrigated wheat.

But all this came at the cost of the future. 'There was enough ground water available to exploit and double crop the area until 1984–85', according to Karan Singh Pawar, the erstwhile Raja of Dhar (in the Malwa region). But, progressively, the water table began to fall. In 2006, water levels were beyond 450 feet and tubewells were being dug to reach 1,000 feet.

The shallow, accessible surface aquifers, which are within the soil strata and pervious rock used to get replenished within the first two weeks of the mon-

42   Ibid.

soons through the practice of kharif fallows. But the entry of soy had stopped this practice. Without giving it a chance to percolate and fill these surface aquifers completely, farmers start draining the water out of the fields to save the soy. By the 1990s, wells had dried up and the shallow water was all gone.

Deeper tube wells enabled farmers to access ancient water embedded in volcanic rock making up the Malwa plateau. Water was trapped in crevices or fracture zones of igneous rocks as various lava flows cooled and hardened one on top of the other in what is known as the Deccan trap. These large hard rock aquifers (accumulated over several thousand years), were now being emptied out through tubewell technology which made this 'primary water' accessible.[43] Unfortunately, these aquifers are not rechargeable through surface water or rain water because the rock is impervious.

The consequence has been reduced water availability throughout the year and unbearable scarcity in the summers as most tubewells stop functioning after February or March (and some even in December). By the time April comes around, a handful of tube wells far away in the fields supply a dribble of water to service the everyday cooking, cleaning and washing needs of vil-lagers. In the monsoons, on the other hand, the run-off from the fields adds to the drainage from the increasing built-up area in the region, contributing to recurring floods.

Farmers recognised the ecological fragility of sustaining this cropping pattern year after year, especially in the face of a failed year of rain when the soil aquifers did not get recharged. Yet, they did not want to shift from water intensive varieties of wheat because, even with less irrigation, they yielded more than unirrigated varieties. Despite a ban on boring new tubewells in most districts of Malwa, farmers continued to dig them in the hope of striking water.

Leaving land fallow during the soy season to promote recharge of surface aquifers also caused immediate economic loss. The value of land is much higher in terms of productive potential by growing soy or by selling it as real estate. Converting part of it into a farm pond is considered to have too great an economic opportunity cost. Unfortunately, the value of water can only be imputed today by proxy calculations such as the cost of extraction or piping. There is no intrinsic value given to it otherwise. Unless the creation of water bodies is equally valued, the water crisis will continue to spiral out of control.

---

43   P.S. Vijay Shankar, 'Four decades of agricultural development in MP', *Economic and Political Weekly* **40** (48) (2005): 5014–24, at 5020.

Dryland regions across India have had a long history of valuing water and creating and maintaining perennial water bodies. But this practice was abandoned with the coming of colonialism in many parts of British India.[44] Since Malwa was never directly ruled by the British, its ponds and lakes survived – until the onslaught of technologies that used up water very quickly and those that enabled frenzied water extraction, along with those which financed the process.

Deepening Distress: The Reproduction Squeeze

In India, growing soy monocultures using industrial inputs has meant riding a technological treadmill that is threatening to push people off the land. Despite the growing remuneration from soy over the years, independent family farmers in central India have found it increasingly difficult to reproduce themselves year after year. The constancy of crisis has meant that they are able to tide over in the years when the remuneration is good enough to cover the costs, but in other years, they find themselves under mountains of debt – squeezed between the inevitable risks of monoculture production and the debilitating risks of a global market.

In years when soybean was highly remunerative (2004, 2007, 2008), they recalled the investment in farming and increase in living standards it had engendered. But in other years (2006, 2009), when remuneration barely covered the ever-rising cost of production, farmers pointed out the damage to the land, soils, rising price of more potent pesticides and herbicides leading to indebtedness and everything else that is wrong with the pursuit of single-crop productivity.

A further irony is that there is no direct connection between a farmer being highly productive and that same farmer obtaining a higher market price. In 2009, farmers in Ranipura village broke all previous records of soybean productivity. Farms produced 36 quintals a hectare! The ordinary ones produced 20 or 24 quintals a hectare. But income? The harvest of 2009 saw some of the lowest price points in the last four years. Ultimately, farmers remained where they were.

However, even with its reducing productivity and increasing cost, soy was still a better crop compared to other possible substitutes. Farmer Omkar Singh Chauhan explained that, although local varieties of sorghum gave a

---

44   M. Davis, *Late Victorian Holocausts: El Niño Famines and the Making of the Third World* (New York: Verso, 2002).

good market rate, it took five to six months to grow, delaying the planting of the rabi crop. Hybrid Shankar *jowar* (sorghum) grew in 100–120 days but the rate was less and it was not very tasty – hence few farmers chose to grow it. Soy required less manual labour to harvest, thresh, grade and store the crop as compared to corn and sorghum.

Moreover, in the Indian context, prices can never precipitously fall because demand from the local processing industry is three times the total production. Liberal government support for establishing industries in backward districts, and entrepreneurs enticed by the prospects of profit in the export of deoiled cake, led to the creation of excess and idle capacity. These plants were eager to purchase soy to keep the unit functioning for as many days in the year as possible.[45] Government policies disallowing the import of soybean seeds forced the industry to look at the domestic market for raw material.[46]

In 1984, the processing capacity was nearly 2 million tons against production of 0.99 million tons.[47] Processing capacity increased from 19 million tons in 2006 to 22 million tons in 2014 and 30 million tons in 2017.[48] Production, on the other hand, grew from 6 million tons in 2006 to a high of 12 million tons in 2012 and has been hovering between 8–10 million tons since then.[49] Thus, despite price fluctuations in the international market, domestic demand did not falter. And soybean farmers benefitted from growing a crop where prices did not crash.

At the same time, the yellow revolution continues to promote an agricultural system that is parasitic upon the future by drawing resources from the environment without accounting for the cost of polluting them or their

---

45   In the light of excess capacity, processing factories had unsuccessfully petitioned the government to allow imports of raw soybean (see S. Persaud and E. Dohlman, 'Impacts of soybean imports on Indian processors, farmers and consumers', *Journal of Agribusiness* 24 (2) (2006): 171–86). The capacity utilisation rate of all solvent extraction plants in India was between 30 and 40 per cent according to the Solvent Extractors' Association of India (see Persaud and Landes, 'The role of policy'; S. Persaud, 'Impacts on India's farmers and processors of reducing soybean import barriers, OCS-19J-02', *USDA, Economic Research Service A Report from the Economic Research Service*, Oct. 2019: https://www.ers.usda.gov/publications/pub-details/?pubid=95138).

46   Persaud and Landes, 'The role of policy'.

47   EPW, 'Soybean processing: Ill conceived expansion,' *Economic and Political Weekly* 20 (5) (1985): 176; SOPA, 'India's soyabean production'.

48   M. Ahuja, 'High taxes lead to closure of soya processing units in MP', *Hindustan Times*, 26 Dec. 2014: http://www.hindustantimes.com/indore/high-taxes-lead-to-closure-of-soya-processing-units-in-mp/article1-1300597.aspx; Persaud, 'Impacts on India's farmers.'

49   Data from SOPA: https://www.sopa.org/statistics/world-soybean-production/. Processing capacity in Madhya Pradesh alone was 12.5 million tons (Ahuja, 'High taxes').

unavailability in the future. With a depleting pool of fossil fuels reserves to produce fertiliser and the growing carbon costs of the production of fertilisers, pesticides and herbicides, the future survival of such an agro-industrial complex is in question all over the world. Much of the environmental and health cost is externalised. Poor farmers and labourers pay for it when they spray the pesticides and fall sick. Consumers in urban areas pay for it with their poisoned food. The land pays for it with soil toxicity. And farmers pay for it in the long run when it becomes harder and harder to stay on the treadmill.

Increasingly, farmers are in a market where prices of inputs and output are controlled by a handful of giant agribusinesses. At the same time, they are dependent on a state which is unable to deliver fertiliser, electricity and petroleum products on time and at a reasonable price. Scientists (in the public and private research system) keep the promise of technology alive even though their research on new seeds and agrochemicals is playing catch-up with the ever-evolving pests and genetic vulnerability engendered by single varieties. But ultimately, to farm means to work with nature and its variability and, despite irrigation and modern technology, the vagaries of the weather and the rain, and now climate change, are ever-present. As farmers often said, one can never know how much will ripen until it [soybean, wheat, maize, etc.] is safely [threshed and stored] inside the house. It is no wonder then, that the very existence of farmers in India (and in other parts of the world) is frequently threatened with some form of crisis, with the most common one being debt.[50]

Deepening Distress: The Nutrition Factor

Despite the yellow revolution, India witnessed a growing scarcity of edible oil beginning in the mid-1990s, due to a shift of land away from mixed cropping of oilseeds-cereals-pulses to crops like wheat, rice and sugarcane that were given special incentives and purchase guarantees by the government.[51] In addition, liberalisation of edible oil imports in 1994 paved the way for cheap palm oil from Indonesia and soybean oil from Argentina to flood the domestic market, despite high tariffs of up to 300 per cent, thus further

50   R. Kumar, N. Agarwal, A.R. Vasavi and P.S. Vijayshankar, 'State of Rural and Agrarian India Report 2020', Network of Rural and Agrarian Studies (NRAS), 30 Nov. 2020: http://www.ruralagrarianstudies.org/wp-content/uploads/2020/11/State-of-Rural-and-Agrarian-India-Report-2020.pdf.

51   Import dependence of edible oils was 2–5% from 1961–75 and 36–47% from 1976–87. With the tightening of import restrictions it came down to 4 per cent in 1993 (Sekhar, 'Agricultural price volatility', 4733).

Richa Kumar

disincentivising domestic producers.[52] From 1994 to 1999, imports rose to over 50 per cent of domestic edible oil consumption,[53] from 2000–2009 they ranged from 40–50 per cent[54] and from 2009 to 2016 they went up to nearly 75 per cent.[55]

Companies like Adani Wilmar and General Foods (Ruchi group) set up edible oil refineries at ports in Gujarat and Maharashtra to process imported crude palm oil as well as crude soybean oil in large quantities. Since these companies were already processing soybean meal as an important export product and bringing in valuable foreign exchange, they were able to situate these refineries for imported edible oil in port based Special Economic Zones, thus benefitting immensely from government policies providing subsidised land, energy and water, and tax breaks.[56]

Palm oil and soybean oil became cheaper substitutes for other edible oils, especially in the manufacture of vanaspati (refined, hydrogenated vegetable shortening), a very popular substitute for more expensive ghee (clarified butter). Together, they displaced a multitude of traditional and regional oilseeds and oils from farms and plates across India including mustard, sesame, linseed, coconut and groundnut, as farmers found these crops unremunerative.[57]

Moreover, monocultures of soybean replaced many other rainy season crops including maize, sorghum, groundnut, lentils, sesame and several other

52   E. Dohlman, S. Persaud and R. Landes, *India's Edible Oil Sector: Imports Fill Rising Demand.* Electronic Outlook Report from the Economic Research Service. United States, Department of Agriculture, OCS-0903-01, Nov. 2003; B. Dorin, 'Globalization and self-sufficiency: A tentative revolution in oilseeds', in F. Landy and B. Chaudhuri (eds), *Globalization and Local Development in India: Examining the Spatial Dimension*, pp. 171–98 (New Delhi: Manohar, 2004).

53   Sekhar, 'Agricultural price volatility'; P.V. Shenoi, 'Oilseeds production, processing and trade: A policy framework', *Occasional Paper 26. Department of Economic Analysis and Research, National Bank for Agriculture and Rural Development* (Mumbai, 2003): https://www.nabard.org/demo/auth/writereaddata/File/OC%2026.pdf.

54   A.A. Reddy, 'Policy options for India's edible oil complex', *Economic and Political Weekly* **44** (41) (2009): 22–24.

55   Persaud, 'Impacts on India's farmers.'

56   In 2021, the government incentivised palm monocultures in the forested areas of northeast India and the Andaman and Nicobar Islands in an effort to boost domestic edible oil production notwithstanding the environmental and nutritional problems associated with this policy (First Post Staff, 'Why Centre's plan to expand oil palm cultivation across North East worries environmentalists', *First Post*, 6 Oct. 2021:https://www.firstpost.com/india/why-centres-plan-to-expand-oil-palm-cultivation-across-north-east-worries-environmentalists-10030361.html; M. Ramnath, 'With palm oil expansion, India is blazing a trail to a parched future', *The Wire*, 12 Aug. 2018: https://thewire.in/agriculture/india-palm-oil-cultivation-foreign-currency).

57   Dohlman et al., *India's Edible Oil Sector*; Dorin, 'Globalization and self-sufficiency'.

food and fodder crops in Malwa. This meant a shift in diets away from the nutritious staple foods of the past – maize and sorghum bread, lentil soup and roasted groundnut, sesame and sorghum snacks. Nutritional needs were expected to be fulfilled through income earned from market products of the farm or from wages. Local agricultural production had been de-linked from local food consumption.[58]

The conversion to cash crop monoculture farms compromised the ability of rural people to access a nutritionally wholesome diet.[59] This was compounded by national level shortages of lentils and oilseeds leading to a rise in retail prices of these products after liberalisation in the 1990s. While imports filled the gap, they were mainly supporting middle class consumers in urban areas and the rural elite. These crucial parts of a wholesome balanced diet were unaffordable for the poor.

Finally, soybean processing made the solvent extraction process of separating oil and meal from oilseeds popular in the 1980s and other oilseeds also began to be processed using this method. In addition, modern refining techniques created a homogenised edible oil product by heating the oil, deodorising and decolouring it using chemicals and also through the process of hydrogenation. This was linked to the marketing of a homogenised (blended) oil under the label of 'refined' edible oil as a modern choice for urban consumers.

This was in complete contrast to the past where traditional methods of oil extraction (cold pressing) produced oils which retained their distinct colour, flavour and smell. These characteristics of oil were an integral part of food and culinary practices around the world. Thus, Indian cooking (in the east) was built upon the distinctiveness of mustard oil, in the north and the west it was sesame and groundnut oil and in south India it was coconut oil. Similarly, west African consumers were comfortable with the colour and smell of traditionally processed palm oil. However, making soybean and palm oil palatable to Indian consumers required stripping them of their distinctive aromas and making them part of a generic edible oil.[60] Growing acceptance

58   For a similar argument in South Africa, see R. Patel, *Stuffed and Starved: Markets, Power and the Hidden Battle for the World's Food System* (Toronto: HarperCollins Canada, 2010), p. 239–40.

59   Small quantities of soy are processed into food items but rural families do not eat soybean products such as tofu, soya milk or nuggets. However, many of them use soybean oil for cooking since it is cheaper than other oils, albeit with a slightly altered taste.

60   R. Sharma, 'India's colourless revolution: Replacement of traditional oils by soy', *Independent Science News*, 12 May 2008: https://www.independentsciencenews.org/health/indias-colourless-revolution-replacement-of-traditional-oils-by-soy-and-palm-oils/; B.M. Vyas and M. Kaushik,

of oil from 'nowhere'[61] has led to a shrinkage of diversity of cooking practices, recipes and tastes, further reducing the demand for traditional oils.[62]

However, scientists and nutritionists are calling into question the health implications of modern refining techniques.[63] India is seeing a resurgence of expensive cold-pressed regional oils for an elite, health-conscious consumer. Whereas, in the past, poor households had access to cold-pressed oil for cooking through small scale oil presses at the village level, the scale and scope of modern oilseed production and processing, driven by soybean, changed all that.[64] Ironically, the food that was supposed to help India overcome protein malnutrition amongst the poor has become a scourge for them – it has wiped the nutrition from their plates.

Conclusion

It is a sad irony that soy took off in India as a commercial cash crop and not a food crop. The government's Public Distribution System (PDS) could have adopted soy as a cheap source of protein for the masses[65] as a nutritional supplement to the carbohydrates made available by the green revolution.[66]

'How India was stripped of its Atmanirbharta in the edible oil industry', *The Wire*, 4 Nov.2020: https://thewire.in/political-economy/india-edible-oil-self-sufficiency.

61   P. McMichael, 'A food regime genealogy', *The Journal of Peasant Studies* 36 (1) (2009): 139–69.

62   M. Hiraga, 'Sucked into the global vegetable oil complex: Structural changes in vegetable oil supply chains in China and India, compared with the precedents in Japan', in S. Hongladarom (ed.), *Food Security and Food Safety for the Twenty-first century*, pp. 179–94 (Singapore, Berlin: Springer, 2015).

63   A.H. Lichtenstein, *'Dietary fat: A history'*, *Nutrition Reviews* **57** *(1) (1999): 11–14*; Sharma, 'India's colourless revolution'; S. Bhardwaj, S.J. Passi and M. Anoop, 'Overview of trans fatty acids: Biochemistry and health effects', *Diabetes & Metabolic Syndrome: Clinical Research & Reviews* **5** (3) (2011): 161–64; V. Dhaka et al., 'Trans fats – sources, health risks and alternative approach. A review', *Journal of Food Science and Technology* **48** (2011): 534–41; S.C. Manchanda and S.J. Passi, 'Selecting healthy edible oil in the Indian context', *Indian Heart Journal* 68 (4) (2016): 447–49; S. Gharby, 'Refining vegetable oils: Chemical and physical refining', *The Scientific World Journal* (2022): 6627013.

64   K. Achaya, 'Ghani: A traditional method of oil processing in India', *FAO Food, Nutrition and Agriculture –11 – Edible Fats and Oils'* (1994): http://www.fao.org/docrep/T4660T/t4660t0b.htm.

65   However, questions have been raised regarding the suitability of factory-processed soybean for human consumption. A compilation of research on this is available at the Harvard School of Public Health website (M.H. Menu, 'Straight talk about soy', The Nutrition Source, Harvard School of Public Health, 12 Feb.2014: http://www.hsph.harvard.edu/nutritionsource/2014/02/12/straight-talk-about-soy/). Also see B.T. Hunter, 'The downside of soybean consumption', *American Nutrition Association NOHA NEWS* **36** (4) (2001).

66   The PDS system did not have extensive reach at that time. But even today, the government has not seriously considered the proposition (See S. Gillespie, J. Harris and S. Kadiyala, 'The

But, instead, as Prof. Bhatnagar put it, 'we exported our cheapest source of protein to fatten the cattle of affluent foreign countries', leaving the poor still malnourished. Moreover, the yellow revolution did little to dent the enormous edible oil import bill that India has been saddled with and, instead, ended up pushing unhealthy oils into people's diets.

Soy's continued persistence across farms and plates in India is a result of multiple factors – its physical resilience on the farm across a wide range of natural conditions, excess processing capacity brought on by governmental largesse and possibilities for profit across the global supply chain, import lobbies and transnational agribusiness linkages shaping the edible oil market along with a dominant perspective in policymaking that views agriculture as situated outside of its ecological and social context. By making visible some of the linkages that form the Soyacene, this paper hopes to contribute towards a reframing of the yellow revolution in India – one that recognises the drastic implications of soybean for the future of farmers and farming.

agriculture-nutrition disconnect in India: What do we know?' *IFPRI Discussion Paper 01187*, June 2012. Available at https://www.researchgate.net/publication/254416377_The_Agriculture-Nutrition_Disconnect_in_India_What_Do_We_Know).

CHAPTER 15.

DOMESTICATING THE EXOTIC: SOYA LIVELIHOODS AND ECOLOGICAL IMPACTS IN MODERN ZIMBABWE, 1900–2021

Vimbai Kwashirai

Introduction

'Domesticating the exotic' begins the soya story around 1900 at the Salisbury Experiment Station (SES) where Rhodesian agronomists researched and tested the acclimatisation and adaptation of several foreign and local cultivars to the then Southern Rhodesia's (Zimbabwe) semi-tropical environments and agricultural regions. Throughout the ninety-year British colonial period (1890–1980) soya trials, production and acreages were largely insignificant because of cautious famer attitudes towards the crop and consumer antipathy to the taste of its products. It was not until the late 1970s that Zimbabwe was believed, albeit on a small scale, to be the second largest soya producer in Africa after Egypt. After independence in 1980, post-colonial Zimbabwean farmers continued to embrace soy production on a piecemeal basis, despite recognising its multiple benefits to livelihoods, livestock, employment and environmental regeneration.

This study explores the domestication and expansion of soybeans in colonial and post-colonial Zimbabwe from 1900 to 2021. In these 121 years, over 2,000 varieties of soyabean have been developed, eight of which nodulate promiscuously in most soils of Zimbabwe. The eight are made up of four arid and four irrigation major hybrids bred from genotypes originally collected in East Asia at the turn of the twentieth century.[1] Promiscuity in nodulation enhances soybean cultivation in a range of semi-tropical ecologies where lack of suitable inoculants would otherwise preclude growing the crop.[2] Small-

---

1   D.M. Khojely et. al., 'History, current status, and prospects of soybean production and research in sub-Saharan Africa', *The Crop Journal* **6** (3) (2018): 226–35.

2   Ibid.

THE AGE OF THE SOYBEAN: 312–331 **doi:** 10.3197/63800040695086.ch15
© Vimbai Chaumba Kwashirai

holder farmers require only access to sufficient seed to be able to efficiently farm soy, which promotes diverse livelihood outcomes and better-quality household nutrition from the extraordinary protein and oil content, cash income from sales of the crop and inputs of nitrogen that rejuvenates soil fertility and contributes to the sustainability of rotational cropping systems.[3] Promiscuous soybean varieties therefore signify a highly suitable technology for cultivation by smallholder farmers. Also, the exploitation of varieties with greater yield potential together with rhizobial inoculants has largely been an appropriate technology for commercial production of soy by farmers who have ready access to agricultural inputs. Soy production also saves the country scarce foreign currency.

Colonial Soya: 1900–1980

Soy cultivation probably first reached the east African coast in the late 1800s as a result of long-established trade with eastern China. It spread to South Africa at Cedara in Natal and the Transvaal, spreading to Zimbabwe from about 1900. Rand gold mines used soya to improve mineworkers' diets.[4] Imperial European nations considered African colonies as strategic sources of raw materials including minerals and crops such as cotton, tobacco, maize and soya. British peripheries like South Africa and Zimbabwe led in natural resource extraction and exploitation to benefit the UK metropole. White settlers in Zimbabwe pioneered soya domestication from the 1900s, particularly from 1906 at the SES, largely for purposes of improving soil fertility. Zimbabwe heavily borrowed agricultural and other social, economic and political knowledge from South Africa, including soya taming experiments and production. Generally, colonial Zimbabwe learnt much of its early administrative, agricultural, forestry and mining policies and practices from South Africa.[5] Under the 33 years of British South Africa Company rule (1890–1923), particularly the peak era of land alienation from the indigenous Africans, five soy varieties were cultivated both to regenerate soil quality in Zimbabwe and for fattening livestock with maize-soya blended feeds.[6]

---

3   Khojely et al., 'History, current status and prospects'.

4   National Archives of Zimbabwe, (NAZ), FG209, H. Burkill and C.S. Crosby, 'The flora of Vavau, one of the Tonga Islands', *Journal Botany* **35** (242) (1900): 20–65, at 20; B.-D. Sawer, 'Soyabean', *Rhodesian Agricultural Journal* (*RAJ*) **3** (1906): 19.

5   Annual Report of the Transvaal Department of Agriculture, 1935, p. 7.

6   Ibid.

Post Company domination, robust soya experiments and breeding at the SES took fifteen years, 1925–1940 resulting in the release of several *Hernon* strains.[7] Large scale commercial soy production took off during World War II in 1940, an imperative of increased wartime food and oil demand. In 1941 Arnold, Manager at the SES reported on soya production, tests and first palatability trials for boiled soybeans.[8] He notes that:

> Requiring lengthy cooking, they had a nutty flavor. Soya took too much time and fuel to cook and the taste was not well accepted by many farmers and families. The mines and other employers of native labor may eventually purchase large quantities of them … During the cooking tests several Europeans ate the beans and freely expressed a slight liking for them. Native employees at the station soon came to like the soybeans parched, toasted, or boiled and mixed with kaffir or haricot beans. Shortly after World War II a brief attempt was made, unsuccessfully, to introduce tempeh soya to Zimbabwe.[9]

However, at tortoise pace, it was a decade after the war that soy production gathered a sluggish momentum. For example, in 1954–55, the total acreage for soy cultivation in the country was only 770 hectares (ha), from which 326ha was for seed, yielding 174.2 tonnes (t) of soybeans translating to a harvest of 534kg/ha. In the next farming season 1956–57 the total acreage more than doubled to 1,886ha, 404.8ha for seed, yielding only 230.6t of soybeans or 569.7kg/ha.[10] In the Federation years – 1953–1963 – soybeans were sold direct to the trade and the Grain Marketing Board (GMB) acted as a residual buyer. Soy was planted on few farms at a time when the federal authorities actively sought to achieve food and cash crop self-sufficiency through diversified production away from maize – *chibage*; cotton – *donje* and tobacco – *fodya*. Efforts to introduce soy in food rations among African agricultural and mine workers partly failed because it was considered too tough and requiring too long in cooking. Millers too, preferred not to

---

7   W. Shurtleff and A. Aoyagi, *History of Soybeans and Soyfoods: 1100 B.C. to the 1980s* (Lafayette: Soyinfo Center 2007).

8   Ibid.

9   NAZ, DG112, Salisbury Experiment Station Annual Report 1941. Tempeh is a traditional Javanese soy product that is made from fermented soybeans. It is made by a natural culturing and controlled fermentation process that binds soybeans into a cake form. From the 1960s aid agencies such as FAO and USDA introduced wide-ranging soya products in southern Africa including soyfood, ProNutro, soy flour and corn-soy-milk, corn-soy and wheat-soy blends to the 1980s (Shurtleff and Aoyagi, *History of Soybeans*).

10   NAZ, Agricultural Experiment Station Report, 'Results of Experiments Seasons 1959–60,' *RAJ* **58** (3) (1961): 154–66.

15. Domesticating the Exotic

mix soy flour with wheat flour for bread because of consumer boycotts.[11]

Nonetheless, the Federal government established the Griffith Committee to explore the potential of expanding soy production in Malawi, Zambia and Zimbabwe. The committee noted that low soy uptake and yields were also due to low profits for the majority of soy growers.[12] After the dissolution of Federation in 1963, Rhodesian settlers greatly expanded soy breeding programmes on farms around Salisbury. Yields slightly increased and for the four years 1964–1968 the GMB guaranteed relatively favourable soybean prices. From 1967, production showed signs of rapid expansion in many parts of the country – see Table 1 below. During the liberation struggle for independence in 1974, resilient soy production reached 10,000t, considered by government as the take-off tonnage as the national average yield reached 1,700kg/ha with variety trials reaching 5,500kg/ha.[13] In 1977, Egypt and Zimbabwe were viewed as the top soya producers in Africa. By 1981, total soya production in Africa reached 265,000t, mostly from the four largest producers: Egypt (136,000t), Zimbabwe (97,000t), Nigeria (80,000t) and South Africa (26,000t).[14] However, soybeans and soyfoods still played a very minor role in much of colonial Africa.[15] The resilience of food and cash crop production in Zimbabwe defied the intensification of the armed liberation struggle beginning with the 1966 Chinhoyi battle and generally spreading nationwide and gaining intensity from the mid-1970s onwards. One outstanding feature of the war was the imposition of rural concentration keeps in which emergency powers and curfew law applied. Table 1 below shows the slow growth in kg/ha soy yields by large scale commercial white farmers, medium African commercial growers and peasant cultivators. The increased soy output alongside other crops such as maize, groundnuts and tobacco defied the intensifying and spreading guerrilla warfare. However, the state supported and subsidised white settler producers more than Africans.

11   Ibid.

12   NAZ, F57, Griffith Committee Report, 1960, p. 11.

13   J.R. Tattersfield, 'The role of research in increasing food crop potential in Zimbabwe', *The Zimbabwe Science News* **16** (1) (1982): 6–10 and 24.

14   NAZ, NM230, *Soybean Digest*, Nov. 1964; *FAO Production Yearbook* 1969.

15   Shurtleff and Aoyagi, *History of Soybeans*.

Year	A	B	C	Total
1969	1 310	829	-	2 139
1970	684	292	-	976
1971	1 140	192	319	1 651
1972	1 591	617	450	2 658
1973	1 067	178	330	1 575
1974	1 918	568	374	2 860
1975	1 754	442	500	2 696
1976	1 808	794	324	2 926
1977	1 938	689	650	3 277
1978	2 010	735	780	3 525
1979	2 108	750	750	608

Table 1.

Kg/ha Soya Yields in Colonial Zimbabwe, 1969–1979

A: White commercial farmers; B: African Purchase Area farmers; C: Peasant communal farmers

Sources: K. Muir, 'Crop production statistics 1940–1969', Salisbury, University of Zimbabwe, Department of Land Management, Working Papers, 4/81, May 1981, p. 23; R. Tattersfield, 'The role of research in increasing food crop potential', Salisbury, *Science News*, 1981.

Post-colonial Soya: 1980–2017

At independence, from 1980, soy diffused spontaneously to most smallholder farming areas in the higher rainfall zones of Zimbabwe led by Mashonaland Central Province (MCP), followed by Mashonaland West Province (MWP), Mashonaland East Province (MEP) and the Midlands. Soy production remained insignificant or non-existent in the other six provinces of Zimbabwe – see Figure 1 below and Map 1.[16] A minority of farmers grew soy in Masvingo and Matabeleland under supplementary irrigation. Soy producers were represented by the three main farmers' unions: the Commercial Farmers Union (CFU), Zimbabwe Farmers Union (ZFU) and Zimbabwe National Farmers Union (ZNFU). With the CFU, soy farmers fell under the Commercial Oil Producers Association (COPA). As with the many unions,

16   (MCP); Districts of Glendale, Bindura, Shamva, Mvurwi, Centenary and Mt Darwin, (MEP); Goromonzi, Beatrice and Marondera, (MWP); Norton, Banket, Chinhoyi, Mhangura, Doma, Chegutu, Chakari, Karoi, Tengwe, Kadoma, (Midlands); Kwekwe and Gweru, (Manicaland); Rusape, Odzi and Mutare.

several institutions conducted soya research, including the Zimbabwe Oil Pressers Association (ZOPA), Soya Bean Task Force (SBTF), Agriculture Research Council (ARC) and Zimbabwe National Soya Bean Commodity Association (ZNSCA).

Grain legumes in general and soya in particular, replenish soil nitrogen, providing much needed protein in maize-based diets. In the four major soya belts of Zimbabwe, soybeans (*Glycine max* (L.) Merrill) emerged as a leading crop that regenerates soil fertility and improves income, food and nutritional security among smallholder farmers.[17] Despite such wide-ranging benefits, soya adoption among smallholder farming systems in post-colonial Zimbabwe and sub-Saharan Africa more generally remained very lethargic. The majority of smallholder farmers grew soy on small plots with grain yield barely exceeding 500kg/ha.[18] Low soy uptake was attributed to biophysical, social and economic conditions found in developing countries. Key determinants in soya yields in Zimbabwe were soil types and conditioning, land preparation, timing the weather, varietal choice, weed control and irrigation. During President Mugabe's reign from 1980 to 2017, commercial farmers and ARDA estates grew soy at highly mechanised levels and supplied 65 per cent of the country's soy output.[19]

In contrast to soy, maize has invariably been the dominant staple crop across most of southern Africa- foremost in Zimbabwe such that over eighty per cent of the smallholder farms are maize fields, *minda*. With relative peace in 1980, soya production in Zimbabwe increased to 100,000t from 97,000t in 1977.[20] Thereafter, production declined, stagnating below 100,000t until 1989 when it increased to 125,000t due to exceptionally good weather.[21] In the first decade of independence of 'socialist rhetoric', the total national acreage under soy also remained below 60,000ha except for 1988 and 1989 when it rose to 63,000ha and 70,000ha respectively.[22] This was because, unlike other crops, soy has huge benefits of being a small acreage but large yield crop. In the years 1986–1990, soy production increased due to government support and subsides offered to smallholder farmers: see Figure 2 below.

17   B. Zamasiya, et al. 'Determinants of soybean market participation by smallholder farmers in Zimbabwe', *Journal of Development and Agricultural Economics* **6** (2) (2014): 49–58.

18   Ibid.

19   Ibid.

20   Agritex October 2012. Available at http://www.izimbabwe.co.zw/news/local.

21   SNV, 'Zimbabwe Soya Bean', 16.

22   Ibid.

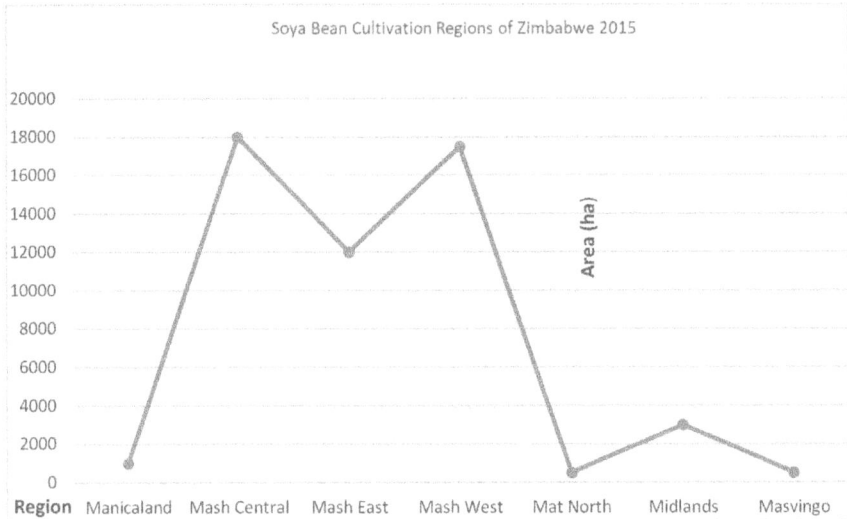

Figure 1.

Soy cultivation regions of Zimbabwe, 2015

Source: Netherlands Development Organization, (SNV), Zimbabwe Soya Bean Sub-Sector

Study, 2012: https://docplayer.net/42638959-Zimbabwe-soya-bean-sub-sector-study-with-support-from.html

During the 1990s World Bank and IMF inspired economic structural adjustment programme, the Zimbabwe Soyabean Promotion Task Force (ZSPTF) launched an initiative to popularise and promote soy as a beneficial smallholder crop. From an initial 55 participating farmers, soy producers rapidly increased within four years to over 10,000 growers by 2000. The ZSPTF initiative assisted large numbers of smallholders to cultivate soy, exploding long-held attitudes and beliefs in Zimbabwe that soy was unsuitable for smallholders.[23] For instance, between 1999–2002 soy production rose because of the continuing promotional drive by the ZSPTF (see Figure 2 below). Soy output reached a peak of 140,000t in 2001. In the 2002/04 seasons, soya beans were selling at Z$2.4 million (US$396) per megaton (mt), while maize, which was subject to strict government controls, fetched Z$700,000 ($116) per mt.[24] In the decade 2002–2012, national

---

23   B. Zamasiya and K. Nyikahadzoi, 'Supporting smallholders in soybean cultivation: the example of Zimbabwe', in *Achieving Sustainable Cultivation of Soybeans*, Vol. 1 (Sawston: Burleigh Dodds, 2018)

24   Ibid.

15. Domesticating the Exotic

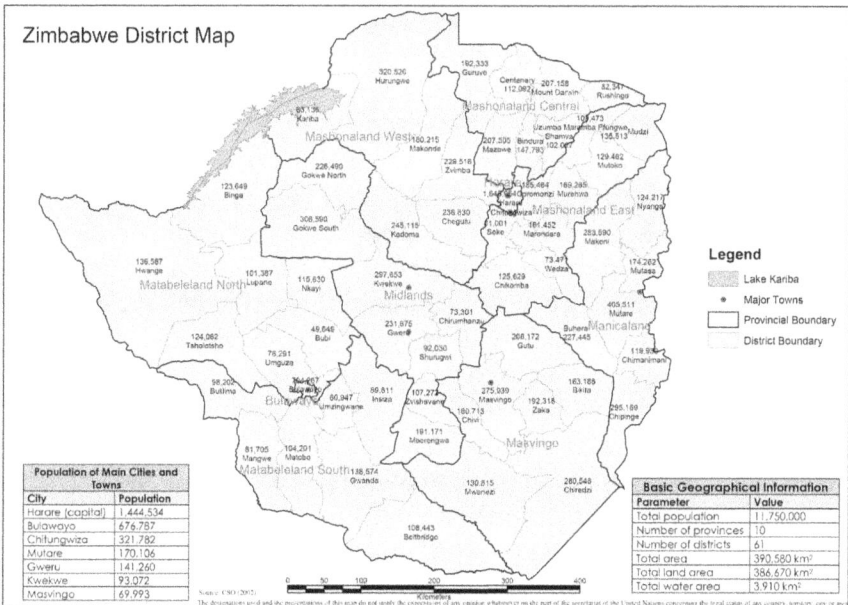

Map 1.

Zimbabwe District Map.

Source: https://reliefweb.int/map/zimbabwe/zimbabwe-district-map-2002

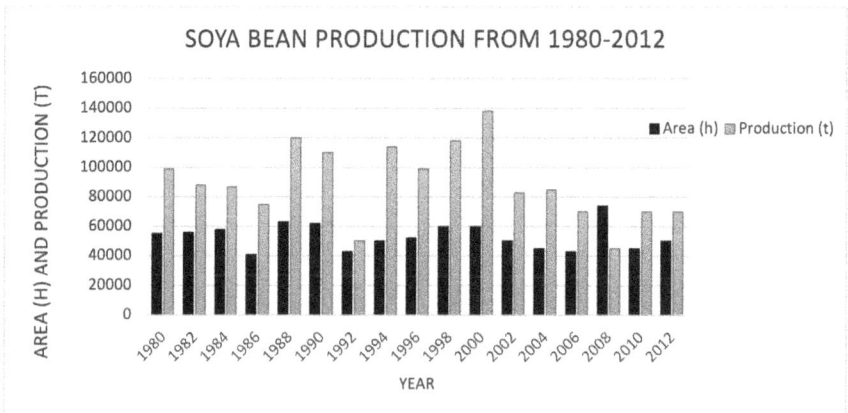

Figure 2.

Soy Production 1980–2012.

Source: Agritex Report October 2012: https://docplayer.net/42638959-Zimbabwe-soya-bean-sub-sector-study-with-support-from.html ; New Ziana, 17 January 2012, cites a figure of 170 000t for the same year.

yields generally declined in comparison to the average in the 1990s, due to poor agronomic practices and policies that resulted in low productivity. Soy production generally decreased notably after 2005 as a consequence of a chaotic Fast Track Land Reform Programme from 2000 which destroyed a vibrant agricultural sector. Efforts to resuscitate the agricultural industry in 2007–2009 by a Reserve of Bank of Zimbabwe funding scheme made little progress to jumpstart and enhance production levels in the country. Cyclical drought impacts also inhibited soy cultivation particularly on a commercial basis, since most production was rain fed agriculture. Also, the average market price of US$260/t after 2002 was much lower than the production costs which averaged US$500/ha, causing losses for farmers.[25]

The levels of soy productivity in Zimbabwe (1.55t/ha) were low compared to high producing countries such as Argentina (3.4t/ha). Commercial farmers' average yield was 1.8t/ha compared to smallholder farmers' produce at 0.5t/ha.[26] The output gap was largely accounted for by the combined effect of decreases in land cultivated and actual productivity. To meaningfully earn a living out of soy production (with a poverty datum line of US$400/month for a family of 6), a Zimbabwean smallholder farmer needed to plant between 5–7ha.[27] Also, the average production levels of 50,000t recorded over the seven-year period 2013–2017 fell short of the national demand of 220,000t per year. To satisfy the soy deficit from production, Zimbabwe imported seventy per cent of its soy requirements, mainly from South Africa, with beans and cake from Zambia and Malawi, depleting scarce foreign currency reserves.[28]

Despite the challenges, soy remained one of the shortest season crops with lucrative returns on investments, capable of sustaining many rural livelihoods at the farm level specifically in Zimbabwe and in Southern Africa more generally. Costs of soy production were consistently very high and prohibitive in the years 2013–2017 and after. The general input expense structure per ha ranged between US$700–$900.[29] At that level, a Zimbabwean soya

25   Agritex Annual Report for 2012; *New Ziana*, 17 Jan. 2012 cites a figure of 170 000t for the same year.

26   J. Basera and H. Mushoriwa, Seed Company Agronomy Report, Soyabean and Legumes, 2011; SNV, 'Zimbabwe Soya', 16.

27   Ibid.

28   Ibid.

29   Confederation of Zimbabwe industries by Zimbabwe economic policy analysis and research unit (Zeparu), *Development of a Competitive Soya Bean Value Chain. Opportunities and Challenges* (Harare: Belvedere, 2017), p. 37.

farmer targeted to surpass the Argentinian productivity level and harvest between 3.5-6t/ha by cultivating Seed Company hybrid varieties. The average seasonal livelihood income earned by such farmers was US$2,000.00.[30] For a minority of farmers it was higher, dependent upon altitude, acreage, variety and productivity levels. This meant that a farmer could realise a gross return of at least US$1,100.00/ha after four to five months. The break-even yield level for soya was about 1.7t/ha at the prevailing prices, averaging US$550 per tonne. A farmer could earn up to US$3.00 per dollar invested.[31] Increased soya production and yield undoubtedly enhanced the livelihood incomes of over 35,000 smallholder households by an average US$138 per annum.[32]

Besides cash incomes, soy production boosted farm cash flows and employment creation for farming communities during peak work periods in April and May, strategically positioning the crop to the nation. Huge potential in soy-related employment creation lay in exploiting the value chain right from input supply, production and processing as well as in downstream effects when stronger linkages were created between the segments of the value chain and when the soya value chain was linked to livestock value chains like poultry, dairy, beef and piggery.[33] The small-scale poultry sector contributed to economic growth and employment generation, since an estimated 65 per cent of poultry production in Zimbabwe was conducted by small scale producers of which seventy per cent were women.[34]

Several female soya producers in Glendale in MCP enumerated the wide-ranging livelihood benefits derived from soy. For example, Susan Mabhande labelled soy a wonder crop – nutritious, cheap and easy to produce.[35] Vimbai Chatikobo from Goromonzi, MEP echoed Mabhande's lived experiences from cultivating soy, noting that soy drastically transformed the lives of many poverty-stricken smallholder farmers, particularly women-headed households without male breadwinners.[36] Among others,

30 Ibid.

31 Ibid.

32 Techno Serve and National Agricultural Marketing Association (NAMC), *Business Solutions to Poverty, Southern Africa Soy Roadmap – Zimbabwe value chain analysis* (Arlington: Techno Serve, Nov. 2010–Feb. 2011), p. 2. See also, D. Garwe et al. *A State of Plant Genetic Resources for Food and Agriculture in Zimbabwe – A Country Report* (Department of Agricultural Research for Development, Ministry of Agriculture, Mechanization and Irrigation Development, 2009).

33 Zeparu, *Development*, p. 34.

34 Ibid., p. 38.

35 Interview with Susan Mabhande, Farmer, Glendale, MCP, 12 Nov., 2020.

36 Interview with Vimbai Chatikobo, Farmer, Goromonzi, MEP, 23 Jan. 2021.

Mabhande and Chatikobo argue that, although seasonal, soya incomes empowered single mothers with essential cash for building better houses, purchasing groceries, clothing, paying education and health fees, as well as buying farm inputs.[37]

Uses of Soya

Testimonies of soya farmers have encouraged joint government and non-governmental organisation (NGO) initiatives and projects that incorporated and benefited over 20,000 soya growers in the producing areas. The farmers acquired new skill-sets in the production, processing and marketing soy-based products. For instance, Peter Moyo states that the soya schemes have greatly improved the living standards of his household in more ways than one, including the ability to purchase five head of cattle and an ex-Japanese half-tonne farm lorry.[38] Other farmers in MWP had similar soya experiences to those of Mabhande, Chatikobo and Moyo. US-based non-profit organisation Africare's provincial project coordinator, James Machikicho agrees with the farmers regarding some benefits resulting from soya stating that:

> ... since 2000 Africare has taught soya production techniques to farmers from the districts of Rushinga, Mt Darwin, Shamva and Bindura. Prior to the initiative, the farmers' options for making a livelihood had been confined to producing maize and cotton, both of which require expensive inputs, such as fertiliser, chemicals, seed and labour while soya beans are cheaper to produce, have higher yields and are less labour intensive. Soya beans need more seed per hectare than maize, but yield more in the same space. Soya has increased social mobility and improved farm livelihoods.[39]

Apart from early household incomes and high yields, soya offered invaluable nutritional benefits to both rural and urban communities because it was one of the richest crops in terms of crude protein ranging between 35–45 per cent and also contains twenty per cent oil.[40] Many farmers, including non-soya growers, received training in the preparation of soya-based products – especially flour, beverages, porridge and confectionery. The education helped to diversify not only on-farm economic activities but

---

37   Ibid.

38   Interview with Peter Moyo, Farmer, Goromonzi, MEP, 22 Dec.2020.

39   Interview with James Machikicho, Farmer, Glendale, MCP, 15 Nov. 2020.

40   Interview with Nelson Banda, Crops Branch, Department of Agricultural Technical and Extension Services, Harare, 10 Feb. 2021.

also income generation. Soya rapidly replaced meat as a main source of protein for villagers and urban dwellers alike. Soya *meatfication* was in the form of a popular relish called chunks/mince. Soyabean products were also ten to twenty per cent cheaper than those derived from fish, beef and poultry sources.[41]

In addition, soya provides valuable on-farm products including soap, cooking oil, milk and margarine. According to Rumbidzai Chigidi, health, income and lifestyles have improved in all directions in several soya farm-ing communities.[42] Through planting soy on just one hectare of land, the Shingai AIDS Support Group in Rushinga was empowered to sustain the long term nutritional security requirements of 23 orphaned children and six adults, three of whom were HIV positive.[43] The group harvested sufficient bags of soya, ensuring secure soyfood livelihoods by providing porridge for breakfast and supper enhanced with multiple soya relishes. Tsitsi Mvemwe, founder of a ten-member local support group claims that: 'The children are fitter and healthier than those with living parents.'[44] In Mt Darwin District, a soya milk-production plant called 'Vita Cow', manned by trained community members sold soya milk and related products to local households and shops in towns such as Mt Darwin and Bindura, as well as in the capital, Harare. Production by Vita Cow rose from an initial sixty litres to 250 litres per week. Silas Moyo argues that: 'Vita Cow production costs are 75 percent cheaper than for dairy milk, but because of soya's nutritive value, its price is comparable to dairy milk.'[45]

Other NGOs helped soya farmers to get the best deals possible by establishing links between their farming associations, the input suppliers and commercial soya buyers to reduce exploitation by middlemen. Mary Tembo notes that:

> Initially, the farmer was getting 50 percent of what their crop was worth, but middle-men are now offering competitive prices because of the linkages. To take advantage of the high demand for soya, farmers are constantly being encouraged to build up their produce and sell it to contracted buyers prepared to provide transport.[46]

---

41 Interview with Nozipo Dube, Department of Agricultural Research for Development, Harare, 10 Feb. 2021.

42 Interview with Rumbidzai Chigidi, Farmer, Glendale, MCP, 20 Nov. 2020.

43 Interview with Silas Zinhu, Farmer, Rushinga, MCP, 22 Nov. 2020.

44 Interview with Tsitsi Mvemve, Farmer, Rushinga, MCP, 26 Nov. 2020.

45 Interview with Silas Moyo, Farmer, Mount Darwin, MCP, 30 Nov. 2020.

46 Interview with Mary Tembo, Chairperson of Dotito Farmers' Union, Dotito, MCP, 24 Nov. 2020.

Some soy farmers set up village loan schemes and seed banks, managed by the farmers themselves, offering savings and credit facilities. Seed bank membership in MCP and MEP was as high as 5,000 in 2016.[47] However, frequent spells of drought and high input costs posed serious production challenges. Like maize, the soy crop was planted in the wet season – generally from November to December – followed by a three-month maturation period. The heavy dependence on rain fed agriculture in Zimbabwe and across sub-Saharan Africa was a huge dilemma for farmers, exacerbated by climate change, erratic rains and frequent acute drought. To climate variability farmers added other woes like seed limitation, high input prices and usurious loans from private companies exploiting vulnerable farmers. Arguably, a combination of these difficulties partly accounts for low soy production.

Nonetheless, farmers' livelihoods were directly and indirectly improved by soy because it remained an affordable source of protein for livestock feed. From colonial times, soy was used for value addition on farms where stockfeed were formulated and mixed. Generally in livestock production systems, feed constituted about seventy per cent of the total cost structure and feed costs among Zimbabwe's soy farmers reduced by 25 per cent, ultimately enhancing the profitability levels of wide-ranging livestock production from chickens to pigs, goats and notably cattle.[48] Soy served as a high-protein livestock feed and, when grown in rotation with maize and cotton, greatly improved soil fertility.[49] Soy products also contributed thirty per cent of the national stock feed raw material requirements, of which an estimated 69.3 per cent was destined for poultry production. This contributed to an increase in poultry production and consumption in rural and urban areas. However, poultry feed production was inadequate to meet local demand. It was estimated that on average a bird consumed three kilograms of feed during the entire six-week period of its production.[50]

From the micro farm livelihoods to the macro national demands, Zimbabwe under Mugabe required over 220,000mt of soya annually for food, feed and other industrial needs. Changes to the agrarian structure following land reform shifted producer configurations from large scale commercial to small and medium sized farms in new resettlement areas. State led farm schemes

47   Ibid.
48   Interview with Nelson Banda.
49   Zeparu, *Development*, p. 38.
50   Ibid.

augmented by private sector participation through contracts became the major drivers in the soya sector. The role of the ZNSCA and other farmer organisations improved farming practices through knowledge sharing, price bargains and marketing, albeit in a struggling and fragile economy operating under cyclical droughts and western sanctions imposed since 2001 for gross human rights violations.[51] While soy's contribution to the national economy (GDP) was small, its significance lay in its import-saving potential with regard to edible oils and stock feeds, since it contributed thirty per cent towards the former.[52] Production and productivity were constrained by a combination of factors related to farmer-level skills and knowledge, the quality of support services (extension and finances) and input supply. The limited availability of small parcels of land of between 5ha-7ha constrained production in the communal and resettlement areas, and land available in A2 farms remained contested lands. The country used less than ten per cent of installed soya processing capacity because of low production and productivity of soya. Regardless, soy contributed significantly to food security and the country projected attaining autarky in soy production in 2030.

Self-sufficiency in soy required addressing capacity challenges amongst the 50,000 smallholder producers connecting them to private sector companies keen on contract farming arrangements and strengthening the support environment. Close to ninety per cent of soy farmers used returned seed and did not apply the recommended agronomic practices such as the use of herbicides and inoculants. This contributed to the low productivity among smallholder farmers (average 0.5t/ha). The provision of quality extension services was very weak. Smallholder farmers did not have access to financial inclusion programmes. As a result, the production of soy was on a declining trend to 2017 caused by these several limitations (see Figure 2 above). After 2010, the area under cultivation increased but the yields remained below the commercial production benchmark. Large scale commercial farmers produced 65 per cent. Estimates showed that about 50,000 smallholder farmers grew soy and accounted for 35 per cent of national output in part because pieces of land devoted to soya were tiny. National output dropped to 50 000t per year since 2017 due to a shrinkage in the producer base and loss of productivity amongst both smallholder and large-scale farmers.

51   Interview with Chipo Size, Agronomy Research Institute, Department of Agricultural Research for Development Harare, 12 Feb. 2021.

52   Zeparu, *Development*, p. 38.

Low output caused considerable shortages of raw materials for cooking oil expressing and stock feeds.

The bulk buying and storage of the soy was done by the GMB and private organisations. The country installed capacity to process 500,000mt per year but capacity utilisation was only ten per cent. The leading processors were Surface Investments and Olivine Industries. Besides low-capacity utilisation, the processing equipment needed refurbishment with some of it over thirty years old. The three processing methods were chemical/solvent extraction, pressing (mechanical extraction) and hydraulic pressing. Soya oil wholesale was handled by major players such as Mohammed Musa and N. Richards. The retailing was done through supermarkets such as OK, TM and Spar. The soya oil subsector had four market outlets which were: own farm consumption, poor urban and rural consumers, high income urban earners and institutional buyers such schools and hospitals. Soy was a high value crop with strong industrial linkages. It supported the industrial processing of value added products such as soya cake, soymilk, soy yoghurts, flour, margarine and soyabean oil. Soy produces thirty per cent of cooking oil nationally, and its oilcake, which is a by-product of oil extraction, is sold to feed manufacturers.[53]

Ecological Impacts of Soy in Zimbabwe

Soy has to a large extent generally generated beneficial ecological impacts, in sharp contrast to many other food and cash crops that are associated with environmental degradation around deforestation and soil erosion. The yield of a cereal crop such as maize following soya in a rotation was usually enhanced because of the important residual nitrogen. Soyabean is a legume which fixes nitrogen.[54] A well-managed soy crop left a residual nitrogen level of up to 90kg/ha, benefiting the next crop in a rotation.[55] Soya has huge potential to address the need for beneficially diversifying cropping systems to assist in overcoming pervading and pervasive soil fertility constraints. Crop diversification away from maize monoculture using soy to improve soil fertility through biological nitrogen fixation was the most important consideration for domesticating and producing the crop, at least in Zimbabwe. Soy

---

53   Ibid.

54   J. Basera and H. Mushoriwa, *Seed Company Agronomy Report, Soyabean and Legumes* (2011).

55   K.E. Giller et al. 'Soyabeans and sustainable agriculture in southern Africa', *International Journal of Agricultural Sustainability* **9** (1) (2011): 50–58.

is quite specific in its requirement for rhizobial strains, meaning that only a few types of rhizobia are able to form nodules on soya roots and actively fix nitrogen. These varieties require inoculation with effective rhizobium strains in order to nodulate and fix nitrogen. Many other tropical legumes, notably cowpea, groundnut and common sugar beans, are more permissive and form nodules and fix nitrogen with a wide range of rhizobia that are present in most tropical soils.

Rhizobia are soil bacteria that adhere to and colonise root cells of leguminous plants including soyabeans and groundnuts. Rhizobia have the ability to reach the centre of root hair cells and together with proliferating plant cells form a nodule. Here, rhizobia fix nitrogen, converting molecular nitrogen (N_2) from the air into ammonia, nitrates, and other nitrogenous compounds to support plant metabolism and growth. Rhizobia are particularly important to plants in nitrogen-deficient soils such as those in natural regions 3 to 5 in Zimbabwe. In return, rhizobia receive from the plant carbon-rich organic compounds, important for their own energy production.[56] Common dry and irrigated land soya varieties in Zimbabwe include the SC hybrids: Serenade, Safari, Status, Sequel and Squire.[57] These are termed 'promiscuous' or 'naturally nodulating' legumes, and they make effective use of the inherent soil biodiversity of rhizobia indigenous to the soils.[58] Other nodulating varieties, Roan, Nyala, Sonata and Solitaire yielded more than a tonne of extra grain with inoculation compared to Magoye and Local, demonstrating the advantage of the breeding programmes for increased yield and disease resistance.[59] The amounts of nitrogen fixed from the atmosphere by soya ranged from 60–130kg N ha21 on smallholder farmers' fields to 160–260kg N ha21 on prime land fertile soils.[60]

Strong rotational benefits were observed by breaking the prevalent maize monoculture in Zimbabwe that has extensively destroyed environments. Soya was also compatible with long and short rotations. An example of a long rotation was maize (summer)-soyabean (summer) which was common in the non-irrigated farming systems of MCP and MWP as well as the Midlands.

56   L. Cegelski, C.L. Smith and S.J. Hultgren, 'Adhesion, microbial', in M. Schaechter (ed.), *Encyclopaedia of Microbiology* (San Diego: San Diego State University, 2009), pp. 1–10.

57   https://www.seedcogroup.com/zw/farmer-s-hub/soya

58   S. Mpepereki et al., 'Soyabeans and sustainable agriculture: Promiscuous soyabeans in southern Africa', *Field Crops Research* **65** (2–3) (2000): 137–49.

59   Ibid.

60   Zamasiya and Nyikahadzoi, 'Supporting smallholders', 1.

Soya (summer)-wheat (winter) was an example of a short rotation (also known as double cropping system) in the irrigated farming systems such as those in Bindura and Mazoe Districts of MCP. Soya rotated particularly well with winter-irrigated wheat as it took less time in the fields even when planted in November or early December. Both rotation setups were beneficial to the farmer and ecology. In rotations, the yield of maize/wheat following soyabeans was generally greater than following maize at both low and high levels of nitrogen application. In Zimbabwe, soya varieties were generally early to medium maturing- 115–135 days from planting to physiological maturity.[61] The rotational benefits often led to a doubling of maize yields in soils where maize had been grown for many years. These benefits were due to the addition of nitrogen to the soil from soy residues that gave a small but significant improvement in soil fertility.[62] The first certified agricultural methodology for reducing carbon dioxide (CO_2) emissions under the United Nations Clean Development Mechanism issued in 2008 was rotation with soya. The reductions in CO_2 emissions were largely due to substitution of soil damaging and expensive artificial fertilisers with nitrogen fertiliser in soya–maize rotations, which was highly relevant for Zimbabwe's climate smart agricultural policy. Smallholder farmers in Zimbabwe used little N fertiliser compared to North American farmers. However, Zimbabwean farmers required both strong technology development and strong institutions in a broad sense, including extension, input and output markets.[63]

Low and declining soil fertility had long been recognised as a major impediment to intensifying agriculture in Zimbabwe. Biological nitrogen fixation in soy was economically and ecologically beneficial in Zimbabwe where most soils were deficient in nitrogen and artificial fertilisers were not affordable to most farmers, owing to generally poor economic conditions.[64] Also, intercropping in Zimbabwe was one of the most valuable practices for improving land use. Using maize–soybean strip intercropping increased grain yields. Mixed farming systems were the major farming systems in Zimbabwe with maize as the most widely grown crop and a staple food for the majority of rural and urban people. The high risk of extensive maize monocropping practices in Zimbabwe had to a large extent been mitigated by diversification,

61   Basera and Mushoriwa, 'Seed Company'.
62   Ibid.
63   Giller, 'Soyabeans and sustainable agriculture'.
64   Interview with Chipo Size.

intensification and efforts at inclusion of grain legumes such as soya. In this context, a maize–promiscuous soybean rotation and/or intercropping provide an alternative approach for enhancing the productivity and sustainability of farming systems in Zimbabwe.[65] Moreover, soybean-maize intercropping suppressed infestation by striga (*Striga hermonthica*), a parasitic weed that infested over sixty per cent of farmland under cultivation. The continued cultivation of maize in the field provided a host for striga. Economic analysis of these systems showed an increase of fifty to seventy per cent in gross income compared to those of farmers still following continuous maize cultivation.

Additional economic and ecological benefits arising from cultivating the soya varieties in Zimbabwe prior to maize included better soil moisture conservation due to the early maturity of soy and the canopy cover it forms guarding against erosion. Resistant to certain diseases, soy was used to control weeds, crop pests and diseases such as red leaf blotch (*Pyrenochaeta glycines*) and frogeye (*Cercospora sojina*), especially in rotation systems with cereal crops.[66] Soy had relatively few pests compared to other grain legumes. However, it was subject to attack by leaf eaters such as semi-loopers and leaf rollers. Nonetheless, soy could tolerate up to thirty per cent defoliation without significant yield loss, above which economic yield loss occurred and insecticides were required. As a cultural practice, rotation was used to manage weeds, pests and diseases in Zimbabwe cropping systems. Frog-eye leaf spot was the only disease of consequence to soy and the problem was solved through the use of new disease-tolerant varieties that did not need spraying with fungicides. However, when soyabean rust appeared, it required the application of expensive fungicides. The SPTF persuaded chemical companies such as Syngenta to package fungicides in 500 ml bottles, thereby improving access for smallholder farmers. New rust-tolerant soy varieties developed by Seed Co. provided a long-term solution to disease problems because they did not require fungicides.[67]

Regardless of its perceived environmental and economic benefits, soy has been shown to have some shortcomings. Soy production losses resulted from the increased frequency and intensity of extreme weather patterns, particularly floods, droughts, heatwaves, and windstorms.[68] The impacts are felt more in

65   Ibid.

66   Basera and Mushoriwa, 'Seed Company'.

67   Ibid.

68   J.R. Porter and M.A. Semenov, 'Crop responses to climatic variation', *Philosophical Transactions of the Royal Society* **360** (2005): 2021–35.

developing countries such as Zimbabwe with weak fall-back strategies in terms of climate stresses. Several studies predict and show that soy yields in Africa are likely decrease under climate crises.[69] The increased vulnerability of soy growers in Zimbabwe has been worsened by high temperatures, low and highly variable rainfall and the country's lop-sided economy, hugely reliant on agriculture, as well as low adoption of modern technology. Global warming might benefit crop and pasture yields in temperate regions but has the opposite effect of reducing crop yields in the semi-arid tropical Zimbabwe. Crop yields are estimated to decline in production by 22, 17, 17, 18, and 8 per cent for maize, sorghum, millet, groundnuts, and cassava, respectively.[70]

The Organization for Economic Co-operation and Development states that climate change adaptation has become important for building ecological systems and resilient socio-economic activities. Zimbabwean farmers respond to these environmental changes and challenges by adopting new farming strategies, such as changing cropping dates or crop varieties, economic diversification, and more efficient use of agricultural inputs. Nonetheless, many barriers to adaptation exist, including lack of information, policy distortions and market failures.[71] Soya, however, appears versatile in Zimbabwe embracing diverse climatic conditions and different types of soil, uncommon in traditional crops like maize, cotton and sorghum. Even though Zimbabwe has conducive environmental conditions for soy production, the crop can only withstand drought to some extent and is weak against flooding stress.[72] Regardless, it still performs well in African climatic systems compared to other parts of the world.

Soy production in Zimbabwe is not linked to the extensive and intensive use of ecologically harmful agrochemicals, some of them like Paraquat or Glyphosate banned in parts of the western world. Many Zimbabwean farmers do not appear to specifically expand crop fields for soy production but for maize, cotton and tobacco. Soya is accommodated on lands cleared and prepared for other crops. The environmental consequences of expanding soy

69   J. Turpie et al., *Economic Impacts of Climate Change in South Africa: A Preliminary Analysis of Unmitigated Damage Costs* (Cape Town: Energy and Development Research Centre (EDRC), 2002).

70   J.M. Makadho, 'Potential effects of climate change on corn production in Zimbabwe', *Climate Research*, Zimbabwe **6** (2) (1996): 147–51.

71   A. Ignaciuk and D. Mason-D'Croz, 'Modelling adaptation to climate change in agriculture', *OECD Food, Agriculture and Fisheries Papers* **70** (2014).

72   F.F. Hou and F.S., Thseng, 'Studies on the flooding tolerance of soybean seed: varietal differences', *Euphytica* **57** (2) (1991): 169–73.

cultivation from the perspective of deforestation and biodiversity loss can only be indirectly linked to soy production. Narratives on the socio-environmental consequences of soy expansion are not directly connected to expanding soy because its uptake by both smallholder and commercial farmers has been very slow in the more than a century of soy history examined in this study. Soy history in Zimbabwe resists conclusions that chime closely with soy experiences in other continents and ecosystems.

Conclusion

This study has explored the history of soy domestication, production and use trends in colonial and post-colonial Zimbabwe. Early twentieth century origins of soya were traceable from China into Zimbabwe through South Africa. A wary attitude and general dislike for its taste constrained the uptake of soy farming in both Rhodesian and Zimbabwean times. Acreages under soy and quantities produced have remained inhibited despite a recognition of the economic and ecological significance of soya in improving livelihoods and the environment, respectively. The environmental impacts of soy in Zimbabwe remain ambiguous but largely beneficial. The extant literature and oral evidence suggest that soy growing rejuvenates soil fertility and benefits smallholder farmers in very diverse ways.

CHAPTER 16.

SOY LANDSCAPES: PRODUCTION, ENVIRONMENT AND QUALITY OF LIFE IN THE PROVINCE OF BUENOS AIRES, ARGENTINA (1996–2020)

José Muzlera

> *It seems to be easier for us today to imagine the thoroughgoing deterioration of the earth and of nature than the breakdown of late capitalism.*
>
> Fredric Jameson, *The Seeds of Time*, 1994

> *El cambio climático es un problema global con graves dimensiones ambientales, sociales, económicas, distributivas y políticas, y plantea uno de los principales desafíos actuales para la Humanidad*
>
> Pope Francis, *Carta Encíclica Laudato Si': sobre el cuidado de la casa común*, 2015

Introduction

Humanity has radically transformed its relationship with food and the environment during the last century. Although the world population grew from 1.6 to 6 billion during the twentieth century, food production is exceeding this, with more overfed than malnourished people (despite distribution flaws). The proportion of income destined to food decreased, as did the number of producers, creating a rural-urban migratory dynamic.[1] This scenario has corresponded to new production technologies and radical landscape transformations.[2]

'Landscape' is a polysemic term that first appeared in Renaissance Europe, associated with paintings and literature describing the environment. During the nineteenth century, the concept was appropriated by the social sciences.

---

1   J.R. McNeill and E.S. Mauldin (ed.), *A Companion to Global Environmental History* (New York: John Wiley & Sons, 2015).

2   A.G. Zarrilli (ed.), *Por una historia ambiental Latinoamericana. Aportes para el estudio de la sociedad y la naturaleza en la era del Antropoceno* (Buenos Aires: TESEO, 2016).

THE AGE OF THE SOYBEAN: 332–366 **doi:** 10.3197/63800040695086.ch16

16. Soybean Landscapes: Production, Environment and Life Quality

In the concept's construction, the aesthetic and the subjective, objectivity and functionality, have always been present.[3] Landscape studies analyse the natural areas transformed by anthropic actions, evolution and transformation. As technologies 'evolve' and the ways of organising production are transformed, territorial uses are modified, affecting natural environments and the human beings inhabiting them. People's resource utilisation corresponds to historical, political and cultural contexts. In this sense, landscapes reflect cultural identities that have more to do with the culture than with the natural environment.[4]

> 'Landscapes' are the symbolic environments created by human acts of conferring meaning to nature and the environment, of giving the environment definition and form from a particular angle of vision and through a special filter of values and beliefs. Every landscape is a symbolic environment.[5]

The theoretical mark brought by the social sciences to the study of the landscape lies in the possibility of understanding changes in the physical aspects of the environment and their interactions with different human societies and dialectical dynamics. In the case of Argentina, since the late 1990s, with the rise of agribusiness, the country has tripled its grain harvests (at the same time as poverty and indigence rates have increased). This productive growth has mainly occurred at the expense of the Pampas region and the partial *pampeanisation* of other productive environments, reinforcing the Pampas-centric logic that has accompanied the country since the late nineteenth century.[6]

Agribusiness is the hegemonic paradigm that has shaped Pampas agricultural reality over the last three decades. It has affected human relations and territorial development, patterns of resource exploitation and environmental balances. In this context, soy – the first transgenic crop launched on the Argentine market in 1996 – represented an iconic product for critics and defenders of this productive model. The three most distinctive features of

---

3   T. Greider and L. Garkovich, 'Landscapes: The social construction of nature and the environment', *Rural Sociology*, **59** (1) (1994): 1–24; D. Corrêa, 'História ambiental e a paisagem', *História Ambiental Latinoamericana y Caribeña HALAC* **2** (1) (2012): 47–69.

4   J.C. Radin and C.M. da Silva, '"Um vasto celeiro": representações da natureza no processo de colonização do oeste catarinense (1916–1950)', *Boletim do Museu Paraense Emílio Goeldi. Ciências Humanas* **13** (2018): 681–97; Zarrilli, *Por una historia ambiental Latinoamericana*.

5   Greider and Garkovich, 'Landscapes', 1.

6   A.G. Zarrilli, '¿Una agriculturización insostenible? La provincia del Chaco, Argentina (1980–2008)', *Historia Agraria* **51** (2010): 143–76; and Zarrilli, *Por una historia ambiental Latinoamericana*.

the agribusiness model are the standardisation of processes, outsourcing of labour and production on third-party lands.[7] As we will see throughout this chapter, these characteristics modify the landscape through depopulation, homogenisation and geometrisation mechanisms.

Two main discourses on agribusiness can be identified. First, agribusiness entrepreneurs and some academics promote this model with arguments such as spillover and attraction of extra-sectorial capital. Essentially, they defend the model's high productive efficiency, the creation of a qualified labour force, the generation of large exportable balances and the benefit for local economies (the metal mechanical industry, professional services for agriculture, computer science, specialised shops, activating local economies).[8] Conversely, critics claim that the agribusiness model and its star crop are the axis of an accumulation pattern based on the over-exploitation of increasingly scarce natural resources and the expansion of the agricultural frontier towards territories previously reserved for other uses and practices (jungles, native forests, mountains, valleys). They also insist on this model's negative consequences in terms of health, environmental deterioration and forced evictions.[9]

7   C. Gras and V. Hernández (eds), *El agro como negocio. Producción, sociedad y territorios en la globalización* (Buenos Aires, Biblos, 2013); J. Muzlera, 'Tipos de productores y uso de la tierra en Balcarce y 25 de Mayo (2010–2015). Tras la herencia de los Mega Pools', *Revista Pilquen. Sección Agronomía* **15** (1) (2016): 1851–2852; and 'Análisis de los vínculos Familia-Empresa en los Contratistas de Maquinaria Agrícola Pampeanos', *Revista La Rivada* **6** (10) (2018): 152–66.

8   See C. Angió, 'La competitividad de la soja en El Tejar. MERCOSOJA 2006', *3° Congreso de Soja del MERCOSUR*, Rosario (Argentina), 27–30 June 2006. Available at http://www.planetasoja. com.ar/trabajos800.php; R. Bisang and B. Kosacoff, 'Las redes de producción en el agro argentino', *XIV Congreso Anual AAPRESID*, 2006; R. Bisang et al., 'Cadenas de valor en la agroindustria en la Argentina ante la nueva internacionalización de la producción. Crisis y oportunidades', in Bernardo Kosacoff and Ruben Mercado (eds), *Libro de la División de Recursos Naturales e Infraestructura de la CEPAL* (Santiago de Chile, Naciones Unidas, 2009); R. Bisang et al., 'Una revolución no tan silenciosa. Claves para repensar el agro en Argentina,' *Desarrollo Económico* **190–191** (48) (2008): 165–207; and H. Maiztegui Martínez, 'Una nueva modalidad asociativa en Argentina: el pool de siembra', *Estudios Agrarios. Procuraduría Agraria* (2009).

9   See A. Girado, 'Minería y conflicto social en la provincia de Buenos Aires. Letras Verdes', *Revista Latinoamericana de Estudios Socioambientales* **14** (2013): 48–68. Also S. Cloquell (ed.), *Familias Rurales, el fin de una historia en el inicio de una nueva agricultura* (Rosario: Editorial Homo Sapiens, 2007); C. Gras and V. Hernandez, *La Argentina rural. De la agricultura familiar a los Agronegocios* (Buenos Aires, Editorial Biblos, 2009); N. López Castro and G. Prividera (eds), *Repensar la agricultura familiar. Aportes para desentrañar la complejidad agraria pampeana* (Buenos Aires: Ciccus, 2011); G. Neiman and M. Lattuada, *El campo argentino: crecimiento con exclusión* (Buenos Aires, Editorial Capital Intelectual, 2005); G. Neiman (ed.), *Trabajo de campo. Producción, tecnología y empleo en el medio rural* (Buenos Aires, Ediciones Ciccus, 2001); G. Neiman and G. Quaranta, 'Los estudios de caso en la investigación sociológica', in V. de Gialdino (ed.), *Estrategias de investigación cualitativa,*

16. Soybean Landscapes: Production, Environment and Life Quality

Although agribusiness proponents sometimes acknowledge its harmful impacts, they argue that socio-environmental consequences associated with soybean monocultures are not the consequence of the crop's characteristics but of its productive model.[10] As agribusiness' flagship crop, soybean constitutes a perfect indicator to measure the progress of this model. Soybean's high oleic and protein content explains the demand for it, while its transgenic characteristics (which make it adaptable to different soils and climates) explain its popularity among producers. Finally, the agribusiness model considers soy as a vector of socio-environmental issues but not as the cause *per se*.

Therefore, the main issue at stake is describing the soybean farming system in its complexity, illustrating its links with nature and its implications for territorial development (wellbeing, health, accumulation, etc.), the winners (at least economically) and the losers. In order to address such topics, this study analyses the evolution of soy farming areas and their productivity. First, it illustrates how different crops and livestock implantation evolved in Buenos Aires (the primary reference of the Pampas Region). Second, it describes changing land tenures, their related local environmental conflicts and their impacts on the quality of life.

Historically, Argentina has been an agro-exporter of products from the Pampas region (mainly beef and grains). The remaining agricultural production has been relatively insignificant, at least for a century. However, with the expansion of modern agricultural capitalism – characterised by high investment and dependence on computer technologies, genetics and telecommunications, the outsourcing of labour and financialisation – the Pampas model has advanced over other ecosystems, displacing traditional cultures, their networks and their products. This analysis will mainly focus on the province of Buenos Aires. By 2020, the province contributed 34.4 per cent of national exports (USD 1.692 billion). About 30.4 per cent of these exports were primary products, and 29.9 per cent were agricultural manufactures. Together, these contributed to about 22.29 per cent of the country's total agricultural exports, with a high proportion concerning soybeans and derivatives.

pp. 213–37 (Buenos Aires: Gedisa, 2006); G. Quaranta, *Reestructuración y organización social del trabajo en producciones agrarias de la región pampeana argentina* (Doctoral Thesis, Universidad de Córdoba, 2007); H. Ratier, *Poblados Bonaerenses, vida y milagros* (Buenos Aires: Ed. La Colmena, 2004); M. Sili, *La Argentina rural. De la crisis de la modernización agraria a la construcción de un nuevo paradigma de desarrollo de los territorios rurales* (Buenos Aires Ediciones INTA, 2005).

10   D. Ferrara, 'Fitosanitarios un alternativa segura y sustentable para aumentar la productividad', *Estudios Rurales* **8** (14) (2018): 16–23.

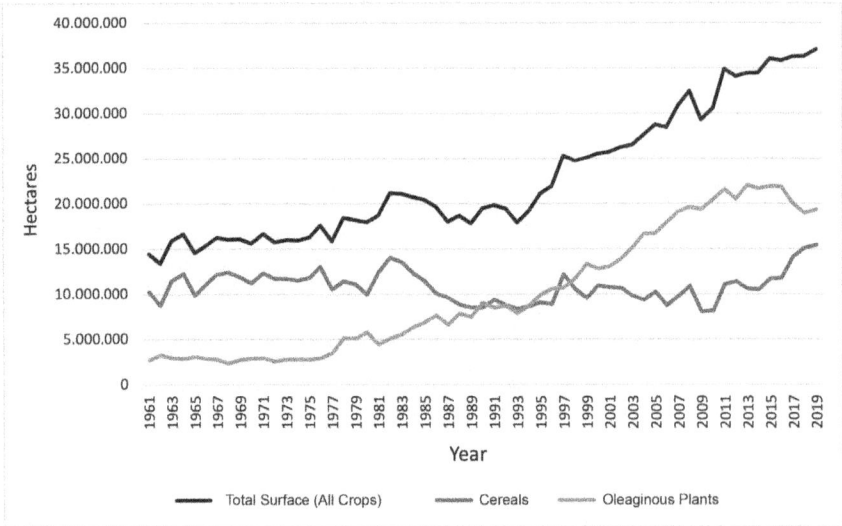

Figure 1.

Surface Harvested in Argentina by Crop Group (1961–2019)

Source: Author's elaboration based on FAO data available at http://www.fao.org/faostat/es/#data/QC

The Advance of Soybeans

Since 1996, when transgenic soybeans resistant to the herbicide glyphosate were released onto the Argentine market, the popularity of this oilseed has had exponential growth, displacing other local crops such as sunflower and even *pampeanizando* other regions in the country.[11]

The graph above shows the weight of cereals and oilseeds in total harvests between 1961 and 2019 (citrus fruits, sugar crops, fibre crops, fruits, vegetables, legumes, roots and tubers, nuts, cereals and oilseeds). From 1961 onwards, cereals and oilseeds always accounted for between 88 and 95 per cent of the country's total cultivated area. Although the surface with oilseeds presented a decrease between 2016 and 2018, it recovered later. This slight fall is explained by the price of futures markets (180 days) that experienced a slowdown compared to the previous decade. Since the graph above does not

11   Ministerio de Economía y Obras y Servicios Públicos. Secretaría de Agricultura, Pesca y Alimentación, *Resolución 115* (1996): https://www.magyp.gob.ar/sitio/areas/biotecnologia/ogm/_archivos/res167-1.pdf.

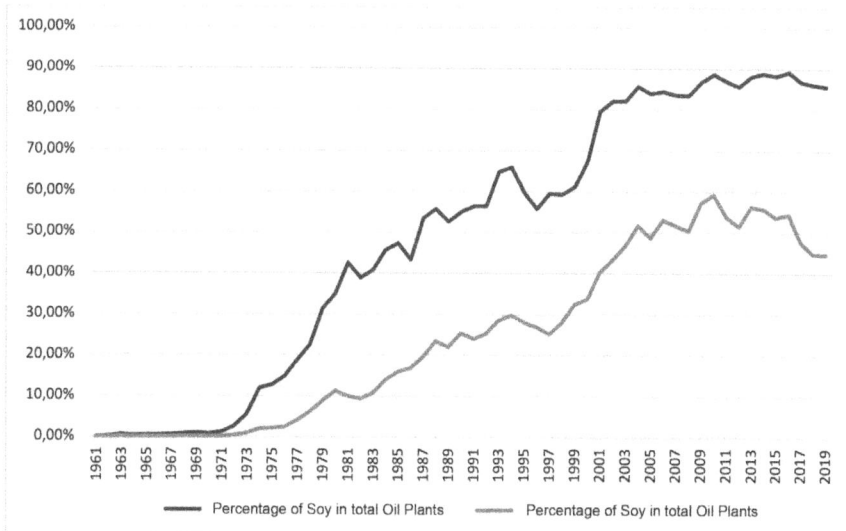

Figure 2.

Percentage of Soybeans in Total Harvested Area and Among Oilseeds (1961–2019).

Source: Author's elaboration based on FAO data available at http://www.fao.org/faostat/es/#data/QC

distinguish soybeans from other oilseeds, the one below seems to confirm that after the release to market of glyphosate-based pesticides, soy cultivation has experienced exponential growth over the last twenty years. Overall, soybean farming amounts to between eighty and ninety per cent of the total land surface sown with oilseeds and fifty to sixty per cent of total cultivated area.

Looking at the cultivated area in the province of Buenos Aires, the soy's significant weight in the total area sown appears evident.

However, analysing the landscape transformation data, the advance of soy displaces other crops and indirectly stimulates *feedlots*. A *feedlot* is a technological process that involves enclosing cattle within fences and feeding them a balanced diet, thus freeing up acreage for agriculture.

Between 2010 and 2019, despite the increase in cultivated area, the province's cattle stock grew by 19.38 per cent. This phenomenon is explained by the increase of double crops, which have already been prevalent for more than two decades, and *feedlots*, increasingly common along regional routes.

José Muzlera

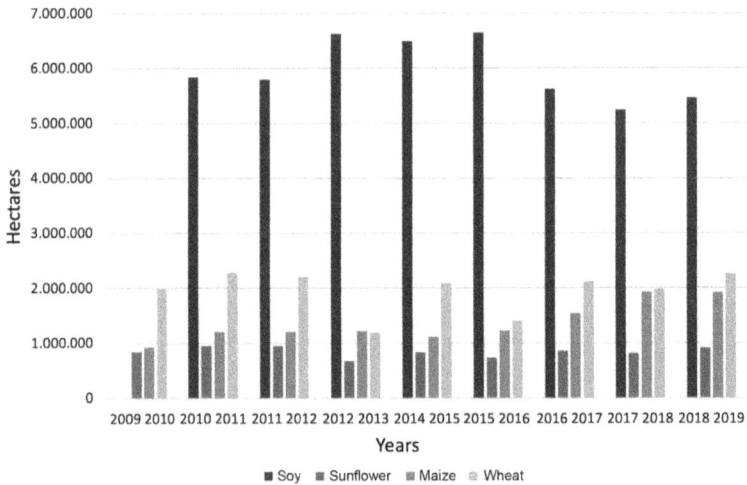

Figure 3.

Surface Occupation (ha) of Main Crops in the Province of Buenos Aires (2009–2019).

Source: Author's elaboration based on data from the Provincial Directorate of Statistics (PBA) available at http://www.estadistica.ec.gba.gov.ar/dpe/index.php/economia/agricultura-ganaderia-y-pesca/estadisticas-agricolas/120-metodologia-estadisticas-agricolas/139-cuadros-estadisticos-ea

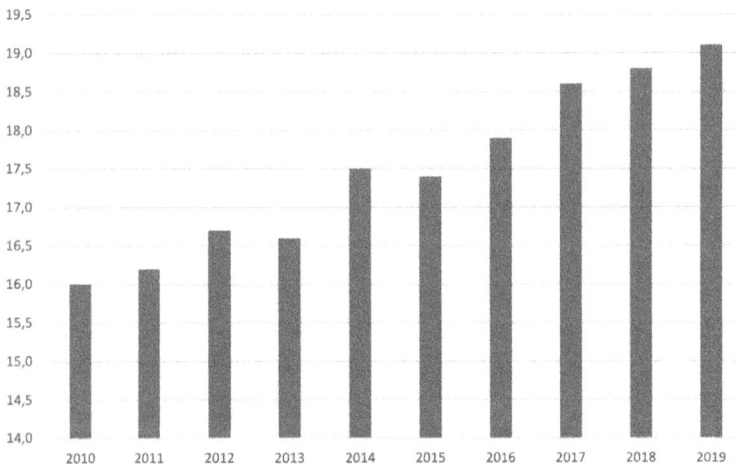

Figure 4.

Millions of Head of Cattle Raised per year. Province of Buenos Aires.

Source: General Directorate of Statistics of the Ministry of Hacienda and Finance of the Government of the Province of Buenos Aires. http://www.estadistica.ec.gba.gov.ar/dpe/index.php/economia/agricultura-ganaderia-y-pesca/estadisticas-ganaderas

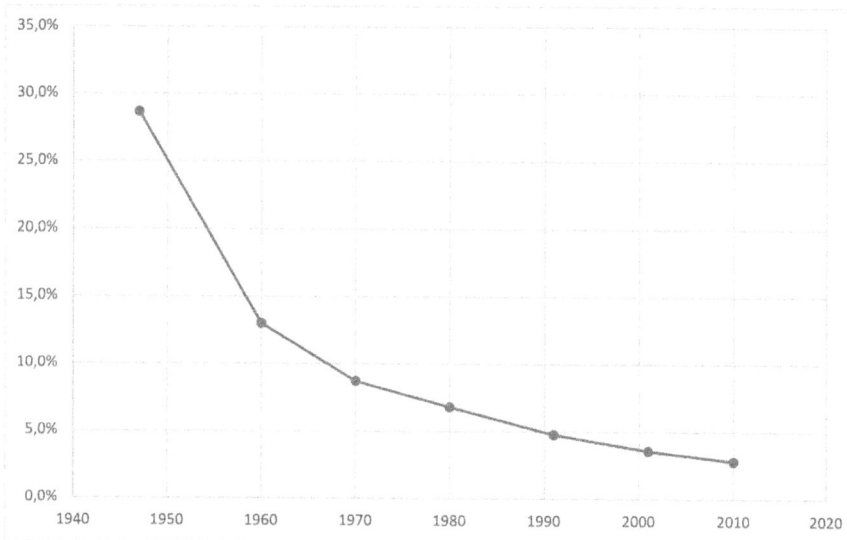

Figure 5.

Evolution of the Rural Population in the Province of Buenos Aires.

Source: Author's elaboration based on the CNPyV – INDEC.

Biophysical Landscape

As explained in the introduction, landscapes are like historical documents recording the relationship between nature and society. Thinking about them implies considering space utilisation and how its resources are affected by the development of capitalism. The modified environment has multiple meanings and symbolic values.[12] The exodus from the countryside to urban centres is a longstanding process, but the current model reinforces it. These processes are no longer only from the countryside to the village but also from the small towns to the intermediate cities. These migrations are no longer explained just by higher comfort levels in terms of livelihoods but because living in productive establishments is more costly and unnecessary. This exodus is part of a process that David Harvey calls 'space-time readjustment', where the capitalist system

---

12   C. Crumley (ed.), *Historical Ecology: Cultural Knowledge and Changing Landscapes* (Santa Fe: School of American Research Press, 1994); A.M. França, 'As Imagens de paisagem como testemunhos de transformação e memória de áreas de conservação', *Boletín De Estudios Geográficos* **112** (2019): 9–45.

340

José Muzlera

puts complex and conflicting logic into operation in using the territory.[13]

One of the ways in which this migratory dynamic impacts landscape is through processes of depopulation and the proliferation of abandoned types of houses: from houses to small palaces or even adobe ranches. 'Green deserts' is an increasingly popular expression to describe this new reality. Transgenic soybeans and extensive production demand less direct labour. Machines – often not owned by producers – possessing greater work capacity and automated processes encourage people not to live in the middle of the countryside. Even crop monitoring can be done through numerous applications such as satellite tracking.

Housing and Sheds

The rationality that the agribusiness model imposes on the Pampas region entails rural depopulation, the transformation of dilapidated houses, the displacement of extensive livestock to fields near routes (becoming intensive productive models) and the expansion of scales.

Photo 1. Tapera.[14]

13   D. Harvey, 'El "nuevo" imperialismo: acumulación por desposesión', *Socialist Register* (2004). CLACSO, Consejo Latinoamericano de Ciencias Sociales, Ciudad Autónoma de Buenos Aires, Argentina: http://bibliotecavirtual.clacso.org.ar/ar/libros/social/harvey.pdf.

14.   All Photographs by the author.

16. Soybean Landscapes: Production, Environment and Life Quality

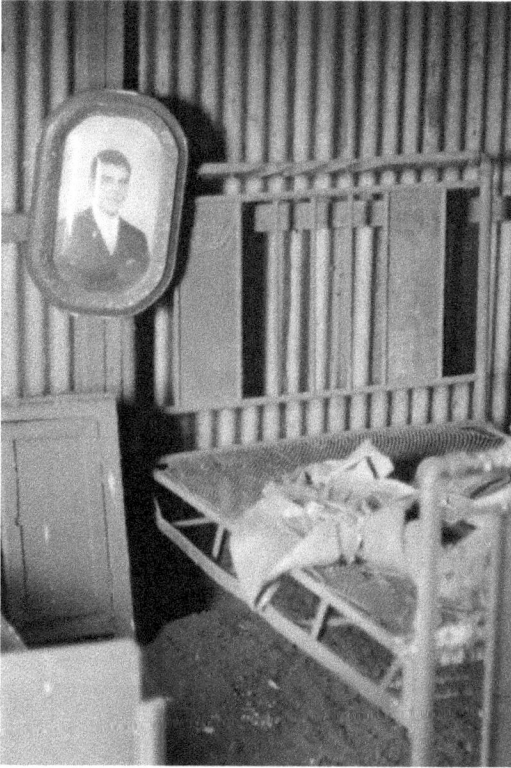

Photo 2. Abandoned ranch interior. Tapera.

Photo 3. Abandoned electricity generator.

Photo 4. Old abandoned house for employees.

Photo 5. Old abandoned house.

Photo 6. Old abandoned house of *estancia* bonaerense*.

*An *estancia* is a capitalist productive unit, which produces on its own land, with its own machinery and wage employment schemes. In general, its productions are extensive and relatively diversified stretching across a large land holding.

Photo 7. Old abandoned *estancia* house.

344

José Muzlera

Photo 8. Old abandoned *estancia* house.

Photo 9. Old abandoned *estancia* house.

Photo 10. Old abandoned *estancia* house.

Photo 11. Old abandoned *estancia* house.

Photo 12. Old abandoned *estancia* house.

José Muzlera

Photo 13. Old abandoned building that served as a ranch warehouse*

* These are parts of farmhouses not intended for productive processes such as sowing or livestock farming (includes the houses, sheds, gardens, etc.).

Photo 14. Old abandoned employee's house.

16. Soybean Landscapes: Production, Environment and Life Quality

Photo 15. Old abandoned
building that served as a ranch
warehouse.

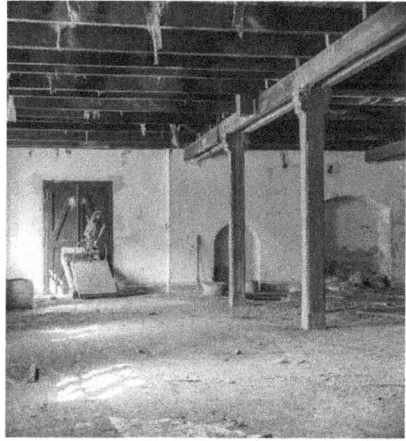

Photo 16. Old abandoned
building that served as a ranch
warehouse.

Photo 17. Old
abandoned building
that served as a ranch
warehouse.

José Muzlera

Photo 18. Old desk in abandoned living room.

Photo 19. Abandoned house.

Photo 20. Barn and bathing shed for pedigree shorthorn bulls, from the early twentieth century.

Photo 21. Butchery (the place where the meat was kept before refrigerators).

Photo 22. Abandoned shed.

José Muzlera

Productive Spaces

Pampean industrial agriculture 'geometrises' the landscape and its territorial planning. Such organisation is embedded in a rationalist conception of the world, which has its roots in modern Europe.[15]

Photo 23. Wheat harvest.

Photo 24. Wheat stubble with soybean field in the background.

---

15   França, 'As Imagens de paisagem'.

16. Soybean Landscapes: Production, Environment and Life Quality

Photo 25. Planting sunflower (front) and corn (behind).

Photo 26. Soybeans, corn and wheat fields.

José Muzlera

Photo 27. Corn plantation.

Photo 28. Sunflower, corn and wheat plantations.

Photo 29. Soybean plantation.

16. Soybean Landscapes: Production, Environment and Life Quality

Photo 30. Second plantation of soybeans on wheat stubble.

Photo 31. Second plantation of soybeans on wheat stubble and abandoned shed.

The combination of technologies and introduced fauna and flora modify landscapes, providing information about the logic of social organisation and cultural specificities. As a territorial pact, spatial organisation configures an image of the natural and/or desired order of natural resources and social values associated with the territory.[16] This tacit 'covenant' epitomises a dispute over resources and the idea of 'natural' and an attempt to justify the manipulations and resource appropriation. As the agribusiness model advances, it produces another meaning of space consistent with its logic. These include the eviction of communities, the extraction of resources and the development of a productive system oriented towards maximising individual monetary gains in the short term, ignoring any collective, long-term perspective of environmental and social sustainability.

Environmental Conflicts and Disputes of Meaning

According to Zinger and Vaquero, environmental conflicts are mainly related to the sensitivity of the environment and the intensity of space utilisation.[17] In other words, environmental conflicts are linked to landscape transformations. In the province of Buenos Aires, environmental conflicts have been of two types during the period analysed. Most of them originated from agrochemicals linked to extensive agriculture – especially the sparying associated with transgenic soybeans. A smaller number were related to mining operations in the mountains of Tandilia.[18] During the last decades, these conflicts have increased considerably in the province of Buenos Aires: 'practically all cities have formed an NGO or simply groups of neighbours linked to public denunciations of the acts of pollution produced in agricultural activity'.[19]

---

16   C. Albadalejo, 'The impossible and necessary coexistence of agricultural development models in the Pampas: the case of Santa Fe province (Argentina)', *Review of Agricultural, Food and Environmental Studies* **101** (2–3) (2020): 213–40.

17   A. Zinger and M. Vaquero, 'Conflictos ambientales y desarrollo turístico en el Sudoeste de la Provincia de Buenos Aires. República Argentina', *Anais do X Encontro de Geógrafos da América Latina*, (São Paulo, 20–26 March 2005).

18   G. García Pascual, *Análisis del establecimiento de Corredores Ambientales en el partido de Magdalena, provincia de Buenos Aires-Argentina* (Master thesis, Universitat Politècnica de València, 2018); A. Guirado, 'Resistencias y conflictos socioambientales en Tandil. Un estudio de caso', *Revista Sociedad y Equidad* **4** (2012): 22–43; M. Sabio, *Conflictos socioambientales por plaguicidas en pergamino. Territorio y poder en disputa* (Master's Thesis, Facultad Latinoamericana de Ciencias Sociales (FLACSO), 2020); H. Villalba et al., 'Conflictos ambientales. El rol del municipio. El caso de Tandil, Buenos Aires. Argentina', *VII Encuentro de Geógrafos de América Latina* (Santiago de Chile, 2001); Zinger and Vaquero, 'Conflictos ambientales'.

19   Sabio, *Conflictos socioambientales*, p. 14.

16. Soybean Landscapes: Production, Environment and Life Quality

The socio-environmental conflicts derived from extensive agriculture with direct sowing and technological packages associated with the industrial agriculture model have a dialectical relationship with local landscapes. On the one hand, they are explained by how space is used and how the environment is affected by replacing human labour with machines (encouraging rural depopulation and concentration) and by increasing the use of herbicides and pesticides that adversely affect human health and that of non-transgenic animal and plant species. On the other hand, this process has a self-reproductive dynamic. As skilled labour becomes scarce, it increasingly relies on artificial technologies. Moreover, as weeds become increasingly resistant to pesticides, more agrochemicals are needed.

The current productive paradigm lacks an integrated vision. The absence of public policies (and generalised social knowledge of how the system works) is governed by market laws, thus ensuring conflicts almost by definition. Focusing on the Pergamino case, Sabio notices how current agricultural production conditions cause environmental degradation and social conflict, affecting people's health and quality of life.[20] This has provoked the mobilisation of local communities. Such a mechanism is at work in almost all the agrocities of the Pampas region. The advancement of industrial agriculture (described above) goes hand in hand with technological packages (not exclusive to transgenic soybeans but all grains) and glyphosate. Direct seeding was crucial for reducing soil erosion processes and moisture conservation. Herbicides contributed to 'clear' weeds, and the package displaced extensive livestock.[21] The result is a greener and more homogeneous landscape, with almost no native flora and enhanced vegetation cover, compared to during the green revolution.[22]

Complementary to this process is the increase in agrochemicals. Until 1996, 30 million litres of agrochemicals per year were applied in soybean plantations, while by 2013, the sum marketed by its affiliates reached 230 million litres.[23] According to IES Consultancy (Sectoral Economic Research), in 2018, the companies included in CASAFE sold for use in Argentina 460

---

20   Sabio, *Conflictos socioambientales.*

21   F. Zorzoli, 'Malezas resistentes y/o tolerantes. (Argentina, 1996–2021)', in A. Salomón and J. Muzlera (eds), *Diccionario del Agro Iberoamericano*, pp. 649–54 (Buenos Aires, Tesseo, 2021).

22   The term agrocity, originating in geography, alludes to predominantly industrial urban spaces, but at the same time considers the number of inhabitants, history, location in the regional and national network, the environment and local imaginaries. See W.P. Umaña, 'Revolución Verde. (Tercer Mundo, 1941–2020)' in Salomón and Muzlera (eds), *Diccionario del Agro Iberoamericano*, pp. 917–22.

23   Sabio, *Conflictos socioambientales.*

million kilolitres of agrochemicals.[24] Looking at the estimates according to the goals set by the Alimentaria Emergency Program, by 2030, the cultivated area will reach 43 million hectares, and agrochemicals will increase to over 600 million kilolitres.[25] As for environmental conflicts, in the case of Pergamino, two main environmentalist groups are active in the field: women from the NGO *Assembly for the Protection of Life, Health and the Environment*, who live in the neighbourhoods of the periphery affected by pesticide drifts, and men belonging to the traditional agricultural society. The conflict mainly revolves around natural resources, health and the right to private property and production.[26] Activists' environmental logic collides with the market and producers' rationality.

> On August 30, 2019, the San Nicolás Federal Court of First Instance of Criminal and Correctional Matters (No. 2), presided by judge Nicolás Villafañe Ruso decided to extend the measures ordered on April 3 and 17 of that year, referring to the provisional suspension of spraying. As a result, he ordered that this prohibition be extended to the entire Parchment Party, setting a limit restricting the use of pesticides and excluding 1,095 metres for land applications and 3,000 metres for aerial applications. Furthermore, for the same reason, the judge ordered the arrest of the owner of the field, the contractor (applicator) and the engineer/tenant who prescribed the pesticide.[27]

The emergence of socio-environmental conflicts conditioned local development and affected the landscape. The consequences were the depletion of local flora and fauna species, demonstrations and civil surveillance to ensure strict compliance with the rules.

Wellbeing

The last part of this chapter will reflect on how soy, or the agribusiness model, affects the quality of life of those who live in agro-territories. The first issue to mention is quality of life. Traditionally, wellbeing has been associated with consumption capacity. Most state measurements to assess the population's quality of life exclusively consider the material conditions of existence. Although the economic is a dimension of great relevance, measuring the capacity of consumption and associating it *vis-a-vis* wellbeing is a typical

24   Cited in Ibid.
25   Ibid.
26   Sabio, *Conflictos socioambientales.*
27   Sabio, *Conflictos socioambientales.*

error of the classical economics paradigm that we intend not to repeat here.[28] Wellbeing has to do with consumption, health, rights, positive freedoms (often guaranteed by the State) such as access to medical treatments and education, and psychological wellbeing that allows people to project a future and adapt satisfactorily to the present.[29] In short, wellbeing is linked to the ability of each individual to build their biography.[30] Psychosocial needs are as central as economic needs to understanding wellbeing. Therefore, different studies identified with this paradigm include variables such as mental health and emotional life as subjective aspects relevant to adequate human development. In order to measure wellbeing, this text will present some data produced within the framework of the First Welfare and Consumption Survey carried out in 2019 in the Balcarce Subdivision (Province of Buenos Aires).[31] One of the survey's objectives was to measure the extent to which agriculture (hegemonised by the agribusiness model) affects the quality of life of these localities.

According to Iscaro's studies in Nicanor Olivera (southeast Buenos Aires), the agribusiness model establishes a dynamic articulation between actors. This is based on the quantities and types of capital and the possibilities of integrating into it by identifying two groups.[32] The first group is composed of those directly associated with the agribusiness model and is integrated centrally (companies supplying inputs, stockpiles, large producers, etc.) and peripherally (rentiers, capitalised contractors, transporters, etc. qualified

28   M. Max-Neef et al., *Desarrollo a Escala Humana. Una opción para el futuro* (Santiago de Chile: Cepaur, 1986); A.K. Sen, 'Capacidad y bienestar', in M.C. Nussbaum and A. Sen (eds), *La calidad de vida*, pp. 54–83 (México: Fondo de Cultura Económica, 1996); A. Salvia (ed.), *Impacto de factores económicos sobre el bienestar subjetivo en población adulta de la argentina urbana* (Buenos Aires: Educa, 2018).

29   Positive freedom alludes to the concrete possibility of choosing, and negative freedom to the absence of prohibition (which does not imply the concrete reality of choosing). As an example, the possibility to travel in a private jet is a negative freedom, if such activity is not forbidden. It would turn into a positive freedom only when the means to travel on a private jet were available.

30   E. Bandrés, 'Amartya Sen Inequality reexamined', *Revista de Economía Aplicada* 6 (2) (1994): 231–40; K. Sen, 'Capacidad y bienestar'; R. Castel and C. Haroche, 'Individuos por carencia', in R. Castel and C. Haroche (eds), *Propiedad Privada, Propiedad Social, Propiedad de sí mismo. Conversaciones sobre la construcción del individuo moderno*, pp. 53–73 (Buenos Aires: Homo Sapiens, 2000).

31   This was held in Balcarce (the subdivision's capital city with almost 38,000 inhabitants), in San Agustín (500 inhabitants), Villa Laguna La Brava (150 inhabitants). The rural population (approx. 3,500 people) was left out of the survey, just like the inhabitants of Los Pinos (400) and Ramos Otero (80 people). The confidence level of the data is 95% and the margins of error +/- 5%

32   M.E. Iscaro, *Territorio y agronegocios. La redefinición de la dimensión económica – profesional de la actividad agropecuaria a partir del avance del modelo de producción de agronegocios. Un estudio de caso. (1990–2016)* (Master's Thesis, Maestría PLIDER, 2020).

rural workers). The second group is linked with a logic of localism (*chacare-ros* producers, decapitalised contractors and informal rural workers, among others). They form a series of juxtaposed territorial pacts at the territorial level that coexist in tension but without manifest conflicts (which does not imply the total lack of conflicts between urban subjects and agricultural producers), resulting in a kind of mosaic landscape where the different groups are interspersed.[33]

Product	Ha. Harvested and number of heads	Gross margin in USD/ ha or USD/head	Estimate of aggre- gate earnings
Corn	34,830	USD 365	USD 12,712,950
Sunflower	51,912	USD 365	USD 18,947,880
Wheat	46,096	USD 390	USD 17,977,440
Soy	71,539	USD 420	USD 30,046,380
Potatoes	7,608	USD 1.114	USD 8,475,312
Bovine Stock (1/3)[34]	99,277	USD 79	USD 7,842,830
Total			USD 96,002,792

Table 1.

Agro-livestock campaign 2018/2019 Balcarce – Rindes.

Sources: Author's elaboration based on data from Governments Province of Buenos Aires (http://www.estadistica.ec.gba.gov.ar/dpe/); F.A. Fillat et al., 'Margen bruto de la producción ganadera bovina de carne de ciclo completo. Economía y sociología, EEA Pergamino', *Informe Técnico n. 5*: https://inta.gob.ar/documentos/margen-bruto-de-la-produccion-ganadera-bovina-de-carne-de-ciclo-completo-julio-2019; S.M. Cabrini and H. Zeballos, 'Margen bruto para principales actividades agrícolas-Campaña 2019/20'(INTA,2019): https://inta.gob.ar/videos/margen-bruto-para-principales-actividades-agricolas-campana-2019-20; Argenpapa, Ing. Agr. Sergio Costantino. The information available is on harvested area and gross margins. For MGs data, average estimated yields were always considered.

33   According to Albadalejo, a territorial pact is A is a coherent and stable set of institutional, organisational, technological and economic arrangements, resulting from a particular national historical stage and the power relations that characterise it, which contributes to establishing particular ways of relating society and territory. In the case of agricultural activities, the term is used to define the forms of its territorial insertion and its transformations (see C. Albadalejo, 'Pacto Territorial. (América Latina, 2000–2021)', in A. Salomón and J. Muzlera, *Diccionario del Agro Iberoamericano*, pp. 769–74 (3rd edition) (Buenos Aires, Teseo, 2021).

34   Stock matchmaker for 2019 according to PBA statistics was 297,829 heads. As the average cycle is three years, this study's estimates use stock/3.

16. Soybean Landscapes: Production, Environment and Life Quality

In attempting to understand how the agribusiness model affects (or does not affect) the economic wellbeing of Balcarceña's population, due to the lack of serial wellbeing data, an ideal distribution estimate can be compared with the real one. The agricultural sector's wealth in 2019 (not counting farms, farms, fruits and vegetables) is just over 96 million USD. The economically active population (EAP) of Balcarce is approximately 15,000 people.[35] According to data from the First Quality of Life and Consumption Survey in Balcarce in 2019, the agricultural sector gives direct income (producers, rentiers, employees, specialised merchants, service providers, etc.) EAP's 13.6 per cent (in contrast to the 26.2 per cent of the state sector, the 45.0 per cent of the non-agricultural private sector; 4.3 per cent work in more than one sector[36] and 10.9 per cent is unemployed). Thus, had more than 96 million dollars been equally distributed among the city's 2,040 people employed in the sector, average incomes in 2019 would range from 47,060.19 per capita (3,921.68 USD per month, about 294,126.20 Argentine Pesos, ARS). If this amount were distributed among the subdivision's whole population (including children, infants, unemployed, etc.), it would give 15,635.23 ARS. According to estimates from the National Institute of Statistics and Census (INDEC), in December 2019, a family of three needed to earn 30,829 USD per year to avoid poverty (12,775 for a single-person household).[37] As a result, if the sole amount produced by the agricultural sector were evenly distributed, each citizen would be comfortably above the poverty line. However, the real data sees 41.2 per cent of households below the poverty line and 14.1 per cent indigent, with only 44.7 per cent of households above the poverty line. In this context, the agribusiness model seems to be more responsible for generating inequalities and evictions than for promoting territorial dynamics of equality and wellbeing. Although it could be objected that the lack of comprehensive data on a series of years omits the evolution of total profits and wealth distribution in Balcarce, unequal wealth distribution is undoubtedly a fact.

However, material inequalities do not directly translate into low rates of psychological wellbeing. As explained by Veenhoven, just as the natural bio-

---

35   An economically active population includes people age between 16 and 65 who actively work or look for employment. People under 16 and over 65 as well as people within this age range who do not work (for whatever reason) and do not actively seek employment do not fall into this category.

36   Almost all of them are municipal employees.

37   Regarding the number of people per household, the average is 2.74, the median is 3.0 and between 1–2 and 3 people per household are 70.6% of households. And as the number of people increases, the relative amount decreases.

logical state of the organism is health, in the psychological area, wellbeing is the 'natural' state.[38] This explains why people living in adverse environmental conditions do not necessarily register worse levels of psychological wellbeing than people living in favourable environmental conditions. However, resilience capacities are exceeded in times of more adverse conditions, and people worsen these indexes. Furthermore, although biological causes affect levels of psychological wellbeing, these are primarily explained by social causes.[39] Therefore, it is legitimate to understand that the discomforts associated with individual causes are random and that psychological wellbeing assessments among different subgroups respond to environmental conditions, softened by the tendency towards wellbeing. In other words, a slight drop in wellbeing would be explained by strongly adverse conditions.

> Research conducted with twins separated at birth indicates that positive and negative attachment has a robust temperamental component. Researchers showed that about 50 per cent of the variation between individuals was attributable to temperamental differences (i.e. congenital, differences); the rest is attributed to environmental factors.[40]

This explains why, despite high levels of poverty, indigence and wealth concentration, indices of psychological wellbeing are usually very good, although they present group differences.

Comparing the different subsectors according to employment, it is observed that among unemployed people, 66.7 per cent have a 'very good' psychological wellbeing index (BIEMPS-J).[41] Such number increases among people linked to the agricultural sector (75.6), to the non-agricultural private sector (85.3) and state employees (88.6). Therefore, among people employed in only one sector, agricultural employees register the worst levels of wellbeing.

---

38   See R. Veenhoven, "Is happiness relative?" *Social Indicators Research* **24** (1991): 1–34 and 'Developments in Satisfaction Research', *Social Indicators Research* **37** (1995): 1–46.

39   E. Durkheim, *El suicidio* (México: Coyoacán, [1897] 1999); R. Mertón, *Teoría y estructura sociales* (México: Fondo de Cultura Económica, 2013).

40   M.M. Casullo (ed.), *Evaluación del bienestar psicológico* (Buenos Aires: Paidós, 2002), p. 14.

41   This psychological wellbeing was measured through a standardised test (BIEMPS-J) that consists of statements to which each person must answer with three options: they agree, neither agree nor disagree or disagree. The final version of the test incorporated four dimensions: a) self-acceptance and control (feeling of control and self-competence, feeling of well-being with oneself); b) autonomy (ability to act independently); c) Psychosocial links (quality in personal relationships); and d) Projects (goals and purposes in life). Responders were awarded a score between 1 and 3 according to the answer. Thus, results varied between 13 and 39 points. Results were grouped into two categories, 'very good' (up to 19 points) and the other three quartiles in order to explicitly target wellbeing. The same criterion was used to group in two categories the evaluation of the 4 dimensions surveyed in the BIEMPS-J (see Casullo, *Evaluación del bienestar psicológico*).

16. Soybean Landscapes: Production, Environment and Life Quality

		Psychological Wellbeing			Total
		Very good	Good, regular and bad	Ns/Nc	
Revenue Sector	Agricultural Sector	75.6%	14.6%	9.8%	100%
	Non-agricultural private sector	85.3%	11.8%	2.9%	100%
	State employee	88.6%	5.1%	6.3%	100%
	Unemployed	66.7%	33.3%		100%
	Does not work	74.1%	16.5%	9.4%	100%
	More than one sector	69.2%	15.4%	15.4%	100%
Total		78.7%	14.4%	6.8%	100%

Table 2.

Psychological Wellbeing Index by sector.

Source: Author's elaboration from the 1st Survey of Quality of Life and Consumption (Balcarce 2019).

	How happy are you with your job?			Total
	Very happy - I enjoy it - It fulfils me	It is indifferent to me - more or less	I suffer from it	
Agricultural Sector	70.7%	29.3%		100 %
Non-agricultural private sector	68.9%	25.2%	5.9%	100%
State employee	87.0%	13.0%		100%
Revenue more than one sector	76.9%	23.1%		100%
Total	74.8%	22.2%	3.0%	100%

Table 3.

Perception of happiness according to employment sector.

Source: Author's elaboration from the 1st Survey of Quality of Life and Consumption. (Balcarce 2019).

The BIEMPS-J is composed of 4 dimensions: a) self-acceptance and control (feeling of control and self-competence, feeling of wellbeing with oneself), b) autonomy (ability to act independently), c) Psychosocial links (quality of personal relationships), and d) projects (goals and purposes in life).[42]

In the first category, only 45.5 per cent of the unemployed declare feeling 'very well'. Conversely, people linked to the agricultural sector total 68.8 per cent, a higher number but still less than people of the non-agricultural private (71.6 per cent) and the state sectors (73.6 per cent). Again, people directly linked to the agricultural sector have the worst levels after the unemployed.[43] Different data emerges when looking at the autonomy dimension, where people directly linked to the agricultural sector are the highest number measuring 'very well' (58.3 per cent), in contrast to unemployed (45.5), non-agricultural and State employees (38.5 and 42.5 per cent, respectively). The data concerning the psychological links dimension differ. Here, agricultural workers are again the worst category after unemployed people (39.4 per cent), with only 60.4 per cent people feeling 'very good', in contrast to employees from the non-agricultural private sector (62.2 per cent) and the State (69 per cent). Obviously, personal relationships are much more helpful for getting a job in the state sector than in the private one, while unemployment is associated with lack of links. In the fourth and last dimension of life goals, agricultural employees seem again to be the better off category, with 58.3 per cent people scoring 'very well', compared to unemployed (33.3), non-agricultural workers (47.3) and state employees (46).

In terms of absolute happiness, state employees are perceived as the happiest, with 87 per cent very happy with their work and none suffering from it. Agricultural sector employees rank in the second position, with almost 71 per cent very happy with their work and none suffering from it, in front of people employed in the non-agricultural private sector, where almost the same amount (69 per cent) declare to be 'very happy' with their work, but 6 per cent suffer from it. Now, let us review how household poverty is linked to unemployment or the respondent's labour insertion sector.

Households directly linked to the agricultural sector through at least one member (respondents to the survey) are the most likely to live above the poverty line (73.1 per cent), slightly more than in households with at least

---

42   Casullo, *Evaluación del bienestar psicológico.*

43   The unemployed were considered people between age 16 and 65 who do not work and are actively looking for work. Those who do not work do not fall into this category, if for whatever reason they are not actively looking for work, nor are people over 65 or under 16.

16. Soybean Landscapes: Production, Environment and Life Quality

	Poor and destitute households			Total
	Destitute	Poor	About poverty line	
Agricultural Sector	3.8%	23.1%	73.1%	100%
Non-agricultural private sector	14.3%	34.3%	51.4%	100%
State employee	1.6%	29.5%	68.9%	100%
Unemployed	25.9%	51.9%	22.2%	100%
Does not work	18.9%	51.5%	29.6%	100%
Revenue more than one sector		30.0%	70.0%	100%
Total	14.1%	41.2%	44.7%	100%

Table 4.

Household poverty level by respondent's sector.

Source: Author's elaboration from the First Survey of Quality of Life and Consumption (Balcarce 2019).

one member belonging to more than one sector (70.0). These are followed by households with a state employee, where 68.9 per cent of the families live above the poverty line. Far behind are those where have at least one member employed in the non-agricultural private sector (51.4). Obviously, households with at least one unemployed member have only a 22.2 per cent chance of not being poor.

In this case, the agricultural sector did not possess the worst indicators. The agribusiness model demands skilled labour, and although the levels of formality present 'a certain laxity', informality does not abound. For example, it is typical for a service provider or a carrier not to enter an establishment without all the papers in order. Moreover, employees are normally legally employed and informally collect production awards. Conversely, workers from the non-agricultural private sector have the highest level of informality.

| | Level of formality | | | Total |
	Formal	Informal	Mixed	
Agricultural Sector	87.8%	9.8%	2.4%	100%
Non-agricultural private sector	64.7%	32.4%	2.9%	100%
State employee	89.7%	3.8%	6.4%	100%
Revenue more than one sector	84.6%		15.4%	100%
Total	76.5%	19.0%	4.5%	100%

Table 5.

Level of formality according to the economic sector.

Source: Author's elaboration from the First Survey of Quality of Life and Consumption (Balcarce 2019).

| | | Revenue Sector | | | | | |
		Agricultural Sector	Non-agricultur-al private sector	State employee	Unoccupied	Revenue more than one sector	Total
Sex	Man	20.4%	39.8%	24.1%	10.2%	5.6%	100%
	Woman	9.8%	47.9%	27.3%	11.3%	3.6%	100%
Total		13.6%	45.0%	26.2%	10.9%	4.3%	100%

Table 6.

Gender by economic sector.

Source: Author's elaboration from the First Survey of Quality of Life and Consumption (Balcarce 2019).

Perhaps unsurprisingly, gender better explains people's employment possibility in each sector, with men having with more than twice women's chances to work in the agricultural sector. Moreover, whenever women managed to get employment in the sector, they did not usually enter either as producers or as direct labour force, but as laboratory technicians, clerks or rentiers.

16. Soybean Landscapes: Production, Environment and Life Quality

	Primary Incomplete	Primary Complete	Secondary Incomplete	Secondary Complete	Tertiary -University Incomplete	Tertiary - University Complete	Postgraduate Course Complete	
Agricultural Sector	9.8%	22.0%	9.8%	31.7%	9.8%	14.6%	2.4%	100%
Non-agricultural private sector	2.9%	18.4%	15.4%	39.0%	5.1%	19.1%		100%
State employee	5.1%	21.5%	8.9%	26.6%	3.8%	34.2%		100%
Unoccupied	6.1%	27.3%	27.3%	36.4%	3.0%			100%
Total	4.6%	20.2%	14.2%	33.4%	5.6%	21.2%	0.7%	100%

Table 7.

Sectoral demand according to the highest educational level of the head of household.

Source: Author's elaboration from the First Survey of Quality of Life and Consumption (Balcarce 2019).

Looking at the highest educational level per head of household, the agricultural sector concentrates its demand at both extremes. It also demands more labour in the educational extremes, with 9.8 per cent of its labour with incomplete primary education. The state sector in second place has almost half this number (5.1 per cent).[44] But, 2.4 per cent of the agricultural sector's employees possess postgraduate degrees (no other sector records similar demand).

Conclusions

The landscape, understood as a conjunction of factors that rationally organise and interpret space, possesses different characteristics depending on each historical moment and socio-productive process. In the agribusiness model, hegemonic in the province of Buenos Aires for at least the last three decades, soy constitutes an iconic crop, occupying fifty per cent of the arable land. Beyond the different ecosystems and historical processes, this model has

---

44   Keep in mind that this survey did not capture the population that lives directly in the countryside, so this percentage is expected to increase since the labour force destined to livestock and as caregivers has lower education levels.

imposed a short-term, mercantilist and extractivist conception of nature. This hegemonic productive paradigm dominates technologies, plant and animal species, people and cultural knowledge. Industrial agricultural and livestock productions, especially soybeans, occupy most of the space, displacing people, damaging flora and fauna species (both native and exotic), homogenising landscapes and occupying those spaces with extractive and productive logic.

The advancement of industrial agriculture goes hand in hand with the techno-agricultural package (not exclusive to transgenic soybeans but covering all grains) and pesticides. Direct sowing was crucial for reducing soil erosion processes and moisture conservation. Herbicides helped 'clear' weeds, and the package displaced extensive livestock farming, largely confining it to *feedlots*. The result is a greener and more homogeneous landscape, with almost no native flora and much more vegetation cover than the period of the green revolution.

A tour of the Buenos Aires countryside would reveal fields practically unpopulated by human beings and with abandoned houses; fields characterised by a vast majority of cultivated plots, geometric formats and uniform coverage, with few livestock herds and a handful of establishments where meat is produced in corrals. Smaller villages have begun to experience the same fate as the fields – depopulation. New inhabitants live in 'dormitory dwellings' (they formally go there, but their leisure and productive life occur primarily in the city centre), although the migratory balance of small towns remains negative. In medium-sized cities, as in the case of Balcarce, the population grows, but so does marginality and the concentration of wealth. Ultimately, although the agribusiness model has managed to protect soils from the growing erosion phenomena brought by the green revolution, it demands the deployment of agrochemicals harmful to health, concentrates wealth, disperses labour force and decreases biodiversity.

INDEX

Index

Index

Index

miso 85, 250
Mississippi River 80, 170, 178, 182
Mitsui 39, 251, 252
mixed cropping 292, 299, 301, 328
modernisation 51, 59, 60, 82, 85, 86, 100,
101, 117–18, 143–44, 157, 158, 238,
248, 255, 260, 264
monoculture 2, 7, 11, 13, 29, 38, 46, 59,
105, 126, 133, 143, 145–46, 152–55,
157–58, 160, 162, 186, 188, 190–93,
202, 220, 254, 260, 292, 299, 300,
302, 305, 308, 309, 326, 327, 335
Monsanto 108, 165, 178, 183, 188, 213,
225, 242, 243, 265, 264
Moore, Jason 133
Morse, William J. 24
Muszyński, Jan 53, 55

N

narrative 3, 9, 13, 65, 66, 92, 138, 188, 201,
267–82, 284–88, 292, 331
Nazi 234–36
Nemecek, T. 164
neoliberalism 79, 80, 82, 93, 95, 106, 108–
09, 11, 112, 264
Netherlands, The 12, 23, 74, 76, 77, 80, 87,
88, 89, 223, 265–88
nitrogen 37, 180, 185–203, 254, 265, 272,
273, 274, 276, 287, 299, 313, 317,
328
cycle 12
fixation 21, 30, 103, 129, 144, 253,
257, 326–28
crisis *see* crisis
non-governmental organization (NGO)
95, 279, 281, 282, 322, 323, 354, 356
Norberg, Matilda Baraibar 8, 11, 91–114
no-till farming 82, 108, 152, 154, 156, 158,
181, 189
nutrification 180
nutrition *see also* malnutrition 5, 10, 12, 24,
39, 54, 57, 58, 73, 143, 164–66, 221,
227, 232, 234, 235, 237, 250, 252,
263, 292, 299, 307, 309, 310, 313,
317, 322, 323
crisis *see* crisis
soil *see* soil

O

oil, vegetable 160, 207, 210, 231, 294–96,
298, 299, 307–11
palm 210, 238, 307–09
soy 6, 19, 21, 24, 29, 32, 33, 35–39,
42–44, 46, 51, 54, 58, 59, 67, 68,
73, 74, 81, 85, 90, 122, 169, 187,
209–11, 222, 225, 235, 236, 238,
239, 245, 250, 251–52, 255, 266,
267, 284, 291–96, 307–08, 313, 314,
323, 326
Oilfed 293, 294, 296, 300
oilification 73, 81, 237
oilseed 57, 80, 86, 126, 128, 209, 210, 212,
231, 233–36, 238, 244, 291, 292,
295, 299, 308–10, 336, 337
Osborne, Thomas 24
ownership *see* landownership

P

Pampa (Brazil) 115, 150, 151, 153–55, 162
Pampas (Argentina) 12, 26, 150, 185–203,
333, 335, 340, 355
Paraguay 7, 8, 9, 12, 76, 91, 92, 94, 95, 97–
99, 101, 103–04, 106–08, 112, 113,
135, 137, 148, 150, 155–58, 242,
247–64, 270, 282, 284
Paraná (Brazil) 25, 48, 130, 150, 157, 158,
159, 170, 174, 177, 180, 254–56,
258–63,
River 282, 284
paste, soybean 34, 35, 37, 39, 229
pasture 103, 110, 116, 152, 330
peasant 2, 48, 51, 54, 78, 98, 101, 105, 106,
109–10, 147, 200, 253, 315
Pelotas (Brazil) 50
periphery 2, 7, 65, 94, 148, 159, 237, 313,
356
Peru 76, 88, 134, 256
Pesticide 3, 8, 26, 102, 105, 170–71, 174–80,
182–84, 196, 200, 209, 213, 220,
263, 300, 301, 302, 305, 307, 337,
355–56, 366
pests *see also* insect 57, 60, 213, 255, 263,
295, 300–02, 307, 329
Piauí 118, 121, 122, 126, 132, 133
pig 21, 35, 45, 54, 85, 86, 88, 135, 139, 187,

Index

Index